Spectrum of Modern Geography
Essays in Memory of Prof. Mohammad Anas

Spectrum of Modern Geography
Essays in Memory of Prof. Mohammad Anas

Mohammad Shafi
Mehdi Raza

CONCEPT PUBLISHING COMPANY, NEW DELHI-110 059

ISBN: 81-7022-285-0

First Published 1989

Published and Printed by
Ashok Kumar Mittal
Concept Publishing Company
A/15-16, Commercial Block, Mohan Garden
New Delhi-110 059 (INDIA)
Phones : 25351460, 25351794 *Fax* : 091-11-25357103
Email : publishing@conceptpub.com

Prof. Mohammad Anas : In Memorium

PROFESSOR Mohammad Anas was an eminent geographer, a gifted teacher and a man with wide-ranging interests. His colleagues and those who worked with him held him in affection and realized the depth and breadth of his learning. He will be remembered by generations of students as a gentle-hearted person, who was as good in the field as in the lecture room. He was an all-rounder, trained in all aspects of geography. As a man he was modest, with few pretensions. In fact he was generous to a fault with both money and time. His remarkable work, his brilliance in his subject, his broad vision, his persuasive power to settle differences between contending groups, his respect for his elders and affection for his juniors, and a balanced attitude towards his colleagues are the traits for which Prof. Anas will always be remembered.

Prof. Anas was born on 13 July 1925 in Ghazipur, a town in eastern Uttar Pradesh. After receiving his early education in Ghazipur, he joined the Aligarh Muslim University in 1939 and passed his Bachelor of Arts and Master of Arts examinations with distinction from this university. He was appointed Secretary of the Aligarh Muslim University Geographical Society and subsequently its vice-president (the president of the society is the head/chairman of the department). He joined the teachnig staff of the university in 1949 and in course of time became in succession, Reader and Professor of Geography.

Prof. Mohammad Anas was awarded a scholarship by the Australian National University to work for his doctoral thesis on Australian New Guinea. Ae conducted his field work in a very rigorous environment, and under the supervision of Prof. Spate wrote a thesis which could be envy of a geographer anywhere in the world. A paper based on this thesis was published in the *Geographical Review*, U.S.A.

In 1965 he was invited by the University of Susex, Brighton, England to give a course on the Geography of South Asia and the Pacific Islands. The lectures that he delivered are still fresh in the memory of those who attended them. In 1967 he revisited Australia at the invitation of the Australian National University, Canberra. The purpose of this visit was to collect more recent data to be incorporated in his thesis which was to be published. He was a perfectionist and would not publish anything unless he was thoroughly satisfied with it. He had an equal command over both English and Urdu. His book *The Earth We Live on* and his articles published in the *Geographical Review* of the Association of American Geographers and in *The Geographer* are eloquent testimony to his grasp over language and the subject.

Prof. Anas was a man of very wide interests. He was a teacher and a keen researcher and took great interest in raising the academic standards of the university. He was particularly interested in the organization of the education of children. As Manager of the University Schools, he took personal interest in the re-organization of the teaching in the schools. In fact, whatever work he undertook, the attainment of perfection was his moto. As provost of a residential hall of students in the university, he would not remain content with only the administrative work of the hall, but worked hard for the betterment of the academic and cultural life of the students.

Prof. Anas was equally interested in the education of young children, so much so that he became virtually the life-secretary of the *Bal Biradri* (Brotherhood of Children) and would arrange excursion of groups of young children to hill stations in the summer holidays.

Prof. Mohammad Anas was a fluent speaker and a fine conversationist. While on an excursion, either with students or with older groups, he would not remain content until he had explained in detail the historical background of the area, and had given a lively description of the entire feature.

He was a member of the Geography Terminology Committee set up to coin Urdu geographical terms. He had a fair knowledge of Persian and some knowledge of Arabic. He would not easily give his consent to the use of equivalent Urdu words of English geographical terms, unless he was fully satisfied.

It was not difficult for him to speak on any subject of interest before an audience, and his exposition of the subject and marshalling of arguments would strike one with his depth of knowledge.

On 19 March, 1983 he left for New Delhi to participate in a ymposium of the IGU Working Group on Resource Management in Drylands. While at New Delhi, he extended on invitation to Prof. H.G. Mensching, chairman of the Working Group to visit Aligarh. On 30 March Prof. Mensching came to Aligarh and delivered a lecture at which Prof. Anas requested to introduce the distinguished speaker.

It was a fateful day. Nobody, thought that this would be Prof. Anas last public appearance. That evening he suffered a massive heart attack and breathed his last at about midnight on 30 March 1983.

MOHAMMAD SHAFI

PREFACE

THE PRESENT volume has been published to commemorate the memory of a distinguished geographer and a fine human being. The topics of the essays were selected by the contributors themselves. The editors deliberately refrained from imposing any restrictions on the authors except that of space. The idea was to a low individual authors to pay their tribute to the memory of Prof. Mohammad Anas in their own field of specialization. The wide variety of topics covered, therefore, should be taken as a fitting tribute to the memory of a person whose involvement in geography was very varied and who had a singularly wide range of interests, knowledge and experience. Although the essays present a great diversity of subject matter, it has been found convenient to arrange them in eight sections : Geographical Thought, Urban Geography, Geography and Social Problems, Planning and Development, Agricultural Geography and Resource Management, Physical Geography, Environmental Pollution and Management and Political Geography.

The first section dealing with Geographical Thought, includes two essays. Academician I.P. Gerasimov presents a resume of modern constructive and integrational trends of thought in the general theory of Soviet geography. The author makes detailed comments on the articles included in *Soviet Geography Today*, published in three volumes in English and containing sixty articles by leading Soviet geographers on the main themes of the XXV International Geography Congress, 1984.

Prof. C.D. Deshpande in the other article of this section examines the possibilities of geography serving India through the Gandhian perspective of man-environment interaction.

The section on Urban Geography contains eigth essays. These essays examine the various aspects of urbanization. Prof. Dirk Bronger makes a comparative analysis of the process of metropolization in India and China in the demographic perspective. He also focuses his attention on the emergence of pronounced regional disparities in the developing countries.

Prof. Ronald Fuchs discusses the urban and population distribution policies in the Asian—Pacific region. He has made an incisive analysis of urban-related policies adopted by Asian and Pacific nations in response to the perceived problem of population maldistribution and examines the apparent policy deficiences.

Prof. R.L. Dwivedi's paper discusses the evaluation and demographic character of five important towns of Uttar Pradesh, India. After discussing in detail the evalution and growth of each of the Kawal towns, he makes a comparative study of the trends of population growth, population densities and functional character of the towns.

Prof. A.K. Dutt and Mohdudul Huq examine the validity of Christaller's Centre Place (CP) model in the context of the research work that has been done during the last fifty years in the theoretical and pragmatic dimensions of the central place concept. He concludes that the CP model has stood the test of time, continues to be viable and will in future continue to interest geographers and regional scientists.

Prof. R.P. Misra discusses the urbanization process in a cross-cultural perspective. He traces the history of human settlements with reference to their pattern of growth and spatial organization. Misra points out that that the challanges faced by the urban centres of developing countries are different from those faced by their counterparts in the developed world. In the Third World Countries rapid urbanization has occurred after the Second World War without the accompanying industrialization, and by the end of the century more of the large metropolitan centres will be in the developing countries of Asia and Latin America.

Prof. C.R. Pathak highlights the relationship, from an historical perspective, between industrialization and the urbanization process in India. He discusses various components that make up India's urban and industrial pattern at a sub-national level and then examines the quantitative relationship between these components at the sub-national and aggregate level. He concludes that the urbanization process in India can hardly be explained by the process and pattern of industrialization.

Prof. N.B.K. Reddy and Prof. Uttara Prabhu examine the components and characteristics of cities and towns of Tamil Nadu, India. They examine all the 241 urban agglomerations and towns listed in the 1971 Census and observe that larger towns, owing to their locational importance and accessibility, have a high magnitude of population and, therefore, a reduced per capita availability of basic socio-economic facilities. Consequently, a differential development policy is required for providing infrastructural facilities for towns of different size.

Amer Singh's article examines the nature of urbanization in the National Capital Region and cames to the conclusion that in terms of his empirical data the process is more urban accretion that urbanization.

The section on Geography and Social Problems includes four essays. These essays generally are concerned with monitoring the social state of the nation in terms of selected indicators. Prof. Moonis Raza and Y.P. Aggarwal examine the regional dimensions of inequities in the level of literacy in India. The findings of this study have serious policy implications for educational policy and evidently point to the need for a different strategy for educational development. The analysis highlights the fact that universalization of education is intrinsically linked with the development process as a whole, and that a narrow sectoral approach to educational planning would not achieve the national objectives.

Prof. Mehdi Raza and Farkhunda Burke analyse the regional dimensions of income disparities in Uttar Pradesh, India and attempt to identify the factors responsible for the marked spatial disparity in the distribution of per capita income in the region.

Prof. Aijazuddin Ahmad's essay examines the socio-cultural changes in a developing society, taking India as a case study. He observes that prosperity and poverty exist side by side. This underlines the changes that have occurred in India as a result of the introduction of new technology and the resulting changes in social institutions. The author also emphasizes the constraints imposed by the existing social structure and finally refers to the role of education in social transformation.

Prof. Mohammad Shafi, in his essay "Economic Development and Social Tension", attempts to analyse how rapid economic development generates tensions in a society where sharp economic disparities already exist. He concludes that if technology is not applied judiciously, social tensions and cultural conflicts are bound to arise.

The section on Planning and Development includes five essays. Prof. Satyesh Chakraborty has attempted to delineate a structure for decentralized regional planning in India which could revive public commitment for guided social change in such a manner as to achieve the identified objectives and ensure the overall welfare of the nation—state, with the added advantage of dealing positively with micro-level features and resource attributes.

In a highly industrialized country such as Britain, planning for recreation is of the utmost importance for the health of the people. Prof. J.T. Coppock writes on planning for countryside recreation in Great Britain. He shows how planning for countryside recreation has evolved in Britain. Coppok mentions its characteristics and evaluates its effectiveness in the light of current trends.

Prof. A.L. Mabogunje's essay deals with local and regional development in the Third World taking Africa as a special case. He critically examines the problems of implementing local and regional development plans and identifies four major problems that have vitiated all efforts towards greater achievements.

Prof. Sunil Kumar Munshi examines the spatial organization of the Indian economy since the start of the period of colonial rule in terms of the concept of enclaves and growth poles. He comes to the conclusion that the growth poles of today do resemble the enclaves of yesterday. The basic ideology persists with the net result that the regional imbalances between developed and backward regions are increasing.

Prof. O.H.K. Spate's paper on "Road to Waigani" is of special interest. Prof. Spate was not only the teacher of late Prof. Anas but also a close personal friend. In view of Prof. Anas's own exhaustive and masterly work on Papua New Gunea, Prof. Spate's essay on the siting of the university of Papua New Gunea is most appropriate.

The section on Agricultural Geography and Resource Management includes three essays. S.S. Dhillon et al. suggests the classification of Indian agriculture into major systems so as to render easier, the understanding of agricultural distribution at all levels.

Prof. Jerzy Kostrowicki, who was the Chairman of the I.G.U. Commission on Agricultural Typology for 1964-76, describes in his paper the efforts of individual European countries to map agricultural types based on a set of selected variables representing social, operational, productional and structural attributes of agriculture. Kostrowicki also refers to the draft of the map showing agricultural types of Europe exhibited at the XXV, I.G.U. Congress held in Paris in 1984.

Prof. H.G. Mensching is the Chairman of the I.G.U. Working Group on Resource Management in Drylands. In his paper Mensching describes the academic discussions of the working group in its two meetings held in Egypt in 1982 and India in 1983. He points out that the most important issue in the drylands is the use and misuse of land.

The section on Physical Geography has only two essays. Prof. A.B. Mukerji in his paper "Landslide and Man" analyses the regional and temporal pattern of landslides and avalanches in the western Himalaya and the role of man in aiding and abetting landslides and avalanches.

Prof. N. Ahmad and S. Rais in his paper on Himalayan glaciation try to disprove the classical fourfold glacial pattern as established by De Terra and others. In his opinion the Himalaya witnessed only three glacial episodes during the Pleistocene era.

The section on Problems of Environmental Degradation and management again includes two essays. Prof. A.G. Noble examines the geography of noise in the city of Kandy, Sri Lanka. He has concluded that the daily profile of noise at each of his recording sites varies greatly, reflecting both the land-use of the surrounding areas and the relative location of the site of Kandy.

Prof. S.L. Kayastha and V.K. Kumra discuss the quality of the environment in Kanpur city, India. They attribute the imbalance in the ecosystem to overcrowding, uncontrolled growth of city population, lack of resources, inadequacy of the sewage disposal system and high traffic density.

The last section of the volume is on Political Geography and contains two essays. Prof. Friedrich Stang analyses the political and economic geography of the partitioned island of Cyprus and brings out the supreme importance of the geographical location of the island in influencing its political and economic developments. Stang traces the history of the island from early times to the present and shows how geographical factors have been behind the problem of economic development of the island.

Prof. R.C. Sharma in his paper on Shatt-al-Arab discusses the present Iran-Iraq conflict from a geopolitical point of view.

He is hopeful of a solution of the problem provided sincere efforts are made by all concerned.

The editors are grateful to all the contributors for their cooperation and enthusiasm at all stages of the preparation of this volume. In spite of the short notice, all contributors readily agreed to join us in paying this tribute to a distinguished, departed colleague and fellow geographer.

CONTENTS

LIST OF CONTRIBUTORS

Y.P. Aggarwal
Head EDPR Unit, National Institute of Educational Planning and Administration, New Delhi, 110016—INDIA

Aijazuddin Ahmad
Professor and Chairman, Centre for the Study of Regional Development J.N., University, New Delhi, INDIA

N. Ahmed
Professor of Geology, Aligarh Muslim University, Aligarh—INDIA

Dirk Bronger
Professor of Geography, Institute of Geography, Ruhr University, Postfach 102 48, 4603 Bochum 1, West Germany

Farkhunda Burke
Research Scholar, Department of Geography, Aligarh Muslim University, Aligarh (INDIA)

Satyesh C. Chakraborty
Professor at the Indian Institute of Management, Calcutta, INDIA

J.T. Coppock
Professor of Geography, University of Edinburgh, Edinburg EHI INR, Scotland, U.K.

C.D. Deshpande
Professor of Geography and Editor, Annals of the National Association of Geographers, India 1205/12, Shivaji Nagar, Pune, INDIA

Sher Singh Dhillon
Lecturer, Department of Geography, Kurukshetra University Kurukshetra, INDIA

Ashok K. Dutt
Professor of Geography and Urban Studies, University Akron, Akron Ohio, 44325, U.S.A.

R.L. Dwivedi
Professor of Geography, Allahabad University, Allahabad, INDIA

Ronald J. Fuchs
Professor, Deptt. of Geography, University of Hawaii at Manoa Honoluln, Hawaii 96822, U.S.A.

I.P. Gerasimov
Director, Institute of Geography, Soviet Academy of Sciences Storomonetny 29, Moscow 109017, U.S.S.R.

Mohdudul Huq
Research Associate, Department of Urban Studies, University of Akron, U.S.A.

S.L. Kayastha
Professor of Geography, Banaras Hindu University, Varanasi, INDIA

V.K. Kumra
Lecturer in Geography, Banaras Hindu University, Varanasi, INDIA

Jerzy Kostrowicki
Director, Institute of Geography, Polish Academy of Sciences, Warsaw, Poland

Horst G. Mensching
Professor and Chairman, Institute of Geography, University of Hamburg, Hamburg, West Germany

R.P. Misra
Vice, Chancellor, Allahabad University, Allahabad, INDIA

Akin L. Mabogunje
Pro, Chancellor, Ogun State University, Ago—Iwaya, Nigeria

A.B. Mukerjee
Professor and Head, Department of Geography, Punjab University, Chandigarh, INDIA

Sunil K. Munshi
Professor of Geography, Centre for the Studies in Social Sciences, Calcutta 700029—INDIA

Allen G. Noble
Professor and Head, Deptt. of Geography, University of Akron, Akron, Ohio 44325—U.S.A.

C.R. Pathak
Professor and Head, Department of Arch. and Regional Planning, I.I.T., Kharagpur, INDIA

M. Uttara Prabhu
Professor of Geography, Sri Meenakshi College, Madurai 625002—INDIA

S. Rais
Research Scholar, Department of Geography, Aligarh Muslim University, Aligarh, INDIA

Moonis Raza
Vice, Chancellor, University of Delhi, DELHI 110007

Mehdi Raza
Professor, Department of Geography, Aligarh Muslim University, Aligarh, INDIA

N.B.K. Reddy
Professor and Head, Department of Geography, Sri Venkateshwara University, Trupati—517502, INDIA

Mohammad Shafi
Professor of Geography, Aligarh Muslim University, Aligarh—INDIA

R.C. Sharma
Professor, School of International Studies, J.N. University New Delhi—INDIA

R.L. Sharma
Lecturer, Deptt. of Geography, Kurukshetra University, Kurukshetra, INDIA

Jasbir Singh
Professor of Geography, Kurukshetra University, Kurukshetra, INDIA

Amar Singh
Head, Department of Geography, D.S. College, Aligarh, INDIA

O.H.K. Spate
Professor, at the Research Institute of Pacific Studies, The Australian National University, Canberra, Australia

Friedrich Stang
Director, Institute of Geography, Der Rheinisch—Westfalischen Technischen Hochochule, 51 Achen, West Germany

LIST OF FIGURES

LIST OF TABLES

SECTION ONE

GEOGRAPHICAL THOUGHT

ONE

I.P. GERASIMOV

MODERN CONSTRUCTIVE AND INTEGRATIONAL TRENDS OF THOUGHT IN THE GENERAL THEORY OF SOVIET GEOGRAPHY

ABOUT thirty years ago in 1956, after a long break owing to the First and the Second World Wars, geographers of the U.S.S.R. resumed their participation in the activities of the International Geographical Union. Within four years, that is, in 1960, Soviet geographers prepared a collective review entitled *Soviet Geography, Results and Tasks* for the Stockholm International, Geographical Congress (Moscow, Geografgiz, 1960). It was an important contribution to world geographical literature, and was translated into English. In the opening article of this book—'*Geography in the Soviet Union*',—the following principal concept was put forward :

> Modern geography should be the science of the present twentieth century. It cannot remain as it used to be, mainly a descriptive-cognitive scientific discipline . . . . It should become a science of an experimental-reformational nature—and its major task should not be so much helping in pioneer development of new lands and their natural riches, as a comprehensive scientific service to the global work of mankind for diverse and an ever intensive use of natural resources, for rational modification of nature and effective development of economy.

It can in short be put in the following way (In modern scientific language) : the principle task of our science is fundamental elaboration of the geographical aspects of the current complex interaction of nature and society.

There has been a constructive trend in the thinking of Soviet geographers in elaboration of the above task basing to stress the integrational potential of geographical science. For instance, in the article on 'Constructive Geography ; the Goal, the Methods and the Results' (Izvestiya VGO, 1966, No. 3) it was underlined that :

> if earlier, in the former stages of the development of mankind, relations of society with nature were based on *adaptability* and *spontaneous-consumption* attitude towards natural conditions and natural riches, in the modern era of its existence man should replace such an attitude by a principally new one, which should be called 'constructive-transformational'.

Those modern geographical studies can be called constructive, which are directed towards the investigation of the problems of interaction between nature and society, using advanced scientific methodology. It was suggested that the most important task in such studies should be substantiating and working out new theoretical approaches and practical methods and techniques for solving such problems that are commonly in geography.

Having singled out the most important and the latest methodological tendencies in the current research, it is possible to differentiate constructive geographical studies from more traditional descriptive-cognitive ones.

In a report *Scientific and Technological Progress and Development Tendencies of Geographical Science* presented at the 23rd International Geographical Congress at Moscow in 1976, the concepts mentioned were formulated in a more concrete form. The report emphasized that in the process of modern scientific and technological progress, geography should receive an important and crucial orientation for its further development, to provide for : a fundamental study of the scientific basis of conservation and also the purposeful modification of the environment for the effective spatial organization of society ; the rational use of natural resources ; and the further development of social production and optimizing of the living conditions and activities of the population. These principle tasks of geographical studies in the U.S.S.R., as noted in the above report, would create a new stimulus for the consolidation of Soviet geography, and would strengthen the unity of its system by encouraging a comprehensive interaction of the various scientific disciplines that comprise this system. In other words, it was underlined that the synthesis of scientific knowledge, that is, integration, is particularly indispensable for the development of constructive trends in geographical research.

In 1976 a special article on the integrational potential of modern geographical studies was published. It was addressed not only to the soviet (Integratsionnyi potentsial sovremennykh geograficheskikh issledovanii, *Izv. VGO*, 1976, No. 3), but also to foreign readers ('The contribution of geography to current knowledge', published in the UNESCO Journal *Int. Soc. Sci.*, Vol. XXVII, No. 2, 1975). In this article the general scientific methodology of Soviet constructive works was analysed, emphasizing the five major scientific principles, which, in spite of traditional tendencies towards specialization and differentiation of scientific geographical knowledge, provide for consolidation and preservation of the unity of Soviet geography. These principles are the use of historical, regional, ecological, sociological and anthropocentric approaches in geographical research. Thus, in this article it was underlined that there existed not only a great integrational potential in the geographical science, represented by scientific approaches common for the whole system of geographical sciences, but also an increase in the role of integral geographical studies, in which this potential is most fully and comprehensively realized.

These constructive trends in modern Soviet geographical research based on the integrational potential of Soviet science was also reflected in recently published papers. One is an article entitled 'Scientific Methodology of Soviet Constructive Geography' (Izv. AN SSSR, seriya. geografich., 1981, No. 2), another 'Contribution of Constructive Geography to the Problem of Optimizing the Impact of Society on the Environment' (Izv. *VGO*, 1982, Series 3) and a third, of particular significance, an introductory article to the volume on constructive geographical research in European socialist countries (Translated into Russian) by the 'Progress' Publishers in 1983.

All these articles develop those general concepts which have been mentioned above. For the purposes of this paper it would seem to be more rational to give a short analysis of the most recent issues of the series, *Soviet Geography Today*, published in English for the 25th International Geographical Congress. This series consists of three books : on theoretical problems (1981), on physical geography (1982) and on economic geography (1983). All in all the series include about sixty articles by Soviet geographers. These articles had earlier been published in different geographical periodicals from 1965 to 1981. In a certain sense, these series represents a new Soviet contribution to geographical knowledge, analogous to the collective monograph *Soviet geography : results and tasks* published in 1960.

The scope of topics considered in the selected papers by Soviet Geographers is fairly wide. All of them are included in the series 'Soviet Geography Today'. In the 6th Section of the Congress Programme devoted to 'information, documentation and methods', there is a special topic called : 'Trends of thought and ideologies in geography'. This topic is referred to in general, without any elaboration of the subject. It would have been appropriate, perhaps, to make this topic the main core in the discussion at the Congress, of the general theory of modern geography.

About half of the articles of the series *Soviet Geography Today* could rightly have been included in the sixth Section of the 25th International Geographical Congress Programme devoted to information, documentation and methods under the heading : 'Trends of Thought and Ideologies in Geography'. There is only one, but very important limitation, and that is, the concept of 'ideology'. In their research, Soviet geographers use the only truly scientific ideology, that is, dialectic and historical materialism as well as fundamentals of political economy founded by Marx, Engels and Lenin. This is clearly and definitely proved by all the information that is included in the series, in particular on the unity and integrity of the system of geographical sciences ; dialectic unity of the processes of differentiation and integration in Soviet science ; and a natural evolution of its scientific methodology according to the needs of life and development of society, with strict adherence to the initial ideological positions.

Besides this most important feature of the ideological basis of modern Soviet geographical studies, modern *constructive methodology* based on the *integrational* potential of Soviet science, is a major trend also reflected in the majority of the papers. Naturally, the situation in this respect is more complicated and controversial than in the field of ideology. While, as stated earlier the scientific ideology of all the Soviet geographers is firmly based on materialistic, Marxist philosophy, the rapid progress of the aims and methods of geographical research in the U.S.S.R. naturally has always created and still creates quite a wide range of essentially different theoretical ideas, views and approaches, which sometimes cause heated scientific discussions among Soviet geographers. All this is, to a certain extent, reflected in the contents of *Soviet Geography Today*. However, the main task of this review nevertheless is not to show such individual differences and discussions, but to reveal and substantiate those general and principally new methodological trends in Soviet science which may be termed 'constructive'.

The series *Soviet Geography Today* consists of three volumes. The second and the third of these are devoted to the problems of physical and economic geography, respectively. For Soviet geography such a division is traditional and methodologically justified. Proceeding from their general ideology, Soviet geographers are unanimous in accepting the natural-historical basis of physical geography and the socio-economic one of economic geography. It is clear that the constructive and integrational trends of thought may in equal measure manifest themselves in both. However, we shall start with physical geography (the second volume of *Soviet Geography Today*).

Sources of such modern trends are inherent in the question, what is physical geography ? The most definite answer is given in the article by Grigoriev (1966), in which he suggests that the subject relates to the geographical mantle (or natural environment)—that is, an integral natural formation consisting of the earth's crust, the lower part of the atmosphere, hydrosphere, the vegetation and soil covers and the animal world. Within this mantle substances are found in three aggregate states. Processes in it go on thanks both to the space and telluric sources of energy, and only in this mantle is there life. All the components of the geographical mantle in their structure and development, are so closely and profoundly linked with each other through the exchange of matter and energy, that they form a single indivisible whole.

This important and well-known concept of the outstanding Soviet geographers has never been disputed. On the contrary, it is cited in many articles of the series and, what is most important, it is constructively used for further development of the principles of physical geography. For instance, Grigoriev suggests that the geographical mantle can be broken down into a system of geographical belts, each of which is characterized by a specific correlation of the incoming radiational energy and atmospheric precipitation.

Kalesnik (1970), Isachenko (1975) and Solntsev (1977) mostly pay attention to the substantiation of another, in their opinion, important aspect of physical geography—the natural landscape representing an elementary unit of the geographical mantle. In this connection the article by Isachenko (1975) classifying the types and kinds of natural landscapes in the U.S.S.R. is of special interest. Solntsev also maintains that it is landscape science that comprises the essence of physical geography, and which also provides the basis of ecology. The teaching of 'Single Nature' (science about nature) should develop on this basis.

A further development of the same theme is given in the article by Mil'kov (1972) who introduces the concept of paradynamic landscape complexes, that is, genetic-geographical links of individual natural landscapes that arise owing to the exchange of matter and energy, which are mainly controlled by relief conditions. In fact, the same concepts are developed in the article by Gvozdetsky, where under the general term of 'systems analysis' he unites the following three approaches to the differentiation of natural landscapes : typological (similarity), regional (differences) and functional (integrity).

It should be noted here that the descriptive (morphological) approach and the field mapping of natural landscapes served as a starting point of the papers on Soviet landscape science, this extensive empirical material accumulated by Soviet geography, is particularly emphasized in the paper on classification of the landscapes of the U.S.S.R. by Isachenko. Somewhat methodologically different was the approach suggested by Armand (1975) and Sochava (1975). In Armand's article, along with the term 'natural landscape', there appeared the concept of 'geosystem', which became very popular, which paid due regard to the inner links among the different components of nature. Sochava goes even further. On the whole, he considers physical geography to be a science about natural geosystems, underlining the fact that such an interpretation clearly distinguishes this branch of geography from all the others (sectoral or component), allotting it quite an independent and separate place within the system of geographical sciences. Thus the most important role in this concept is played by the analysis of inner links, phenomena and processes (in particular, flows of matter and energy), which are characteristic of geosystems and which should be studied with the help of models and graphs and with a wide use of mathematical methods. Simultaneously, Sochava also pays a great deal of attention to the spatial interrelations of geosystems. He distinguishes, for this purpose, the concepts of geomeasures (structural) and geohoros (spatial units). Sochava lays special emphasis on the relationship between the ecosystem and the geosystem (he considers the former to be purely biological) ; on the question of the dynamics and evolution of geosystems and the role of co-operation of man with nature in the matter of creating and transforming natural geosystems and on the problems of ecological forecasts. There is no need to emphasize that the constructive orientation of all these theoretical elaborations is evident in the works of outstanding Soviet geographers.

The relation between the concepts of natural landscape and 'biogeocenosis' proposed by Sukachev has been discussed by Gerasimov (1973). The discussion concerns the important methodological question about the relation of ecology to biogeography. A number of other controversial aspects, although indirectly, have been touched upon in the collective article by Preobrazhensky, Zvonkova, Zimina and Gerasimov presented at the Soviet-British Geographical Seminar. Most of these aspects relate to new tendencies in the development of Soviet Physical Geography, such as a wide-integration of geographical knowledge, rapid development of interdisciplinary and frontier scientific trends conditioned by new tasks of geographers in the epoch of the scientific and technological revolution, methodological re-equipment of geography, and a sharp increase of its constructive role.

Many of these modern tendencies in the development of Soviet physical geography started appearing comparatively early, and in fact they have served as a basis for a thorough modernization and transformation of certain traditional scientific trends. The review is a good illustration of this, especially regarding national landscape science. Practical methods of field landscape studies (mainly expeditional and cartographic) ; experimental observations over the flows of matter and energy in natural landscapes (the so-called geophysics and geochemistry of landscapes) ; special attention to the transformational role of anthropogenic factors ; the use of more accurate, quantitative approaches are some of the aspects dealt with in the review.

Such is the general conclusion about the important historical roots of the modern constructive and integrational trends of thought of soviet scientists in the field of physical geography. It becomes still more definite if we get acquainted with theoretical articles of the third volume of the series *Soviet Geography Today*, devoted to economic geography. Here it is possible to consider only a few of the articles from this volume, although a number of theoretical and methodological questions are also considered in some of the others.

The first article that may be mentioned is the one by Knostantinov (1974) which contains both a short review of the development of economic geography in the U.S.S.R., and a more or less complete modern (1973), general structure of this science, reflecting the new tendencies in its development, as it appears (perhaps objectively) to the author. Further, the article by Labrishchev (1979) must be mentioned, which outlines the concepts generally accepted in the U.S.S.R. on the subject, tasks and meaning (theoretical and practical) of economic geography, formulated as far back as thirty years ago, but which sounds quite contemporary. The principal concept is as follows :

> The main tasks of economic-geographical research are the following : the study of the regularities of geographical location of production, of the spatial (geographical) division of labour and of the formation of economic regions ; the economic evaluation of natural conditions and resources : a joint study of the geographical location of industries, agriculture, transport and regions on the basis of a comprehensive study of all of them ; the study of population geography.

(From the Resolution of the Second Cangress of the Geographical Society of the U.S.S.R. in 1955).

Saushkin and Nemmik (1980), discussing the current contents of Soviet social (or to be more precise, socio-economic) geography, believe that, although in the papers of the founders of Soviet economic geography, and especially in those of Baransky, much attention has been paid to the comprehensive study of population and settlement. Nevertheless, in the recent past in the works of Soviet economic geographers this aspect appears to have been neglected. Only very recently this serious gap in Soviet economic geography being filled up is owing mainly to the influence of the Communist Party's decisions and to progress in science and technology. In this connection, quite interesting is the enumeration by Saushkin, of those socio-economic aspects of the 'human happiness' which should be the subject matter for research of a constructive-geographical character. These aspects include : a highly effective employment, an active social life, a happy personal life, good living conditions, trips about the country, good acquaintances and friendship with interesting people.

In his paper, Probst (1975) discusses the controversial question of the relationship between regional economics ('regional science') and economic geography. Recently, this question was the subject of many scientific discussions among Soviet economists and economic geographers. Probst's conclusion probably is at present the most acceptable one. He maintains that regional economics (and not the regional science, which does not have a methodological basis for becoming independent) itself is a part of economic geography.

The general conceptual and methodological conclusion which can be drawn from the above articles on Soviet economic geography is that Soviet economic geographers consider their discipline to be strictly a scientific one because socio-economic regularities are determined not by social regulations but by the laws governing the development of nature.

Lavrov (1979) in his article dealing with a definite ideological and methodological comparison of Soviet and western economic geographies concludes :

> The development of the bourgeois economic geography is characterized by revision of some traditional theoretical

> concepts, by a broad introduction into their interpretation—and, naturally, within their methodological approach—of those concepts which have been elaborated by the Soviet economic geography and the practice of planning ; by the search for new theoretical schemes ; by the links with the newest trends of the bourgeois political economy, sociology, the work of futurologists, and the birth of the 'regional science' ; the transfer from micro-economic to macro-economic and to 'global plans' and weakness and ambiguity of the theoretical principles of the bourgeois economic geography itself.

Let us now turn to the first volume of *Soviet Geography Today*, which is devoted to the questions of the general theory of Soviet geography. This issue contains 12 articles. Their principal topics are : the 'integrity' of geography as a science or a system of sciences ; the 'two-fold' (nature-society) or the 'three-fold' (nature-society-population) subject of its general theory ; the relation of physical geography to socio-economic geography ; and the practical and the theoretical tasks of geography. As seen from the foregoing the most fundamental problems of the general theory of the modern Soviet geography are discussed here.

This volume opens with an article devoted to the historical sources of Russian and Soviet geography (Gerasimov 1976).

The first volume also contains a series of articles which are devoted to the general subject of geography and to different methodological approaches to its study. The orthodox position of Soviet geography in this respect is dwelt upon in the article by Kalesnik (1972). Its main ideas are as follows : natural-production-territorial complexes (regional physical and economic geographies) are the objects of geographical research. Nowadays geography is represented by a system of sciences consisting of the following four groups—the natural-geographic (natural), the socio-geographic (economic geography and others), cartography, and a group of united geographical disciplines (science about countries regions, etc.). Kalesnik is critical of the concept of the 'monistic' or, as he calls it, 'single' geography, especially in respect to the methodological unification of physical and economic geography. He also criticises the views of Anuchin (1975), and Saushkin (1976), expressed in some of their earlier articles. However, this criticism is now mainly of historical interest, as an indication of the acuteness of scientific-discussions in Soviet geography. Later it will be seen that the more recent views of the authors criticized by Kalesnik have undergone considerable modification.

Similarly, Markov's (1978) article, is also very interesting. It presents the author's opinion about the state and the fate of Soviet geography, and devotes considerable space to the problem of unity of geography as a science, a topic that has only recently become the concern of many Soviet geographers.

However, from the view point of geography as a unified discipline, even more important the article by V.A. Anuchin, published in 'Vaprosy Filosofii' in 1975. Although the article is mainly a critique of dichotomy in geography (mainly between physical and economic geography), it also suggests a methodology for uniting the various sub-disciplines of geography into a single system. This aim cannot be achieved in terms of uniting the natural and social laws which help us understand the natural and socio-economic phenomena respectively. The answer, according to Anuchin, lies in conceptualizing a special geographical environment which manifests itself both in physio-geographical and socio-geographical factors and processes.

It seems to me that the conceptualization of the 'geographical environment' by V.A. Anuchin as—'a complex system of natural and social phenomena and processes' (and at the same time, as the subject of geographical science), although disputable, is essentially different from his earlier concept. His earlier view was that the 'natural' and 'social' phenomena and processes characteristic of the 'geographical environment' preserve their 'methodological' independence ; but here he means their interaction and mutual influence, in the process of which their own, specific regularities are born. But such position seems to sidetrack the whole problem of the 'single' geography and clearly comes closer to the elaboration of the problem of interaction of nature and society, which has long been a major concern of Soviet geography. The latter approach has been well exemplified in the works of Fedorov (1974), Gerasimov (1977) and a number of other articles published in *Soviet Geography Today*. These works are all devoted to an appraisal of the modern form of nature-society interaction in the present day complex scientific and technological context. The most important aspect of their appraisal has been a comprehensive consideration of the complex global, regional and local forms of interaction between natural environment and society. For many Soviet geographers such an approach constitutes the general theoretical framework of geography.

High-lighting the clear interdisciplinary character of the Complex Nature—society interaction, the present author (Gerasimov 1976) stressed the evident necessity of putting forward of a definite scientific methodology capable of formulating the problem in its entirety and organizing comprehensive studies that could integrate the results of multidisciplinary work dealing with various facets of the complex man-environment interaction.

A similar interpretation of the subject of geography was suggested by Slanikashvili who wrote that 'the subject of the geographical science (modern physical geography), is a universal process of spatial relations of universal phenomena of energy, matter and information exchange taking place in the anthropogenic landscape.

The present author is also in general agreement with the ideas about the development of theoretical geography expressed by Gokhman and Rodoman (1976). Two of their concept specially need mention : one regarding the principal role of geography in the study of the spatial structures of the 'mesoworld' as against that of the macroworld and the other regarding the role of comprehensive geographical modelling in the study of the mesoworld structures. The latter concept has been formulated as follows : 'So long as natural factors interact with anthropogenic factors there is a need to have a discipline which would unite both in one and the same model.

In concluding the review of theoretical issues raised in the *Soviet Geography Today* series ; it is necessary to mention three other articles. The article by Preobrazhensky (1972) emphasizes the dialectical basis of geography. The dialectical approach is not only necessary for a rational understanding of the geographical structure but also for the coordination of the typological and regional approaches.

The article by Saushkin (1976) is also very constructive. His concept of socio-economic geography is quite interesting which he defines as the science which deals with structure and development of spatial systems which emerge as a result of nature-society interaction. The rational and scientific management of these spatial systems for optimum benefit of society also lies within the perview of this definition.

Finally, of undoubted importance is the article by Isachenko (1976, 1982) on determinism and interdeterminism in foreign geography. The article examines urgent questions of foreign political geography and the ideologies underlying them : Environmentalism, possibilism, cultural and social determinism and different manifestations of interdeterminism is very useful. First, it is useful for readers from other countries as an indication of the great interest of Soviet geographers in the ideology and methodology of foreign geography (there are very few such reviews concerning Soviet geography in the west), and second, it is useful for Soviet readers as a source of reliable information and a serious methodological analysis of many theoretical trends of foreign geography, alien to the Soviet ideology.

In conclusion it is pertinent to note that it may not be an easy task to select articles by Soviet geographers to be presented to the 'trends and ideologies' section of future International Geography Congresses. This is so because it is a cardinal principle of Soviet geographers to base their work on Marxist scientific methodology which advocates integral link between theory and practice and believes in the development of theory by generalizing the empirical and experimental data and in the verification of theoretical models by practical experience. Therefore, if by 'trends of thought' one means certain scientific methodology of research, then most works by Soviet geographer may qualify for such a selection as is evidenced in the articles included in *The Soviet Geography Today*.

It may be said, by way of summation, that the three volumes of *The Soviet Geography Today* (in English), containing 60 articles covering all major themes of the 25th International Geography Congress, serve an extremely useful purpose because they highlight and bring to the notice of foreign geographers the recent constructive and integrational trends in Soviet geography.

REFERENCES

The First Issue : 'Aspects of Theory' (1981)

Anuchin, V.A., 'The Relationship Between Society and Nature in the Geographical Environment and Philosophical Aspects of Geography', *Voprosy Filosofii*, No. 4, 1975.

Fedorov, E.K., 'From the Description of Nature to its planning', —*Dokiady Institutua Geografii Sibiri i Dal'nego Vostoka*, No. 43, Novosibirsk, 1974.

Gerasimov, I.P., '*The Historical Background to Russian and Soviet Geography*', Progress, Moscow, (1976).

Gerasimov, I.P., 'The Technological Revolution and Trends of Development in Geographical Research, *Voprosy Filosofii*', No. 5 (1977)

Gokhman, V.M., Rodoman, B.O., 'Certain Trends of Development in Theoretical Geography in the USSR', '*Voprosy geografii*, No. 100., 'Mysl', Moscow 1976.

Isachenko, A.G., 'Determinism and Indeterminism in the Work of Foreign Geographers', *Vestnik Leningradskogo Universiteta*, No. 2, No. 6, 1976, 1982.

Kalesnik, S.V., 'The Subject of Geographical Sciences. Their System and Classification' —(with co-authors). *Theoretical Problems of Physical and Economic Geography*, Vol. 1, Moscow (1972).

Markov, K.K., 'General Aspects of Geography Today'. From the book : *Markov K.K. Geografiya. Dva aspekta*, Moscow, Mysl, 1978.

Preobrazhensky, V.S., 'The Continuity and Discreteness of the Geographical Sphere' (with co-authors). *Aktul'nye voprosy sovetskoi geograficheskoi nauki*, Moscow, 1972.

Saushkin, Yu. G., 'Prospects of the Development of Geographical Science : Chapter from the book *The History and Methodology of Geographical Science*, Moscow, 1976.

Slanikashvili, A.F., 'The Subject of Geographical Congnition', *Izvestija AN SSSR, ser. geograficheskaja*, No. 2, 1978.

The Second Issue 'Physical Geography' (1982)

Armand, D.L., *Landscape and Geosystem.*

Gerasimov, I.P., Preobrazhensky, V.S., Zvonkova, T.V., and Zimina, R.P. 'Modern Tendencies in the Development of Soviet Physical Geography'. *Materials of the Soviet-British Seminar*, Moscow, Progress, 1980.

Grigoriev, A.A., *Theoretical Problems of Modern Physical Geography*, 1966.

Gvozdetsky, N.A., An Introductory Article. 'Soviet Physical Geography in the Modern Stage of its Development.'

Gvozdetsky, N.A., *On the Systems Analysis in Physical Geography*, 1979.

Isachenko, A.G., *Classification of Landscapes of the USSR*, 1975.

Kalesnik, S.V., *Some Ideas on the Modern Ways of Development of Physical Geography and its Role*, 1970.

Mil'kov, F.N., *The Principle of contrast in Landscape Geography*, 1975.

Sochava, V.B., *The Teaching about Geosystems. A brochure*, 1975.

Solntsev, N.A., *Natural Geography, Landscape Science and Natural Science*, 1977.

The Third Issue 'Economic Geography' (1983)

Konstantinov, O.K., 'The Modern System of Economic-Geographical Knowledge in the USSR', *Izvestiya AN SSSR, seriya geografich.* No. 6, 1974.

Labrishchev, A., 'On the Subject of Economic Geography of the USSR and on its Present-Day Tasks',—*Communist,* No. 15, 1979.

Lavrov, S.B., 'An Analysis of Some Modern Concepts of the West Economic-Geographical Research', *Voprosy geografii,* No. 112, Moscow, 'Mysl' 1979.

Nymmik, S., 'On Methodology of Socio-Economic Geography', *Voprosy geografii,* No. 115, Moscow, 'Mysl', 1980.

Probst, A., Regional Economy and Economic Geography' (with co-authors). *Teoreticheskiya aspekty ekonomicheskoi geografii,* Leningrad, 1975.

Saushkin, Yu. G., 'The Aspects of Soviet Social Geography', *Voprosy geografii,* No. 115, Moscow, 'Mysl', 1980.

TWO

C.D. DESHPANDE

TOWARDS A GANDHIAN GEOGRAPHY
A Preliminary Essay

In my informal discussions with Prof. Anas at the Bombay Congress of the NAGI, we talked about Vedantism and Sufism and the thoughts of St. Thomas Aquinas, Ignatius Loyola, Thoreau, Tolstoy and Gandhiji inevitably and often came in. Marxist philosophy did not escape our attention. Little did I realise then that it was, alas, the last of our meetings. The present essay is dedicated to the memory of the keen and sensitive geographer that Anas was.

THE CURRENT TRENDS IN GEOGRAPHICAL RESEARCH

A review of contemporary literature is enough to show the stance that modern geography has taken and the methodologies and techniques it seeks to adopt in search of theory and for acquiring the attributes of science as defined by recent philosophers. Undoubtedly, notable achievements have been recorded during the last two decades. Yet all this has not proved satisfying as can be seen from the emergence of 'social' or 'welfare' geography (Smith 1977). Mainly, perhaps, because the search for spatial patterns and interaction represents an external point of view and does not sufficiently look into the many facets of human interaction with nature and interaction within human groups, particularly the growing inequality between the rich and the poor, and the advanced regions and the backwards regions. It is, therefore, time that we took a look at Gandhian philosophy in so far as it concerns the terrain, the environment and the people.

RESEARCH PROBLEMS AND THE REALITY

In our context, a little elaboration on the research themes and their bearing on our problems is indicated. No one has ever claimed that geography holds the answer to all our problems, but few would deny that our research needs to have some relevance to our contemporary scene. This is more true of 'human' and 'social' geography. Our studies on 'environmental influence on man' tend to be too general and naive. Land-use surveys do not pay enough attention to the current economic forces operating on land such as ownership and prevalent crop prices. Morphological, structural and functional analysis in urban geography does not evoke meaningful results 'settlement hierarchies', 'location of service centres', 'regional planning and developmental inputs' follow the trend set by economists. Such themes are many. A common feature of such exercises is that there is almost a ritualistic emphasis on methodology, techniques, data generation and processing, and the conclusions are too general and all too well known. This seems to be largely owing to our zeal in applying concepts derived from the West to an Indian situation and not the other way round, that is, first identifying a problem and then applying relevant techniques to find an answer. It is in this context that the Gandhian perspective becomes important.

THE GANDHIAN PERSPECTIVE

The Gandhian principle of *ahimsa* connotes several ideas that are important in understanding man-environment relations. First of all it implies that man is a part of the larger ecosystem and must play a role that is proper to his wisdom and understanding. In doing so he must live with minimum needs without causing a heavy strain on his environment and thereby degrading it. He should enrich the environment through the wise use and promotion of yields from the soil and domestic animals. In short, he must live in harmony with nature.

Man-Man Relationship

Gandhiji was even more concerned with man-man relationship. Throughout his references to the rich and the poor, the aged and the young, male and female, caste differentiation, discrimination on the basis of race and religion, there was a running thread of argument in favour of promoting respect, mutual help and understanding among human groups. Self help and mutual help were the basis of his Basic Education Programme. Westernization and industrialization had shaken, if not destroyed, the roots of stable societies, sweeping away the good aspects along with many bad ones, Gandhiji was keen on finding out how these new influences on mankind could be stemmed and turned to good advantage.

GANDHIAN IDEAS IN THEIR TERRITORIAL FRAME

It is exciting to see how these basic ideas express themselves with reference to the terrain—the geographer's main concern.

The Village and its Area

Gandhiji's concept of *gramarajya* is too well known to bear elaboration. We may concern ourselves only with its relevant aspects. In visualizing India as 'a land of self-sufficient village communities', Gandhiji was arguing for maintaining the balance in a local ecosystem and a local energy cycle in which man was the key figure. With proper education, some standard of living and mutual help, man can generate a healthy and beneficent eco-system. He must aim at a proper and maximum use of local resources and skills, and try to meet the local needs in food, shelter and clothing to the largest extent possible. It will be seen that landuse (cultivated land, pastures, forests and water etc.) and production by handicrafts (clothing, oil, and leather, iron, wood products) will be governed by this major consideration.

Self-sufficiency and the Energy Cycle

As Gandhiji wrote :

> This concept of self-sufficiency stems from our traditional living pattern in a village ; the farmer produces agricultural and dairy products ; oxen are used, principally for farm activities, and also for transport. Trees on the farm and grain stalks are used for his huts ; animal manure goes back to enrich the soil. This is a largely self-sufficient energy cycle in which external technological and economic forces are

now making severe inroads, more to degrade than enrich village life.

A Model

Since models have now become a familiar technique in geographical research, it would be desirable to conceptualize and develop a model of economic production and distribution in a Gandhian village, that is, by taking Gandhiji's ideas as basic assumptions, testing it with empirical studies, and extending this test from a simple tribal village to the most advanced one as in Punjab or the Andhra delta where rural electricity, irrigation, commercial farming and transport have made heavy dents in the village economy. An analysis of the various kinds of external impacts should enable us to screen out the undesirable ones (e.g., textiles) and retain the more desirable ones (e.g., electricity) which have a potential for increasing village productivity and improving the quality of life.

The Approach

The approach in such a study is radically different from the one we adopt today in our villages—settlement, land-use and other studies in their special frame. We start with the families : their needs, their economic production, level of skills and abilities for achieving better results, and the local sources that are available ; also, their inadequacies and deficiencies in these aspects. It should be possible to construct an input-output model within the economic framework of the village and its land, progressively generate models in which external inputs are introduced, and observe their impact. Such an approach to village study has many exciting possibilities as further extensions in research. Studies in environmental perception should help us better to understand how mental barriers between caste groups could be weakened if not totally removed. Innovations, and identifying barriers to their acceptance and adoption, offer a fruitful field for promoting production and distribution as well as social cohesion. This could be within the parameters of Gandhian philosophy and Gandhiji himself did not rule out modern technology if it were truly to serve the Indian villager.

The Gandhian Regional Landscape

It would be interesting and rewarding to conceptualize the Gandhian landscape of a region. It it possible that Christaller's hexagons may have an expression, though in a faint pattern, for, more than economic forces, the settlement interaction would be generated by social movements such as marriage contacts, religious fairs and the like. The service centres would be perhaps a better organized form of traditional peripatetic weekly bazaars and small market towns. Although the area commanded by a 'central place' would be the same as Christaller postulates, 'the field of influence' in terms of services and goods would be weak. Perhaps service in the form of technical guidance, educational and health facilities and administrative aspects could be more important than exchange and distribution of commodities.

The Gandhian Urbanism

Gandhiji was not against urban settlements. He felt that they were necessary for essential functions such as administration and higher social services. But he was against proliferating bloated cities created by industrialization and overweighted commercial and service functions all leading to over-urbanization and its associated formidable problems. In the Gandhian landscape the town would be small, clean and manageable, and would exercise a beneficent influence on the countryside, supporting it and giving it a cohesion, instead of exploiting the region and drawing its manpower to an unknown future in the urban slums.

The Urban Role

It would be interesting to conceptualize the Gandhian town, its structure, its functions and its relations with the lower and higher order settlements. In structure the town would be more modest in buildings ; in functions it would serve the countryside in goods and services which are desirable but not produced in the villages. It needs further study of his writings to understand how he visualized the relationship between small towns and large cities, and the role of modern industry as an inescapable development.

Regional Planning and Development

With the doctrine of self-sufficiency, running throughout his argument, it would appear that Gandhiji would have considered man first in any planned inputs. He would emphasize the individual's ability to acquire skills and the capability to accept innovations. From his insistence on self-help and mutual help, it would appear that the external inputs from governmental agencies need to be routed through the community according to its needs and scale of preferences. The official machinery would stand by to guide, aid and promote. In this perspective it is difficult to see how 'central places' or 'service centres' currently advocated by geographers and regional scientists, would generate regional ecomomy. It is perhaps the other way round : if the rural community raises its economic capacity, then alone can it support and possibly stimulate service centres in their functions, and they would in their turn promote further regional development. Much has been said for the grass root approach in regional development (Whyte 1981), but little has been thought about the need to study the regional community, their perception as regards regional development, and their ability to motivate themselves to a better performance so as to achieve their needs according to what they want and what they should desirably have.

MODERN TECHNOLOGY AND DISTORTED LANDSCAPES : THEIR IMPACT ON MAN

In the context of Gandhian thought it is relevent to review the way modern technology tends to distort landscapes—both natural and human. We are aware of the degradation that the local ecosystem suffers with the introduction of tractors leading to a shortage of natural manure, apart from the fact that it leads to heavy dependence on the supply of diesel oil (e.g., the Punjab rural economy). Modern industrial technology being point based (which is often a city) is strongly centripetal in drawing resources and manpower to industrial location. Transport lines feed them and it is no wonder that there is not only exploitation of the poor by the affluent classes, but also of the backward regions by the more advanced ones. In a free market economy this has proved to be inevitable and has given rise to regionalism not only in our country but in advanced countries such as Great Britain as can be seen from the rising regional sentiment in Wales and Scotland.

SMALL IS BEAUTIFUL

While Schumaker's argument, 'Intermediate technology', 'work experience in school education' are approvingly talked about these days, Gandhiji had been, over the last fifty years, urging us to put them into practice. A recent visit to Great Britain revealed a transformation of British society. There are no poor in the country. Except for a few very rich, British society has become, by and large, almost a one class society. There is no perceptible difference between a city and a village in terms of services and amenities. The countryside is environmentally richer than what it used to be in 1930s and the cities have shed their over-crowding and pollution. More interesting is the marked preference for small individual units, sophisticated though they are, especially in new towns such as Telford. There is a greater accent on job creation in spite of the microchip revolution. One cannot resist the feeling that Britain is on its way to socialism in a civilized manner in the way that Gandhiji would have liked, except for the apron-string commitments of the government 'to go nuclear' in its NATO commitments. All this shows that what the communist philosophy seeks to achieve through violent means can be achieved by civilized and humane means. There is thus no reason to believe that Gandhian thought is merely idealistic and has no practical utility. His approach has a far greater relevance to poorer countries.

METHODOLOGY AND TECHNIQUES

It is quite possible to use the current methodology and techniques, including sophisticated quantitative techniques. Only the point of view must be radically changed. It is man and his environmental problem that needs to be the starting point of enquiry. Just as the recent theoreticians in geography start with assumptions such as those of Christaller as regards a plane, surface perfect economic free play over distance, modes and networks, it is necessary to assume that a human group in an area has the motivation according to Gandhian philosophy, that is, his wants are minimal, he is as self-reliant as possible, and he is willing to participate in co-operative effort. With these assumptions it would be interesting to see how the human group reacts within itself, with neighbouring groups and with the natural environment, and to see what kind of external inputs would be needed by the group and what products and services could be sent out.

Kind of Inputs Needed

On developing such a conceptual and quantitative model representing the simplest situation, it should be possible progressively to introduce external inputs such as electric power, innovation in agriculture, energy supply and small-scale handicraft production, and assess their role in improving community life. The right kind of education, imparting literacy, skills and social awareness would be also included as an input. It should be possible progressively to compare these models with real situations as empirical testing. The Gandhian programme in Uruli Kanchan (District Pune) and Anandvan (District Chandrapur) could be studied and compared. It would possibly explain the shortcomings of the earlier Wardha District Plan. It should at least enable us to know how far and where the present reality departs from Gandhiji's concept of India to be. This indeed, would be no mean contribution of geography in the service of India.

REFERENCES

D.M. Smith, *Human Geography, a Welfare Approach*, London : 1977.
R.O. Whyte, *A Spatial Geography of Rural Economics*, New Delhi : 1982.

SECTION TWO

URBAN GEOGRAPHY

THREE

DIRK BRONGER

METROPOLIZATION IN INDIA AND CHINA
A Comparative Analysis and the Demographic Perspective

PURPOSE OF THE STUDY

THE PAST four decades (1940-80) were characterized by the most significant population increase ever seen in the world's history, almost a doubling of the population (from 2.3 billion to 4.5 billion) in just forty years. This rapid increase mostly affected the developing countries (DCs) : their share in the world's population rose from sixty-five per cent to nearly seventy-five per cent within the same period. A specific feature of this demographic process is to be seen in the fast growing concentration of the population within the large metropolises and capital cities. Again, this process has struck first and foremost the DCs : whereas the metropolitan population (places>1 Million$^+$) of the industrialized countries (ICs)[1] increased 3.4 times within this forty year period it increased fifteen times (!) in respect of the DCs. While in 1940 one-eighth (16.3%) of the urban population (20,000 and over) stayed in metropolitan cities, this proportion rose to an almost incredible 46.4 per cent (IC : 36.3 %) in 1980, that is, nearly every second urban dweller already lives in a metropolitan city (Table 3.1).[2]

Two phenomena in their complex causal interplay, the metropolization process together with the emergence of pronounced regional disparities, have to be viewed as the essential *spatial* feature of the DCs, whereas metropolization in most of the ICs has taken place as a continuous process starting

TABLE 3.1 : **Metropolization process in developing countries (DC) and industrialized countries (IC) : 1940-80**

No.		1940		1960		1980	
		DC	IC	DC	IC	DC	IC
I	Metropolitan growth (absolute figures in mill.)	23.1	69.4	110.9	146.9	351.9	235.7
II	Metropolitan : urban population ratio (%)	16.3	24.3	35.4	33.4	46.4	36.3
III	Growth of metropolitan population (1940=100)	100	100	480	212	1526	339

with the beginning of large-scale industrialization in the nineteenth century. In the Third World this process began more or less only after the Second World War and furthermore ahead of the industrialization process.

The target of this study is limited to the selection of one aspect of the phenomenon 'metropolization' : the determination of the demographic development of the metropolitan cities in connection with the extension of their metropolitan area. The starting point of the following considerations is the fact that the information regarding the population data especially of the larger cities—and this refers by no means to the DCs alone—varies significantly not only in the innumerable encyclopedias but also in the available professional literature. To determine the present and past population of the concerned cities as accurately as possible seems, however, to be the precondition for any study on urbanization-metropolization including the process as well as its consequences. The difficulty in carrying out this apparently simple task is that the number of inhabitants of a particular metropolis may be seen on different levels :[3]

1. Cities are usually demarcated by purely political boundaries with the result that their extent is arbitrary from a demographic—statistical as well as geographic—functional point of view.

2. The manner of delimiting cities varies significantly from country to country. Up to the present the units are not at all standardized.

3. For statistical, administrative or whatsoever reasons different types of delimitations, especially in respect of the larger cities in most of the countries, came into existence. Again, these units have not only been variously named ('city area', 'municipal area', 'metropolitan area', 'urban agglomeration', 'conurbation', 'urbanized area', etc.) but what is more important, they are seldom accurately defined.

4. The definition and territorial extent of such cities vary not only from country to country but also within the same place. According to Davis[4] 'underbounded' and 'overbounded' cities exist side by side,[5] making a comparison all the more difficult.

5. To complete this confusion, even within the same city

(i) different delimitations coexist (Bangkok Metropolis : 1,562 km^2 ; Greater Bangkok : 3,106 km^2, or
(ii) have been repeatedly changed even in recent years.[6]

6. Last but not least, in the majority of the DCs the population figures are based only on estimates (although many specialists consider them as more reliable than a great number of census data).

From all these deficiencies we can conclude that population figures without simultaneous information about the territorial extent of the present as well as the past on which the population is based is of little use. Regrettably, also, this information is not often available.[7]

An attempt at a comparative analysis regarding the metropolization process in India and China seems appropriate and at the same time challenging because :

1. As is generally known, we are dealing here with the two by far the most populous countries (together 37% of the world's population) with a correspondingly large number of metropolitan cities (45 out of 81 in Asia).[8]

2. The two countries, India and China. possess a very rich cultural heritage of several millenia. This includes a long-standing agricultural as well as urban, even metropolitan tradition.[9] As a result, in both countries we can find an historically differentiated hierarchical system of urban centres.[10]

3. Probably in conjunction with these two facts, India and China are countries in which the metropolization process did not proceed in such an impetuous and disorderly manner as in the vast majority of the remaining countries of the Third World, including the two other sub-continental states of Indonesia and Brazil.

COMPARATIVE ANALYSIS : BASIC CONSTRAINTS

In this chapter we shall show that a comparative regional analysis of the phenomenon 'urbanization'-'metropolization' already reveals a number of basic constraints which make a quantitative comparison between the two countries, India and China, quite difficult and a worldwide comparison—as undertaken in a large number of urbanization studies—hardly sensible.[11] In concrete terms these constraints refer to :

1. The data basis in general : In India the census has been conducted regularly within a 10-year period since 1881 and provides reasonably accurate and detailed information. In contrast to this there are no complete records of the population of Chinese cities and towns of any period before the first census ever taken, that is, the Census of 1953 and again of the year 1982.[12] All other figures are admittedly estimates. Moreover, according to Murphey, "Chinese population statistics are so poor, fragmentary and often self-contradictory, that one can derive no precise picture of the dimensions of this accomplishment, even for the single largest city of Shanghai."[13] In an earlier study the same author speaks of a 'statistical desert'.[14] Therefore, caution has to be the rule in the use of any statistics pertaining to China. Under these circumstances the figures regarding the population development of the metropolitan cities in China (Table 3.8) not only had to be compiled from a great variety of sources but they are by no means definite.

2. What makes the distinction of the metropolitan population in China and accordingly a comparative analysis still more difficult is the fact that the delimitation of the metropolitan area is quite problematic. In this connection a specific feature has mostly been overlooked : Especially in respect of the larger cities, the population figure do not refer only to a more or less limited urban area as in India but normally incorporate one or several countries (*xian*)—in population size comparable to 3-4 Indian *talukas*—that is, an often considerable agricultural umland.[15] To be precise, in order to compute the population of a Chinese city we have to distinguish between three different area levels :

— *shiqu* : urban district or urban area (city proper)
— *jiaoqu* : suburban district (suburbs)
— *xian* : county (under city administration).

The capital of Beijing may serve as an illustration. In 1981 the metropolitan administration was divided into four urban districts (*shiqu* : 62 km^2 : equal to the old 'walled city'), six suburban districts (*jiaoqu* : 2,639 km^2) and nine counties (*xian* : 14,106 km^2).[16]

Whereas the city area amounts to only sixty per cent of Calcutta Municipal Corporation (M.C.), the total municipal area of Beijing—16,807 km^2—is thirty times larger than the Calcutta Metropolitan area, twenty-eight times larger than Greater Bombay and even more than eleven times larger than Delhi Union Territory.

In this connection it must be admitted that in India, too, one can distinguish between several administrative urban areas : 'City', 'Municipal Corporation' (MC), 'Urban Agglomeration' (UA) and sometimes even 'Metropolitan Region' (Bombay), 'National Capital Region' (Delhi). However, even the Urban Agglomeration area in most of the twelve Indian metropolitan cities although very vaguely defined[17] correspond approximately to Davis 'truebounded' city.

3. Growth and distinction of the metropolitan population will be all the more problematic if we take a third component into consideration which is the frequent and sometimes, quite extensive changes in urban areas. As mentioned before these significant extensions of the municipal boundaries refer to almost all of the Chinese metropolitan cities owing to the Communist regime's attempt to make the capital a virtually self-sufficient economic and administrative unit, 'to promote mutual support between industry and agriculture, and to facilitate the assignment of manpower'.[18] These extensive territorial enlargements which occurred mainly in 1958 and 1959 are demonstrated for Shanghai and Beijing. China's largest metropolises in Table 3.2 and may be compared with those of the twelve Indian metropolitan cities shown in Table 3.3.[19]

The spatial data compiled in Tables 3.2 and 3.3 reveal that in contrast to the Chinese metropolis, the territorial extensions of the Indian metropolitan cities, in the past thirty years, even in the case of UA areas, were relatively limited.

A further difficulty in getting comparable figures is the regrettable fact that the Chinese statistical sources as well as the concerned authors' information regarding the city resp. municipal boundaries significantly differ—despite often identical population figures for the different years concerned.[20] The compilation of area and population data for the metropolitan city of Shenyang (Table 3.4) demonstrates the difficulty in getting a consistent picture of the territorial and population development of a

TABLE 3.2 : Territorial development in Beijing and Shanghai since 1928

Beijing year	1 Shiqu		1+2 Shiqu+Jiaoqu		1+2+Xian	
	Area (km²)	Pop. (000)	Area (km²)	Pop. (000)	Area (km²)	Pop. (000)
1928[12]	62[1]	1,369[2]	707	?		
Before October 1959[3]	62	?			8,770	4,184
After October 1959[4]	62	?	ca. 1,300[5]	?	16,807	6,800
Shanghai						
1928[6]			893	3.122[7]		
17 January 1958		6,770[8]			1,756[9]	6,977[10]
12 December 1958					5,910[9]	10,280[11]

Note : [1]'Walled city' ; [2]Trewartha, 1951, p. 338 ; [3]Chang, 1965, p. 319 ; [4]Chang, 1974, p. 6 ; [5]Calculated from : Chang, 1965, p. 320 ; [6]Feetham Report, 1931, p. 17 ; [7]1930 ; [8]Ash, 1980, p. 219 (= 'urban population') ; [9]Fung, 1980, p. 265 ; [10]Tien, 1973, p. 338 ; [11]Zukang, 1982, p. 5 ; [12]Chinese Yearbook, 1935/36, p. 1712.

TABLE 3.3 : **Territorial development of metropolitan cities in India : 1951-81**

		1951		1961		1971		1981	
		MC	UA	MC	UA	MC	UA	MC	UA
1.	Calcutta	84		103	518	104	569	104	852
2.	Bombay	69[1]	235[2]	69	438	69	603	69	603
3.	Delhi	102[3]		327	1485[4]	446[5]	1485	541	1485
4.	Madras	127		127	226	128	531	170	572
5.	Bangalore	67		134	464[6]	175	499	228	366
6.	Hyderabad	135		187	220	169	299	171	375[8]
7.	Ahmedabad	52		93	124	93	124	98	222[8]
8.	Kanpur	45		262	297	270	299	262	299
9.	Pune	129		130[7]	282	139	325	139	344
10.	Nagpur	53		218	239	218	239	218	237
11.	Lucknow	47		104	135	96	128	104	146
12.	Jaipur	65		65	104	206	259	181	210

Source : Census of India 1951—1981.

Note : [1]City of Bombay ; [2]Greater Bombay ; [3]Delhi and New Delhi ; [4]Union Territory ; [5]UA ; [6]according to the same census : 501.21 km² ('Bangalore Metropolitan Town Group') ; [7]"Poona Town Group" : 170.03 km² ; [8]estimated figures.

TABLE 3.4 : Area and population development of Shenyang, 1900—82

Year	'Built-up Area'		1		1+2		1+2+3	
	Area (km²)	Population (000)	Area (km²)	Population (000)	Area (km²)	Population (000)	Area (km²)	Population (000)
1900[1]	11.0	100	11.0	100				
1917[2]	15.9	250+						
1930[5]	15.3	245++						
1936[1]	31.5	527	63.0	527				
1945[2]			121.7	1,465				
1948[1]	78.6	1,021	262.0	1,021				
1950[4]					3,099	1,551		
1953[7]						2,300		
1958[1]	124.5	1,923	249.0	1,923	3,099	2,423		
1972[3]	106.0	2,800						
1973[2]	156.5	2,867						
1974[6]							8,000	4,200+++
1981[8]			164.0	2,940	3,495	3,918	8,515	5,029
1982[9]			173.0[10]	2,905	3,495	4,020	8,515	5,140

Source : [1]Schinz, 1983, p. 208 ; [2]Lo' Pannell and Welch, 1977, p. 275 ; [3]Lo and Welch, 1977, p. 250 ; [4]Cressey, 1955, p. 38 ; [5]Chang, 1970, p. 70 ; [6]Kuchler, 1976, p. 144 ; [7]Ullman, 1961 ; [8]Statistical Handbook of China, 1981, p. 39 ; [9]Statistical Handbook of China, 1983, p. 59 ; [10]China Official Annual Report, 1982-83, p. 30.

Note : + 1922 (Trewartha, 1951, p. 344) ; ++ 1928 (ibid.) ; +++ 1975 (Ma, 1981, p. 224.)

Chinese metropolis—a precondition for any comparability with other countries in respect of extent and growth of population.

4. As a fourth category of constraints—and now the whole problem of comparability becomes a real puzzle—we have to consider the use of the term 'urban'. More precisely it is not merely the fact that the definition of what is urban differs between India and China.[21] Unlike in India the rural population in China within the municipal boundaries is counted separately, however, not deducted from the official 1953 census figure and the later estimates. Beijing, for example, with a total population of 2,768,149 in 1953 includes a population of 708,000 classified as rural,[22] that is, 25.6 per cent of the total.[23] In 1981, that is, after the significant enlargement of its municipal boundaries (see Table 3.2) the rural population's share amounted to thirty-five per cent.[24]

To make the confusion almost complete in recent times Chinese sources distinguish between total and urban, urban and rural and agricultural and non-agricultural population. As far as the example of Beijing is concerned the agricultural population's share aggregates to forty-two per cent of the total,[25] and indeed according to another source, to 48.3 per cent.[26] In Table 3.5 are compiled the urban population figures so far available for the years 1953, 1964 and 1982.

It has so far not been clarified whether the remarkable increase of the urban population to 211.54 million for the year ending 1982[27] is owing to a new urban definition assumed by Aird.[28] The official explanation of urban for this figure, referring to population living in districts under the administration of cities and towns indicates that the *xian* population of the counties coming under the city's administration is not included.[29] However, the corresponding figure given for Beijing—5.96 million—does neither correspond with the urban population ('living in urban districts'—Beijing : 5.55 million)[30] nor with the non-agricultural population (4.77 million).[31]

To sum up, even a somewhat accurate computation of the present Chinese metropolitan population comparable to that of the Indian metropolises may be considered as highly problematic, mainly because of pronounced differences in what constitutes the metropolitan *area*. This is all the more true of the distinction

TABLE 3.5 : **China : Urban population figures, 1953—1982**

Size of Urban Place	1953 Census Total Population : 582,603,417				1964 Census Total Population ; 694,581,759		1982 (Year-End Figure) Total Population : 1,015,410,000			
	Total population Living in "Urban Places"		Urban Population Living in "Urban Places"		Urban Population		Total Population Living in "Urban Places"		Urban ("Non-Agricultural") Population Living in "Urban Places"	
					Old Definition	New Definition				
	Number %	Percentage	Number %	Percentage	Number %	Number %	Number %	Percentage	Number %	Percentage
Total	87,669 (15.0)	100.0	77,257 (13.3)	100.0	97,910 (14.1)	127,103 (18.3)	211,540 (20.8)	100.0		
>1 Mill.	21,020 (3.6)	24.0	17,474 (3.0)	22.6			75,170 (7.4)	35.5	42,050 (4.1)	
>100,000	26,972 (4.6)	30.7	22,423 (3.8)	29.0>			68,690 (6.8)	32.5	51,420 (5.1)	
> 50,000	5,497 (0.9)	6.3	4,570 (0.8)	5.9>						
> 20,000	8,236 (1.4)	9.4	6.846 (1.2)	8.9>			67,680	32.0		
< 20,000	25,944 (4.5)	29.6	25,944 (4.5)	33.6>						

Sources : 1953 : Ullman, 1958, p. 93
1964 : AIRD, 1983, p. 614
1982 : Statistical Yearbook of China, 1983 : p. 103, p. 107 } author's calculations

(and thus the comparability) of the metropolitan population growth because even the 1953 census (and this is naturally all the more true with the estimates after 1953 up to 1981), does not distinguish between *shiqu*, *jiaoqu* and *xian* population. The same applies to the distinction between total and urban, population. Obviously there are different definitions of 'urban, or, what is more, such definitions came into existence after 1955. The extension of the municipal boundaries after 1953 took place in different years for the cities concerned. In addition, these territorial changes occurred several times for each and the same municipality. Moreover, the figures presented by the authors as well as by Chinese sources quite often contradict each other. Finally, there is another factor relevant also to Indian metropolitan cities which should be mentioned : the population figures do not normally include the large number of unauthorized immigrants.[32] All these factors make a sensible comparison difficult.

As far as the present situation in the two countries is concerned, the available data for the forty-five metropolitan cities are compiled in Tables 3.6 and 3.7 regarding the different area levels (China : col. 3, 6, 10 ; India : 3, 6 and—in comparison to the xian-figures—9) as well as the resp. population figures (col. 4, 7, 9, 11 ; 4, 7, 10). The comparison allows at first two conclusions : on one hand that in respect of the city itself (shiqu) almost all the metropolises are 'underbounded'. On the other hand that the *xian* area corresponds approximately—as proposed by the official Chinese policy—to the immediate zone of influence called its 'umland'[33] or 'city region'.[34]

In fact a detailed survey of Bangalore revealed that the umland of this South-Indian metropolis is almost equivalent to the *xian* area of Shenyang, that is, of a city of comparable size not only in area and population but also in respect of its highly industrial structure.[35] Naturally, no generally accepted rule can be derived from such a result, because the *xian* areas (col. 10) differ too much from each other. The computation of the Indian metropolitan population on xian basis (Table 3.7, col. 9—12) reveals a significant change of the metropolitan proportions. Calcutta would approximately amount to one-and-a-half, Bombay at least to the same size as China's largest metropolis : Shanghai. A comparison of the two capitals—Beijing's and Delhi's—

TABLE 3.6 : **Area and population of metropolitan cities in China : 1981**

No.	Metropolis	1 Shiqu			1+2 Jiaoqu				1+2+3 Xian			Population 1 July 1982	
		Area (km^2)	Population (000)	Density/km^2	Area (km^2)	Population (000)	Density/km^2	Population : urban[1] (000)	Area (km^2)	Population (000)	Density/km^2	1+2	3
1	2	3	4	5	6	7	8	9	10	11	12	13	14
1.	Shanghai	141	5,840[2]	41,418	223	6.134	27,507	6,086	6,186	11,628	1,880	6,321	11,860
2.	Beijing	62			2,701	5.430	2,010	4.665	16.807	9.019	537	5,598	9,231
3.	Tianjin		3,450		4,276	5.023	1,175	3.829	11.305	7.628	675	5,143	7,764
4.	Shenyang	164	2,940	17,927	3,495	3.918	1,121	2.937	8.515	5.029	591		
5.	Wuhan	172			1,557	3.157	2.028	2.662	4.480	4.101	915		
6.	Guangzhou				1.345	3.077	2.288	2.338	11.757	5.537	471		
7.	Chongqing				1.521	2.597	1.707	1.900	9.848	6.434	653		
8.	Harbin	156	2.090	13,397	1.637	2.460	1.503	2.094	1.637	2.460	1.503		
9.	Chengdu	40	1.210	30,250	1.447	2.428	1.678	1.376	3.861	3.957	1.025		
10.	Zibo					2.192		623	7.000+				
11.	Xi'an	157			861	2.145	2.491	1.580	2.441	2.888	1.183		
12.	Nanjing	41[3]			867	2.087	2.407	1.702	4.718	3.680	780		
13.	Taiyuan	168	1.176	7.000	3.044	1.702	559	1.239	6.988	2.149	308		
14.	Changchun				1.116	1.696	1.520	1.309	18.881	5.689	301		
15.	Dalian				1.003	1.452	1.448	1.208	12.574	4.645	369		
16.	Kunming	22	560	25.455		1.399		997	6.593	1.970	299		

TABLE 3.6 : (*Contd.*)

1	2	3	4	5	6	7	8	9	10	11	12	13	14
17.	Lanzhou				2.122	1.381	651	1.075	14.414	2.330	162		
18.	Zhengzhou	80	840	10.500		1.381		859	1.748	1.910	1.093		
19.	Guiyang	54	722	13.370		1.296		827	2.090	1.296	648		
20.	Tangshan		1.293			1.293		865	16.000+				
21.	Jinan	78	845	10.833	483	1.290	2.671	1.010	4.875	3.296	676		
22.	Qiqihar					1.193		899	5,000+				
23.	Anshan					1.177		996	4.000+				
24.	Fushun					1.163		1.012	10.500+				
25.	Hangzhou				430	1.156	2.688	905	16.600	5.207	314		
26.	Qingdao				244	1.153	4.725	1.013	5.966	4.260	714		
27.	Fuzhou	34	617	17.980		1.098		690	3.121	1.623	520		
28.	Jilin					1.049		815	20.000+				
29.	Changsha	53	810	15.283		1.047		835	3.926	2.400	611		
30.	Nanchang	117	810	6.923	561	1.033	1.841	815	4.791	1.030 (?)			
31	Shijiazhuang	322	1.030	3.199		1.031		809	15.000+				
32.	Baotou					1.026		846	10.500+				
33.	Huainan					1.015		539	900+				
34.	Hefei				457	795	1,740	539	2.857	1,559	546		
35.	Hohhot	65	720	11,077					6.079	1.175	193		
36.	Urumqi				1'200	942	785	880	9.094	1.101	121		

Sources : Zhongguo baike nianjian, 1982 ; Statistical Yearbook of China, 1981 (author's calculations).

[1] 'excluding agricultural population residing in urban areas'

[2] 31-12-1980 (ZUKANG 1982) ; according to other Chinese sources ; 150 km^2—6.01 mill. pop. (ESCAP, p. 112)

[3] SKINNER, 1978, p. 761 (for 1974)

+ Zhonghua renmin gongheguo fen sheng dituji (Provincial Atlas of China), 1977 (computed by the author).

TABLE 3.7 : **Area and population of metropolitan cities in India—1981**

No.	Metropolis	Municipal 1 Corporation (M.C.)			Urban 1+2 Agglomeration (U.A.)			1+2+3 Neighbouring Districts			
		Area (km²)	Population (000)	Density (km²)	Area (km²)	Population (000)	Density (km²)	Area (km²)	Population (000)	Density (km²)	Description
1	2	3	4	5	6	7	8	9	10	11	12
1.	Calcutta	104	3,305	31,779	852	9,194	10,788	18,904	20.520	1,086	Haora, Hugli, 24 Parganas
2.	Bombay	691[1]	3,196	46,304	603	8,243	13.671	3,860	10.600+	2,746	Bombay Metropolitan Region
3.	Delhi	541[2]	5,729	10,595	1.485[3]	6,220	4.189	8,085	8.364	1,034	Rohtak, Gurgaon
4.	Madras	170	3,277	19.276	572	4,289	7.499	8,028	6.878	857	Chengalpattu
5.	Bangalore	228	2.629	11.556	366	2.922	7.991	8,003	4.922	615	Bangalore
6.	Hyderabad	171	2.093	12.233	375	2.546	6.787	7,707	3.820	496	Rangareddy
7.	Ahmedabad	98	2.060	20.988	222	2.548	11.471	8,700	3.840	441	Ahmedabad

TABLE 3.7 : (*Contd.*)

1	2	3	4	5	6	7	8	9	10	11	12
8.	Kanpur	262	1.482	5.665	299	1.639	5.482	6,200	3.791	614	Kanpur
9.	Pune	139	1.203	8.670	344	1.686	4.899	15,600	4.162	266	Pune
10.	Nagpur	218	1.219	5.603	237	1.302	5.495	9,900	2.582	260	Nagpur
11.	Lucknow	104	896	8.649	146	1.008	6.907	2,500	2.017	798	Lucknow
12.	Jaipur	181	977	5.391	210	1.015	4.831	14,100	3.406	242	Jaipur

Source : Census of India 1971, Series 1—India, Part II—A (i) : General Population Totals, New Delhi, 1975, pp. 239 ff.
Paper 1 of 1972 : Final Population, New Delhi, 1972, pp. 5ff.
Census of India 1981, Series 1—India, Paper 2 of 1981 : Provisional Population Totals, Rural-Urban Distribution, New Delhi, 1981, pp. 83 ff.
Paper 3 of 1981 : Provisional Population Total, Workers and Non-Worker, New Delhi, 1981, pp. 124 ff.
Part II—B (i) : Primary Census Abstract, General Population, New Delhi, 1983.

Note : [1]City, [2]U. A., [3]Union Territory,
+ approx. figure.

municipal areas directly disclose the impracticability of a comparison at this level (Fig. 3.1).[36]

The first conclusion leads us to the principal question, how can we determine the metropolitan city's population ? What criteria should be applied which would be acceptable for an interregional or, if possible, worldwide comparison ? A detailed discussion would be beyond the framework of this study, but an analysis of the data on the metropolitan cities has revealed the necessity to take into account the concerned area on which a population figure is based. However, this simple statement inmediately discloses number of difficulties. First, the

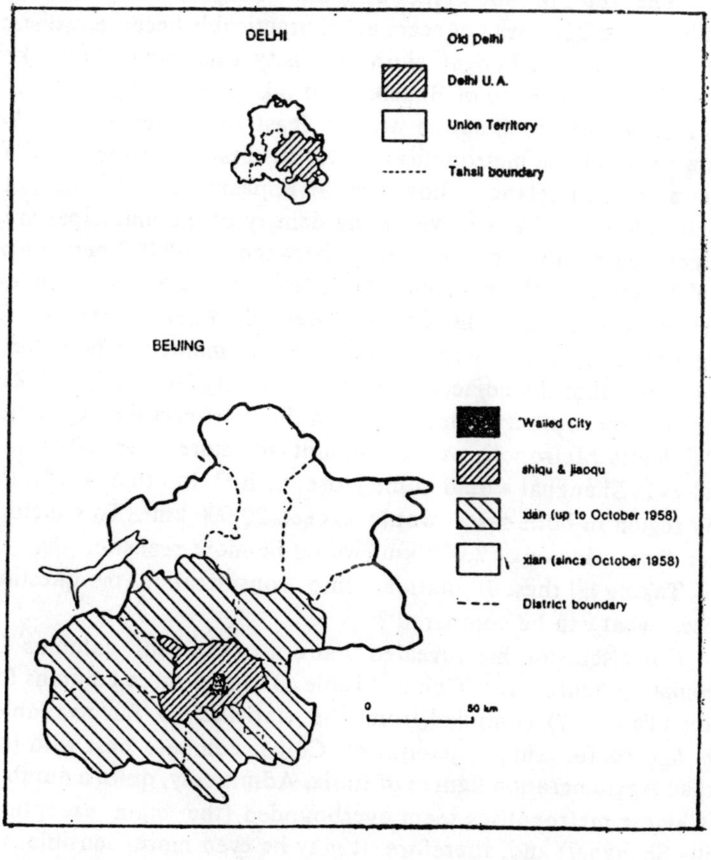

FIG. 3.1 : *Municipal areas of Delhi and Beijing*

regrettable fact that quite often detailed statistics are not available. This appears evident from the gaps regarding the statistics so far available for China (Table 3.6, col. 3 and 6). Second, the area/ population ratio (and consequently the density) differs so considerably even between the metropolises within the same country (e.g., Shanghai—Taiyuan ; Ahmedabad—Jaipur) that each comparison becomes questionable. For this reason the fixing of an area population ratio valid for a world-wide comparison of the metropolitan population is necessary and should be worked out. In the following paragraphs some suggestions are given.

The rule of 400 inhabitants/km^2, adopted by the Census of India (Note 21) does not seem at all practicable because two states (Kerala and West Bengal) show a density exceeding 600 per km^2. The recommendation of Bousted[37] to take a minimum population of 1 million in combination with at least 1,000 persons per km^2 as a basis for the metropolitan area at first sight seems valid.[38] At a second glance, however, it appears suitable only for metropolises of the ICs, where the density of the municipal areas concerned numbers on an average between 1 to 5,000 persons per km^2.[39] As far as the metropolitan cities in India and China as well as in most of the DCs are concerned, where there is a far higher density not only in respect of the municipal boundaries but also within the adjacent umland, a density factor of 1,000 km^2 does not seem practicable. Based on these criteria the population of Calcutta Metropolis would amount to more than 20 million and even Shanghai would come close to it.[40] In other words, the city region in both cases would exceed 20,000 km^2. To conclude therefore, a factor of 2,000/km^2 would be more realistic.

Taking all these limitations into consideration the question arises, what can be compared ?

Our discussion has revealed that out of the four columns of population figures for China (Table 3.6) and three columns for India (Table 3.7) compiled from the statistics so far available, the figures for shiqu+jiaoqu of China compare best with the Urban Agglomeration figures of India. Admittedly, quite a number of Chinese metropolises seem overbounded (the main exception being Shanghai) and, therefore, it may be even more sensible to compare the urban population figure (Table 3.6, col. 9). The disadvantage is that no comparable urban figures for the previous years are available, making it difficult to trace historical metropolitan growth even over the past three decades. Besides, the combined shiqu+jiaoqu area corresponds in most cases with Indias 1953 municipal area ; or is at least comparable to the latter. As matters stand the results derived from the available data can be summarized as follows :

METROPOLIZATION : URBANIZATION—RATIO

Detailed data with regard to the different urban size groups valid for a long period are available only for India. Again, it is difficult to compare the 1953 Chinese urban figures with those of 1982 (see Table 3.5). At least in the case of India, the compiled data (given in Figure 3.1) demonstrate the rapid increase of the metropolitan population's share of the urban population as a whole.[41] When the urbanization—metropolization ratios of 1901 and 1981 are compared a radically different picture emerges in 1981 the metropolitan population accounting for a major share in the total urban population. For example taking into account only the communities of 20,000 and more, the metropolitan share would be 31per cent. Similar picture of the urbanization : metropolization ratio can be observed in China. Though it should not be overlooked that on both sides the urban population's share (> 20.000) at present, and this does not even amount to 20 per cent of the total.

METROPOLIZATION QUOTA

There are a number of similarities in the spatial dimension of metropolizations in the two countries as is borne out by an analysis of regional data. In both countries more than half the states/provinces are without any metropolitan city and some are far away from even the developing ones. On the other hand, a number of regions, though limited, with already a pronounced metropolitan quota of close to or even above 20 per cent have come up in both countries (Maharashtra, West Bengal, Haryana-Delhi in India and Liaoning and Hebei in China).[43] But what is important to emphasize is that the spatial pattern of metropolization in the two countries is due to similar causes which are rooted in this colonial past. To serve the economic interest of

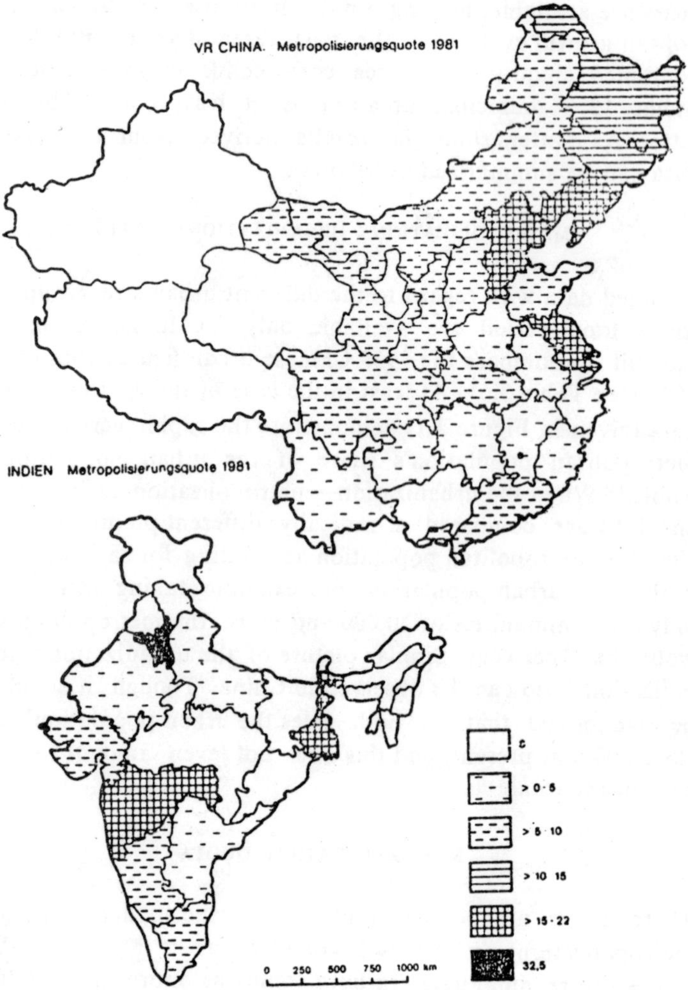

FIG. 3.2 : *Metropolization quota in two countries*

the colonial masters, industrialization had an early start in coastal areas and their hinterland in both India and China. Excluding Delhi-Haryana region which constitutes a special case, Calcutta, Bombay and Madras can be cited as examples of this trend in India. Similarly the "treaty ports" exemplify the trend in China.[44] On the other hand, the regions lacking metropolitan cities are more or less peripherially located with reference to the coastal metropolized zones. The regions lacking metropolitan centres are, however, not necessarily at a lower level of development (Kerala and Punjab !). Viewing the metropolization quota of China (6.7%) and India (6.2%) both subcontinental states on the whole still rank at the lower end within the Asian scene.[42]

GROWTH OF METROPOLITAN CITIES

Since one has to be very cautious with regard to the population figures of the Chinese metropolises before 1953, the following analysis will, concentrate on the development during the last three decades. To begin with, from the data compiled (Tables 3.8 and 3.9) some common characteristics are noticeable. First of all there is a pronounced heterogeneity in metropolitan growth in both India and China. On one hand one finds several cities with a comparatively low growth rate (Calcutta, Lucknow : Shenyang, Fushun, etc.). With a few exceptions this seems unique in the Third World, at least if one excludes the temporary dwellers in the metropolises. On the other hand, other metropolises show a decidedly dynamic growth, even in comparison to the fast growing metropolitan cities of the Far East such as Bangkok, Djakarta, Manila and Seoul.[45] These cities are Delhi, Bangalore and Jaipur in India, and in China, the interior provincial capitals of Lanzhou, Guiyang, Baotou, Xi'an and Chengdu, also Qiqihar in the North.

All in all, however, the differences predominate. First of all the growth of the larger metropolises (Fig. 3.3) is considerably slower in China, especially with regard to the metropolises along the coastal areas. Surprisingly, even the capital of Beijing shows a more moderate growth.[46] As indicated by a number of authors, this is owing mainly to the fact that the rapid growth of metropolitan cities all over China,[47] up to 1958, because of heavy immigration, has considerably slowed down in the majority of cities. This means that the official policy to contain the growth of the largest cities (Shanghai, Beijing, Tianjin, Shenyang, Guangzhou, Chongqing) and 'to redirect the main focus on urban development to newer cities of the interior appears to have been largely successful.'[48] This apparently unquestionable success of 'urban (metropolitan) decentralization'[49] should, however, be viewed with caution because a more detailed view reveals : First that at

TABLE 3.8 : **Population growth of metropolitan cities in China, 1900–82**

	1900	1920	1930	1938	1948	1953	1958	1970	1982	Growth 1953-82 China : 100
1. Shanghai	950	1.539	3.122	3.595	5.020	6.204	6.977	5.802	6.321[15]	59
2. Beijing	728[1]	1.181	1.369[5]	1.574	1.672[12]	2.768	4.148	5.000	5.598[15]	116
3. Tianjin		839	1.392	1.223	1.708[12]	2.694	3.278	3.600	5.143[15]	110
4. Shenyang	100	250[3]	527[6]	772	1.121	2.300	2.423	2.800	4.020	100
5. Wuhan		750[3]	1.584[7]	1.242	1.062	1.427	2.226	2.560	3.230	130
6. Guangzhou		—	830[5]	1.022	1.413	1.599	1.867	2.500	3.120	112
7. Chonqing		351*	298[8]	528	1.003[12]	1.773	2.165	2.400	2.650	86
8. Harbin		200[3]	320[5]	468	760[13]	1.163	1.595	1.670	2.550	126
9. Chengdu		423*	441[8]	458	727	857	1.135	1.250	2.470	166
10. Zibo		—	—	—	—	184	806**	850	2.234**	698
11. Xian		—	188[9]	218	503	787	1.368	1.600	2.180	159
12. Nanjing		300[3]	522[5]	440	1.137	1.092	1.455	1.750	2.130	112
13. Taiyurn		80[3]	139[8]	177	252[12]	721	1.053	1.350	1.750	139
14. Changchun		70[3]	—	360	630	855	988	1.200	1.740	117
15. Dalian		237[4]	586[6]	504	723[12]	892	1.590	1.650	1.480	95
16. Kunming		—	170[4]	184	300	699	900	1.100	1.430	118
17. Lanzhou		110[3]	103[8]	122	204	397	732	1.450	1.430	207
18. Zhengzhou		35[3]	80[10]	197	150	595	785	1.050	1.424	138
19. Tangshan		76	100[5]	146	137	693	812	950	1.333	111
20. Jinan		300[3]	437[9]	472	591[12]	680	882	1.100	1.320	112
21. Guiyans		—	117[9]	145	263[12]	271	530	660	1.314	279
22. Qiqihar		43	76	97	175	345	704	760	1.220	203

23. Anshan		—	—	120	166	549	833	1.050	1.210	127
24. Fushun		181*	118[6]	215	513	679	1 019	1.080	1.190	101
25. Hangzhou		—	468[6]	575	616[14]	697	734	960	1.180	97
26. Qingdao		131	318[4]	592	788	917	1.144	1.300	1.180	74
27. Fushou		—	379[5]	343	331	553	623	680	1.122	117
28. Changsha		180[3]	312[5]	464	422[12]	651	709	825	1.072	95
29. Jilin	100[2]	230[3]	143[6]	132	247	435	583	720	1.071	141
30. Shijiazhuang		—	—	194	190	373	623	800	1.070	165
31. Nanchang		—	206[5]	275	267	398	520	675	1.046	151
32. Baotou		—	67[9]	70	82	149	397	920	1.042	402
33. Huainan		—	—	—	—	287	280	600	1.036	207

[1]1913 ; [2]1907 ; [3]1922 ; [4]1926 : Daiten only ; [5]1929 ; [6]1936 ; [7]1927 ; [8]1934 ; [9]1935 ; [10]1931 ; [11]1928 ; [12]1946 ; [13]1947 ; [14]1944 ; [15]Census : July 1, 1982.

* Xian administrative area ; ** extension of city area ; —no data available.

Sources : 1900 : Murphey, 1953 (Shanghai) ; Chang, 1965 (Beijing) ; —Chinese yearbook 1923 (filin).
1920 : Chinese yearbook 1923 ; Trewartha, 1951.
1930 : Chinese yearbook 1943 ; Trewartha, 1951.
1938 : Ullman, 1961.
1948 : Ullman, 1961 ; Trewartha, 1951.
1953 : Shiger, 1953.
1958 : Ullman, 1961 ; Chen, 1966.
1970 : Chen, 1973.
1982 : Statistical yearbook of China—1983.

TABLE 3.9 : **Population growth of metropolitan cities in India 1901—81**

Metropolis Municipal Corporation	1901	1911	1921	1931	1941	1951	1961	1971	1981	Growth 1901-81 India= 100	Growth 1951-81 India= 100
1. Calcutta U.A.	1,488	1,718	1,851	2,106	3,578	4,589	5,737	7,031	9.194	215	106
Calcutta	934	1,016	1,053	1,221	2.167	2.698	2.927	3.149	3.305		
2. Greater Bombay	928	1.139	1.380	1.398	1.801	2.994	4.152	5.971	8.243	309	145
Bombay City	776	979	1,176	1.161	1.490	2.329	2.772	3.070	3.195		
3. Delhi U.T.	406	414	488	636	918	1.744	2.659	4.066	6.220		
Delhi U.A.	214	238	304	447	696	1.437	2.359	3.647	5.729	931	210
Delhi	214	238	304	374	579	1.120	2.062	3.288	4.884+		
4. Madras U.A.	594	604	628	775	930	1.542	1.945	3.170	4.289	251	147
Madras	503	575	592	734	881	1.416	1.729	2.469	3.277		
5. Bangalore U.A.	159	189	237	306	407	779	1.200	1.654	2.922	639	198
Bangalore	159	189	237	306	406	779	1.141	1.541	2.629		
6. Hyderabad U.A.	448	502	406	467	739	1.128	1.249	1.796	2.546	198	119
Hyderabad	448	502	406	447	720	1.026	1.119	1.607	2.093		
7. Ahmedabad U.A.	186	217	274	314	595	877	1.206	1.742	2.548	477	153
Ahmedabad	186	217	274	310	591	837	1,150	1.592	2.060		
8. Kanpur U.A.	203	179	216	244	487	705	971	1.275	1.639	281	123
Kanpur	203	179	216	219	452	649	895	1.158	1.482		
9. Pune U.A.	164	173	199	250	324	606	791	1,135	1,686	358	147

Pune	153	159	199	198	258	481	607	856	1.203		
10. Nagpur U.A.	167	119	165	242	329	485	690	930	1.302	271	142
Nagpur	128	101	145	215	302	449	644	866	1.219		
11. Lucknow U.A.	256	252	241	275	387	497	656	826	1.008	137	107
Lucknow	256	252	241	251	355	445	595	749	896		
12. Jaipur U.A.	160	137	120	144	176	291	410	637	1,015	221	184
Jaipur						291	403	615	977		

Sources : Census of India 1901—81.
+according to the Census figures : 5,351.

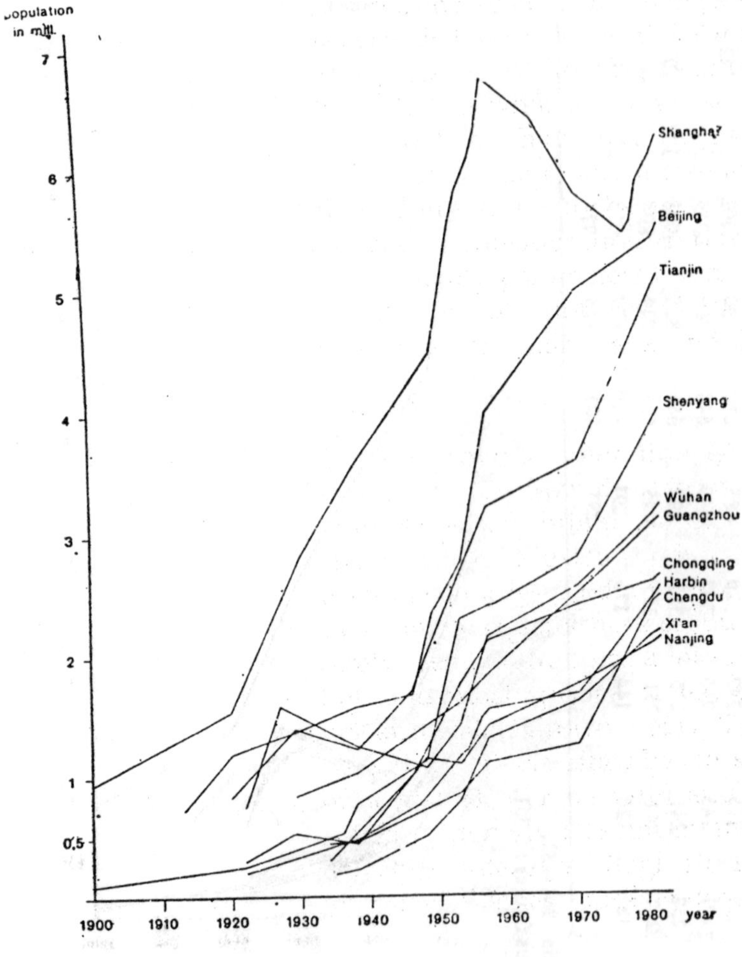

FIG. 3.3 : *Population growth of metropolitan cities in China (1900—1982)*

least half the number of Chinese metropolises show a still noticeable increase of 20 per cent and more above the national average. Second, some coastal metropolises have shown an increasing metropolitan growth since the second half of the seventies. Taking Shanghai as an example—because the most detailed (though not undisputed[50] figures are available for this city it may

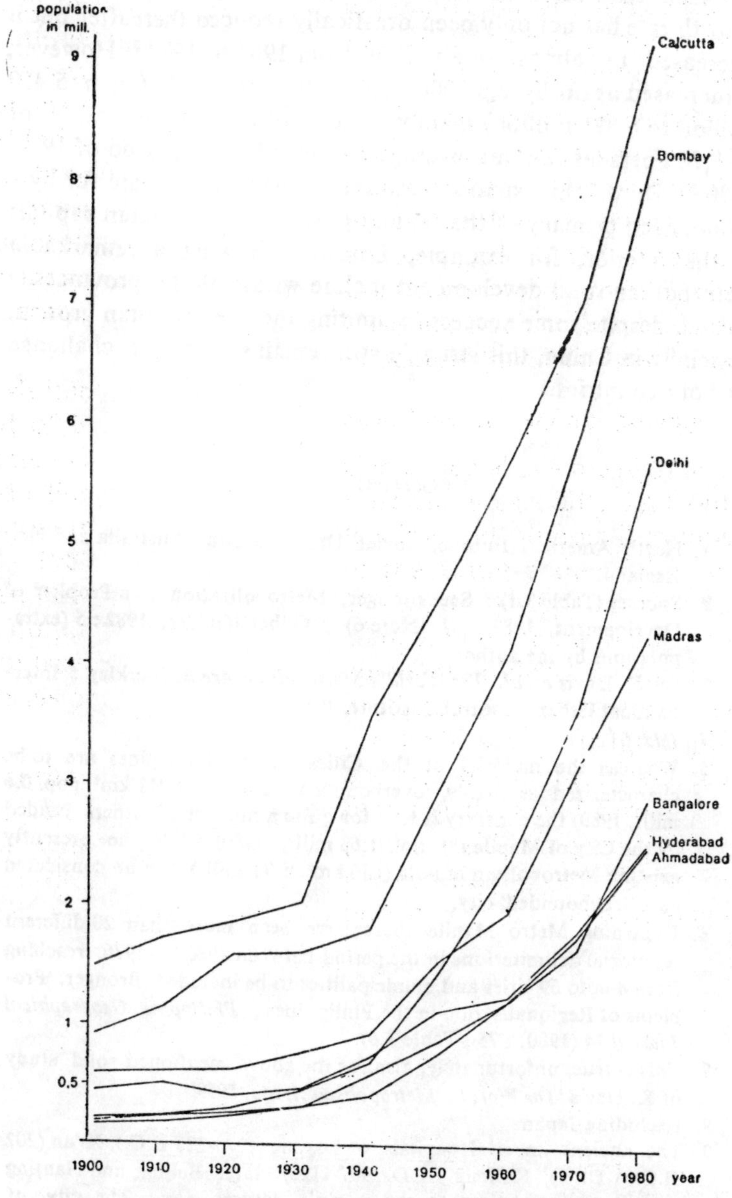

FIG. 3.4 : *Population growth of metropolitan cities in India (1901—1981)*

be seen that after an impetuous growth up to 1960 the city's growth rate has not only been drastically reduced thereafter but it decreased by almost one million from 1965 to 1977.[51] However, it increased again by 8,51,000 from 1977 to 1982 (from 5,470 million to 6,321 million) mainly because of a net immigration of 1,38,200 persons on an average per year for the period of 1978-1980.[52] Third, this spatial redirection had to be paid for by a pronounced primacy of the rapidly growing metropolitan capitals of the interior, for example, Lanzhou showing a remarkable metropolitan-rural development incline within these provinces.[53] Indeed, despite some success in limiting the metropolitan growth, especially in China, this struggle still remains a major challenge for both countries.

FOOTNOTES

1. North America, Europe, Soviet Union, Japan, Australia and New Zealand.
2. Sources (Table 3.1) : See Bronger, Metropolization as a Problem of Development, 1982, 151 (Note 6) ; Gilbert/Gugler, 1982 : 5 (extrapolations by the author).
3. See K. Davis *et. al.*, *The World's Metropolitan Areas* (Berkley : International Urban Research Institute, 1959).
4. *Ibid*, 6f.
5. Whereas the majority of the Cities in the Philippines are to be characterized as highly 'overbounded' (Davao, 2.211 km², pop. 0.6 mill., 1980) the contrary is true for quite a number of others headed by the City of Manila (38 km², 1.63 mill., 1980) while the presently existing Metropolitan Manila (636 km², 5.93 mill.) can be considered as a 'truebounded' city.
6. Regarding Metro Manila there have been more than 20 different territorial delimitations in the period between 1960 to 1976, reaching from 8 upto 69 cities and municipalities to be included (Bronger, 'Problems of Regionalisation in the Philippines', *Philippine Geographical Journal* 24 (1980) : 75 , Table 3.6).
7. This is true, unfortunately, also for the above mentioned solid study of K. Davis '*The World's Metropolitan Areas*, 1959.
8. Excluding Japan.
9. The ancient imperial capitals of Loyang (770-225 B.C.), Xi'an (202 B.C.-A.D. 9), Kaifeng (A.D. 907-1126), later Beijing and Nanjing ranked at their time as the world's largest cities. The city of Ch'angan (Xi'an) was a well known metropolis all over Asia in T'ang times (A.D. 618-906) and with a reported population of about one

million within the city walls S.D. Chang, most likely the largest city all over the world. 'Some observation on the Morphology of Chinese Walled Cities', *Annal's of the Association of American Geographers* 60 (1970) : p. 7.

10. Number and size class of towns in India (1971) and China (1973) were as follows :

	>1 mill.	>1,00,000	>50,000	>10,000	Total
India	9	133	198	1.548	1.888
China	21	111	114	1.263	1.509

Sources : China—C.S. Chen, C.W. Skinner ; India—Census 1971.

11. Regarding the concept of 'comparative regional research', see D. Bronger, 'The Concept of Comparative Regional Research, *Urban Planning and Research* 20 (1977) : p. 146-75.
12. Except for the three cities of provincial status (Beijing, Tianjin, Shanghai) the citywise population figures of the Census undertaken in 1964 have never been published (see J.S. Aird, New York, London), 1978).
13. R. Murphey, *The Fading of the Maoist Vision* (1980), p. 48. Regarding Shanghai see the very detailed compilation in : Bannister 'Mortality, Fertility and Contraceptive use in Shanghai', *The China Quarterly*, 70 (1977) : 259-63, using more than 50 different sources for the period from 1941-75.
14. R. Murphey, *Sanghai* : *Key to Modern China*, Cambridge, Maso : 1953, p. 125.
15. In 1981 there were 2,127 Xian (including 57 'banners') compared to 377 districts and approx. 4,500 *talukas* in India. On a statistical average nearly 5,00,000 people are allotted to one Xian compare to ca. 1,50,000 to each *tahsil*.
16. China, *Official Annual Report 1982*, Vol. 83, p. 28.
17. According to the latest Census an urban agglomeration is defined as 'the continuous urban spread consisting of a core town and its adjoining urban outgrowths which may be either urban in their own right or rural.' Census of India 1981, Series 1—India, Paper 2 of 1981 : 23).
18. *New China News Agency*, 17 September, 1959, cited in S.D. Chang, Peking : 'The Growing Metropolis of Communist China', *The Geographical Review* 55 (1965) : 319.
19. Only the major changes are indicated.
20. See the compiled area data for five key cities (Shanghai, Beijing, Tianjin, Guangzhou and Nanjing) in D. Bronger, 1984. In Table 3.2 only the—in the author's opinion—most reliable figures are given.
21. In China, according to a state council resolution passed in November 1955, urban areas (cities and towns) have to meet any one of the following three criteria :
    - —seat on a municipal people council or a people's council of the xian (county) level or above, regardless of population size,
    - —minimum resident population of 2,000, at least 50% of which is engaged in non-agricultural work,

—places of 1,000 to 2,000 population of whom 75% were employed in non-agricultural work—(R. Chen, *Chinese Economic Statistics*, 1966; 7 *et al.*

—In India, an urban place has to fulfil the following three criteria (this distinction has remained almost unchanged since 1951):

—minimum population of 5,000,

—at least 75% of male working population is non-agricultural,

—density of population of at least 400 per km^2 (Census of India, 1971).

22. Ullman, 1961-1969, p. 87.
23. *Ibid.*, Tianjin : 13.6% ; Nanjing : 24.7% ; Lanzhou : 15.6%.
24. 3.149 mill. out of a total number of 9.008 mill.
25. *Ibid.*, 3.782 out of 9.008 mill.
26. See : Table 3.6, col. 9, 11, 1982 : 48.1% author's culculations (*Statistical Yearbook of China, 1983*, Vol. 35, p. 108.
27. *Ibid.* : 106. For the 1982 Census a figure of 210.83 mill. is given. Another official source states a population figure of 206.6 mill. 'of cities and towns excluding the people in the country which is under the city administration' (1982 Population Census : Vol. 16).
28. Aird, J.S., 1983 : 615. According to this new definition the criteria are :—places with 3,000 or more inhabitants of whom at least 70% are non-agricultural,

    —places with 2,500 to 3,000 inhabitants of whom at least 85% are non-agricultural.

    For a critical assessment of the 1953 urban figures see Orleans, (refer bibliography for details).
29. *Statistical Yearbook of China*, 1983, 103.
30. *Ibid.*, p. 35.
31. *Ibid.*, p. 108.
32. For Shanghai, to cite an example, an estimated figure of 8,00,000 'temporary and illicit migrants' were officially reported for the year 1957 (Howe, 1971, 229, Aird, 1978, 26. Ullman, 1961.
33. K.H. Hottes, 1950, p. 44 ; P. Schöller, 1953, 175.
34. R.E. Dickinson, 1956, p. 165.
35. Singh, 1964 ; 80.
36. The upper figure in Map 3.1 refers to Delhi Union Territory, e.g., by far the largest area among all India metropolises (Table 3.7, col. 6) and includes the neighbouring districts of Gurgaon, Rohtak, Sonepat, Meerut and Ghaziabad (=16.875 km^2), Metropolitan Delhi's population would number 13.841 million, e.g., exceed that of Beijing by 50% ! The Delhi National Capital Region defined a total area of 32.500 km^2 with an approximate population of 19.5 mill. in 1981 (Misra, 1981 ; calculated by the author).
37. Boustedt, 1970, p. 25 (reference details in bibliography).
38. The one million-limit is also adopted by Indian scientists. See : Bhattacharya, 1976, p. 113 ; Raza *et al.*, 1981, p. 85 (reference details in bibliography).
39. As far as the three largest metropolitan cities in West Germany are

concerned, the population density : West Berlin 3.900, Hamburg 2.200 and Munich 4.100 per km².

40. Including the adjacent Nadia district, almost 23 mill.
41. Figures exclude Assam and Jammu and Kashmir.
42. More in detail : Bronger, 1982, p. 154.
43. Including Beijing and Tianjin ; both are entirely surrounded by Hebei province.
44. See the instructive survey of China's space-polity in Whitney, 1970, pp. 26-72.
45. The growth rates are as follows (counted as in Tables 3.8 and 3.9) :
    Djakarta 1961-1980 : 143
    Bangkok 1947-1980 : 156
    Manila 1948-1980 : 152
    Seoul 1960-1980 : 228.
46. In contrast to Chang, 1974, p. 1.
47. See bibliography for the following references : Tien, 1973 ; further : Chen, 1973 ; Küchler, 1976 ; Chang, 1976 ; Pannell, 1981 ; Buck, 1981 ; Kwok, 1981 ; Fung, 1981.
48. Ma, 1981, p. 112.
49. Essentially see : Buck, 1981, p. 116 ; Küchler, 1976.
50. See Note 13.
51. Zukang, 1982, p. 3.
52. Ibid., p. 5.
53. Bronger, 1984, Table 3.5.

REFERENCES

AIRD, J.S. : "Recent Provincial Population Figures", *The China Quarterly*, 73, (1978), 1-44.

AIRD, J.S. : "The Preliminary Results of China's 1982 Census", *The China Quarterly*, 96 (1983), 613-640.

ASH, R. : "The Quest for Food Self-Sufficiency", HOWE, *The Development of Shanghai since 1949*, Ch. Cambridge and New York.

BANISTER, J. : "Mortality, Fertility and Contracceptive Use in Snanghai", *The China Quarterly*, 70 (1977), 255-95.

BHATTACHARYA, M. (1976) : Urbanization and Policy Intervention in India, SOCIAL SCIENCE RESEARCH INSTITUTE (Ed.), *Asia Urbanizing*, Tokyo, pp. 88-117.

BOUSTEDT, O. (1970) : Agglomeration, in : *Handwörterbuch der Raumforschung und Raumordung*, 1, Hannover, pp. 20-38.

BOUSTEDT, O. (1975) : "Grundri B. der empirischen Regional-forschung", Teil III : *Siedlungsstrukturen*, Hannover.

BRONGER, D. (1977) : "The Concept of 'Comparative Regional Research", A Methodological Contribution for Empiricial Developing Country-Research", in : *Urban and Rural Planning Thought*, XX, New Delhi, pp. 146-175.

BRONGER, D. (1980) : "Problems of Regionalization in the Philippines".

Aspects of Regional Planning in Developing Countries I, in *Philippine Geographical Journal*, XXIV, No. 2, pp. 65-82.

BRONGER, D. (1982) : "Metropolization as a Problem of Development in Countries of the Third World—The Far Eastern Scence" (Metropolization and Regional Development Planning 1), *Philippine Geographical Journal*, XXVI, 4, pp. 149-159.

BRONGER, D. (1984) : "Metropolization in China ?", in *Geo. Journal*, 8.2 (in print).

BUCK, D.D. (1981) : "Policies Favoring the Growth of Smaller Urban Places in the People's Republic of China, 1949-1979", in : MA. L.J.C./HANTEN, E.W. (Eds.), *Urban Development in Modern China*, Boulder/Col., pp. 114-146.

Census of India 1961 : Vol. 1. Part II-A (i) : *General Population Tables*. Delhi, 1964.

Census of India 1961 ; *Delhi District Census Handbook*, Delhi, 1964.

Census of India 1971 : Series 1, Paper 1, *Final Population*. New Delhi, 1972.

Census of India 1971 : Series 1, Paper II-A (i). *General Population Totals*. New Delhi, 1975.

Census of India 1971 : Paper No. 13. *A Study on Urbanization in India*, New Delhi, 1979.

Census of India 1971 : Series 27, *A Portrait of Population*. Delhi, 1978.

Census of India 1981 : Series 1, Paper 2. *Provisional Population Totals : Rural-Urban Distribution*. New Delhi, 1981.

Census of India 1981 ; Series 1, Paper 3. *Provisional Population Totals : Workers and non-Workers*. New Delhi, 1981.

Census of India 1981 ; Series 1, Paper 1. *Final Population Totals*. New Delhi, 1982.

Census of India 1981 : Series 1, Part II-B (i). *Primary Census Abstract : General Population*. New Delhi, 1983.

CHANG, S.D. (1965) : "Peking : The Growing Metropolis of Communist China", *The Geographical Review*, 55, pp. 313-327.

CHANG, S.D. (1970) : "Some Observations on the Morphology of Chinese Walled Cities", *Annals of the Association of American Geographers*, 60, pp. 63-91.

CHANG, S.D. (1974) : "Peking", in *Encyclopaedia Britannica*, Macropaedia, 14, pp. 1-14.

CHANG, S.D. (1976) : "The Changing System of Chinese Cities", in : *Annals Association of American Geographers*, 66, pp. 398-415.

CHEN, N.R. (1966) : *Chinese Economic Statistics. A Handbook for Mainland China*, Edinburgh.

CHEN, C.S. (1973) : "Population Growth and Urbanization in China", 1953-1970, in *The Geographical Review*, 63, Pp. 55-72, p. 30.

CRESSEY, G.B. (1955) : *Land of the 500 Million. A Geography of China*, New York, Toronto, London.

DAVIS, K. *et al.* (1959) : *The World's Metropolitan Areas*, International Urban Research, Berkeley, Los Angeles (International Urban Research Institute of International Studies University of California, Berkeley).

China Official Annual Report 1982/83, Hong Kong.

DICKINSON, R.E. (1956) : *City Region and Regionalism*, London.

ENOCOMIC INFORMATION & AGENCY (1982) : The 1982 Population Census of China (Major Figures), Hong Kong (cited as : 1982 Population Census).

ESCAP *et al.* (Ed.) (1982) : Physical Profile of Cities in the ESFAP Region, Yokohama (cited as : ESCAP).

FEETHAM, R. (1931) : *Report of the Hon. Mr. Justice Richard Feetham, C.M.G, to the Shanghai Municipal Council*, I-III, Shanghai.

FUNG, K.I. (1981) : "The Spatial Development of Sanghai". HOWE, Ch. (Ed.), *The Development of Sanghai since 1949*, Cambridge and New York, pp. 269-300.

FUNG, K.I. (1281) : "Urban Sprawl in China : Some Causative Factors", in : MA, L.J.C./HANTEN, E.W. (Eds.), *Urban Development in Modern China*, Boulder/Col., pp. 194-221.

GILBERT, A./GUGLER, J. (1982) : *Cities, Poverty and Development Urbanization in the Third World*, Oxford.

HOTTES, K.H. (1950-54) : Die Zentralen Orte im Oberbergishen Lande, Kol n, Remagen (Forschungen zur Deutschen Landeskunde, 69).

HOWE, Ch. (1971) : "The Level and Structure of Employment and the Sources of Labour Supplyin Sanghai", 1949-57, in : LAWIS, J.W, (Ed.), *The City in Communist China*, Stanford/Cal., pp. 215-234,

KUNHLER, J. (1976) : "htadternewrung in der VR China", in : KUNHLER, J. B./MULLER, M./TOMMEL, I., *Stadtendwicklungsprozeb—Stadtenwicklungschancen : Planning in Berlin, Bologna und in der VR China*, *Göttingin*, pp. 137-218 (Geographische Hoch-schulmanuskripte, 3).

KWOK, R.Y.W. (1981) : "Trends in Urban Planning and Development", in : MA, L.J.C./HANTEN, E.W. (Eds.), *Urban Development in Modern China*, Boulder/Col., pp. 147-193.

LO, C.P./PANNEL, C.W./WELCH, R. (1977) : "Land Use Changes and City Planning in Shenyang and Canton", *Geographical Review*, 67, pp. 268-283.

LO, C.P./WELCH, R. (1977) : "Chinese Urban Population Estimates", *Annals Association of American Geographers*, 67, No. 2, pp. 246-253.

MA, L.J.C. (1981) : "Introduction : The City in Modern China", MA, L.J.C./ HANTEN, E.W. (Eds.), *Urban Development in Modern China*, Boulder/ Co'., pp. 1-18.

MA, L.J.C. (1981) : "Urban Housing Supply in the People's Republic of China", in : MA, L.J.C./HANTEN, E.W. (Eds.), *Urban Development in Modern Ghina*, Boulder/Col., pp. 222-259.

MISRA, B. (1981) : Delhi : "Shaping a Metropolitan Capital—Legacy and Future", HONJO, M. (Ed.), *Urbanisation anul Regional Development*, Nagoya, pp. 227-256.

MURPHEY, R. (1953) : *Sanghai : Key to Modern China*, Cambridge, Mass.

MURPHEY, R. (1980) ; *The Fading of the Maoist Vision : City and Country in China's Development*. New York, London, Toronto.

NISSEL, H. (1977) : Bombay. *Untersuchungen zur Struktur und Dynamik einer indischen Metropole*, Berlin (Berliner Geographischen Studien, 1).

ORLEANS, L.C. (1972) : *Every Fifth Child : The Population of China*, London.

PANNEL, Cl. W. (1981) : "Recent Growth and Chahge in China's Urban System", in MA, L.J.C./HANTEN, E.W. (Eds.), *Urban Development in Modern China*, Boulder/Col., pp. 91-113.

RAZA, M./HABIB, A./KUNDU, A./AGGARWAL, Y. (1981) : "India : Urbanization and National Development", HONJO, M., *Urbanization and Regional Development*, Nagoya, pp. 71-96.

SCHINZ, A. (1983) : Fengtian—Mukden—Shanyang. Wandlung einer chinesischen Großstadt seit der Forschungsreise F.V. Richthofens, 1869, Geowissenschaften in unserer Zeit, 1983, 1/6, pp. 198-209.

SCHOLLER, P. (1953) : "Aufgaben und Probleme der Stadtgeographie", *Erdkunde*, 7, pp. 161-184.

SHIGER, A.G. (1957) : *Administrativno—terriorial'noe delenie zarubezhnuikh stran* : spravochnik. Vtoroe, ispravlennoe izdanie (The Administrative-territorial divisions of foreign countries. Second revised edition). Moskva.

SINGH, R.L. (1964) : *Banglore. An Urban Survey*, Varanasi.

SKINNER, G.W. (1978) : "Vegetable Supply and Marketing in Chinese Cities", in : *The China Quarterly*, 76, pp. 733-793.

STATE STATISTICAL BUREAU, PRC (Ed.) (1982) : Statistical Yearbook of China—1981, Hong Kong.

THE COUNCIL OF INTERNATIONAL AFFAIRS, CHUNG-KING (Ed.) (1968-Repr.) : The Chinese Yearbook 1935/36 and 1943 (Sixth Issue), Shanghai, Nendein/Liechtenstein (cited as : Chinese Yearbook 1935/36 and 1943.

TIEN, H. Yuan (1973) : *China's Population Struggle. Demographic Decisions of the People's Republic*, 1949-1969.

TREWARTHA, G.T. (1951) : "Chinese Cities : number and Distribution", *Annals Association of American Geographers*, 41, pp. 331-347.

ULLMAN, M.B. (1961) : *Cities of Mainland China : 1953 and 1958*, Washington, D.C. (International Population Reports, Series, p. 95, No. 59). Reprint in : BREESE, G. (Ed.), *The City in Newly Development Countries : Readings on Urbanism and Urbanization*, Englewood Cliffs, C.J. 1969, pp. 81-103.

WHITNEY, J.B.R, (1970) : *China Area, Administration and Nation Building*, Chicago (University of Chicago, Department of Geography, Research Paper, 123).

WOODHEAD, H.G.W. (Ed.) (1923) : The China Year Book 1923, New York (cited as : Chinese Yearbook 1923).

Zhonqquo Baike Naidjian 1982 (1982 Yearbook of the Encyclopaedia of China), Beijing and Shanghai.

Zhonghua Renmin Gonghogno Fen Sheng Dituji (Hanyu Pinyinban) Provincial Atlas of China), Beijing 1977.

ZUKANG, Zhao (1982) : "Local Authorities and Human Settlements Development in Shanghai", *An Outline of the Development of Shanghai Human Settlements*, Shanghai (mimeogr.).

FOUR

ROLAND J. FUCHS

URBAN AND POPULATION DISTRIBUTION POLICIES IN THE ASIAN-PACIFIC REGION

URBAN policies, traditionally viewed as an aspect of spatial, economic or regional development planning can also be treated as one element within a broader set of policies concerned with population distribution and redistribution. From the standpoint of governments in the developing world, population distribution has become the critical population issue replacing the earlier concern with fertility (UN, 1979, 1980).[1] The nations of the Asian-Pacific region are no exception to this global concern with population distribution as a policy issue (Table 4.1). In thirty nations surveyed by the UN, fifteen considered their distribution to be 'extremely unacceptable' and in need of 'radical intervention'; only two—Singapore and Nauru—both small island states, indicated that their population distribution was 'entirely acceptable' with intervention uncalled for.

Why are so many governments dissatisfied with their current population distribution ? There is no single explanation that holds for all nations ; much depends on national circumstances. In some nations the concerns are not primarily urban but instead rural, political or ethnic : there may be perceived a need to transfer agricultural populations from older settled regions to newer frontier regions, to reduce population pressure on fragile environments and ecosystems, to settle border areas for purposes of national security, or to increase national integration through redistribution of ethnic groups. However, in most Asian nations the unprecedented recent and projected growth of urban

TABLE 4.1 : Perception of population distribution problems by governments and policies adopted by world regions—1978

Region	Perception of overall spatial distribution						Policies	
	Largely unacceptable		Unacceptable to some degree		Acceptable		Decelerate	
	(No.)	(%)	(No.)	(%)	(No.)	(%)	(No.)	(%)
Africa	35	70	15	30	0	0	33	66
Latin America	21	78	5	19	1	4	23	85
Asia and Pacific	15	50	13	43	2	7	15	50
Western Asia	1	8	8	67		25	6	30
Europe and North America	1	3	25	64	13	33	24	62
More developed regions	5	12	24	57	13	31	25	60
Less developed regions	68	59	42	36	6	5	76	66
Total	73	46	66	42	19	12	101	64

Regarding internal migration						Policies to modify					
Reverse		Accelerate		Maintain		Both rural and urban		The rural configuration		The urban configuration only	
(No.)	(%)	(No.)	(%)	(No.)	(%)	(No.)	(%)	(No.)	(%)	(No.)	(%)
6	12	0	0	11	22	19	38	18	36	1	2
0	0	1	4	3	11	13	48	5	19	3	11
8	27	1	3	6	20	12	40	8	27	2	7
1	8	1	8	4	33	6	50	1	0	0	0
5	12	0	0	10	26	7	18	4	10	15	38
6	14	0	0	11	26	6	14	4	10	18	43
14	12	3	3	23	20	51	44	32	28	3	3
20	13	3	2	34	22	57	36	36	23	21	13

Source : Ed. : compiled and modified from *World-Population Trends and Policies : 1979 Monitoring Report, Vol. II, Population Policies*, New York : United Nations, 1980, pp. 49, 116-20.

populations, and the concentration of much of the urban growth in a single metropolitan region, raising problems of urban management and possible political, social or ethnic instability, have led to the adoption of various policies intended to slow the growth of urban population or to affect the configuration of urban settlements.

Some indication of the importance of these factors to nations within the region can be gleaned from a UN survey of the objectives of distribution policies already adopted (Table 4.2).[2] The deceleration or reversal of rural-to-urban migration trends is apparently the most common goal : some sixteen nations have policies to decelerate flows from rural to urban areas. Another six, including several of the Communist nations, have gone further and are attempting to reverse flows. The second major objective is to alter rural population distribution, in many cases through colonization or resettlement schemes (which may also involve rural service centres). This goal was adopted in twenty nations. The third objective, altering urban configuration, usually through controls on primate city growth and development of small and intermediate-size cities, was adopted by thirteen nations. While this objective, specifically concerned with what are commonly considered urban policies, ranks lowest among the objectives cited, it should be kept in mind that the general objective of slowing or reversing urban growth, also an urban policy, is the most common objective of population distribution policies.

If the stated goals of individual Asian nations are examined in greater detail (Table 4.3), it is apparent that they reflect to some extent differences in national size, inherited patterns of settlement, resource endowment and economic activity, as well as differing political perceptions of problems and needs. Commonly, nations have adopted multiple goals involving modification of both rural and urban migration trends and aspects of rural and urban settlement patterns.

This paper will review the urban-related policies already adopted by Asian and Pacific nations in response to their perceived problems of population maldistribution. The first part of the paper provides a very general overview of programmes and their effectiveness. The second part of the paper, considers apparent policy deficiencies and suggests research needs and other

TABLE 4.2 : **Policies regarding spatial distribution of population adopted by Asian and Pacific governments**

Region and country	Policies regarding basic trends in internal migration				Policies regarding modification of rural and urban configuration of settlement	
	Accelerate	No Intervention	Decelerate	Reverse	Alteration of rural configuration	Alteration of urban configuration
1	2	3	4	5	6	7
China				X	X	X
Japan			X		X	X
Other East Asia						
Democratic People's Republic of Korea			X		X	X
Mongolia			X		X	X
Republic of Korea			X			
Eastern South Asia						
Burma			X		X	
Democratic Kampuchea				X	X	X
Indonesia					X	
Lao People's Democratic Republic			X	X	X	X
Malaysia		X			X	
Philippines				X	X	
Singapore		X				

TABLE 4.2 : (*Contd.*)

1	2	3	4	5	6	7
Thailand			X			
Vietnam			X		X	X
Middle South Asia						
Afghanistan		X			X	
Bangladesh				X	X	
Bhutan	X				X	X
India			X		X	X
Iran			X		X	X
Maldives			X			
Pakistan			X		X	X
Nepal			X		X	
Sri Lanka			X		X	
Australia and New Zealand						
Australia				X		X
New Zealand		X				
Melanesia						
Papua New Guinea		X				
Micronesia-Polynesia						
Fiji			X		X	X
Nauru		X				
Tonga			X			
Western Samoa		X				

Source : UN (1979 : 134),

TABLE 4.3 : **Population distribution policies in selected Asian Nations : Goals, programmes and measures**

Country	Goals	Programmes and measures
Centrally Planned		
China (PRC)	Slow rural-to-urban migration ; reduce growth and decongest large cities ; improve balance among regions and in urban hierarchy ; develop agricultural and mineral resources in interior	Family planning ; rustication programme ; development of new towns and growth centres ; rural development programmes ; land colonization
Mongolia	Facilitate rural-to-urban flow ; increase number of cities and lower-order centres	Settlement of nomadic population ; development of lower-order centres as industrial complexes
Vietnam	Reduce size of Ho Chi Minh City ; transfer labour force from urban to rural areas ; improve urban-rural and regional balance	Rustication of urban population ; colonization of New Economic Zones ; development of agricultural green ring around Ho Chi Minh City
Market/Mixed		
China (Taiwan)	Decentralize population within Taipei and Kaohsiung ; improve balance among regions and centres in hierarchy	Satellite towns ; development of growth poles through infrastructure investment and tax incentives ; improvement of social infrastructure in regions of out-migration
Indonesia	Reduce imbalance between Java and outer islands ; reduce primary and uneven urban growth	Transmigration programmes ; rural and urban development in outer islands ; restriction on Jakarta's growth ; distribution of industrial activities
Republic of Korea	Control growth of Seoul ; reduce imbalance in urban hierarchy ; slow rural depopulation	Long-term physical plan for Seoul, including green belt ; legislation for industrial location ; fiscal and tax incentive/disincentives ;

TABLE 4.3 : (*Contd.*)

Country	Goals	Programmes and measures
		construction of satellite cities ; development of five growth poles ; rural development through Saemaul Undong
Malaysia	Equalize regional development and distribution of economic benefits among ethnic groups ; promote growth in Sabah and Sarawak ; slow growth of major towns	Rural development (extension, credit, marketing, fertilizer subsidence, infrastructure) ; colonization and resettlement (FELDA) ; promotion of industrial development in low-income states (location incentives) ; growth centres ; urban development and renewal, housing
Nepal	Reduce regional disparities between rural population and resources ; reduce rural-migration ; promote urban growth outside Kathmandu Valley	Integrated rural development in hill areas ; improved transport and communication systems ; development of small-scale industries in non-urban areas ; growth centres ; land resettlement
Pakistan	Reduce rural-to-urban migration ; mitigate concentration of urbanization in a few, large cities and resulting problems of housing, services, unplanned development	Population planning ; rural development including increased agricultural productivity, agro-industrialization, and services ; more balanced development between rural and urban sectors ; and between towns ; improved urban services
Philippines	Reduce rural-to-urban migration ; reduce concentration in Manila ; promote balanced urban hierarchy and more even regional development	Reduction in population growth ; integrated rural development ; development of small- and medium-sized cities through industrial dispersion and improvement of basic infrastructure and regional planning

TABLE 4.3 : (*Contd.*)

Sri Lanka	Reduce metropolitan growth and rural-to-urban shift ; promote rural resettlement	Rural resettlement ; establishment of industrial and economic activities outside metropolitan area ; welfare and income transfers to reduce rural-urban disparities ; emphasis on provision of social services outside metropolitan core
Thailand	Limit growth of Bangkok ; develop regional growth centres ; promote intraregional rather than interregional migration	Decentralization of growth in Bangkok metropolitan area through new port, industrial, and government dispersal, taxes, land use controls ; growth centre policies ; regional planning ; rural development

Source : UNPD and UNFPA (1979, 1981) ; Pryor (1979) ; Fuces and Street (1979) ; NESDB (1977).

possible avenues of policy improvement.

POLICY PROGRAMMES AND THEIR EFFECTIVENESS

While individual nations have obviously adopted a variety of policies and measures in response to perceived problems of migration, urbanization and population maldistribution, these programmes, which include both those intended to slow or reverse rural-to-urban migration and to change urban configurations, may be categorized as follows (see Table 4.4) :

—Rustication programmes, designed to resettle urban residents in rural areas.

—'Closed city' programmes, designed to constrain metropolitan growth by stopping or slowing down in-migration.

—Programmes designed to accommodate metropolitan growth by improving the urban habitat. There are many such programmes : two with clear distribution implications are programmes to improve housing and living conditions, particularly in slum and squatter settlements, and programmes aimed at increasing urban efficiency by decentralizing growth in metropolitan areas through promotion of dormitory towns, satellite cities, and commuting.

—Various programmes aimed at dispersion of urban growth through expansion of intermediate-size cities and regional centres.

—Rural development programmes directed at retention of rural populations and the growth of rural service centres.

As is evident from the above classification, population redistribution programmes are often subsidiary components of more general spatial programmes dealing with modernization and economic development, with the importance of the population redistribution component varying significantly from programme to programme.

The widespread adoption of spatial population policies in the Asian and Pacific region has not yet been followed by rigorous evaluations of their effectiveness, in part because of the

TABLE 4.4 : Population redistribution programmes, objectives and instruments

Type of programme	Migration/mobility objectives	Instrument commonly employed
Urban constraints		
Closed city	Reduce or slow in-migration to designated cities, including the metropolitan centre	Tax disincentives ; identity cards or internal passports, residence permits, registration of addresses ; limitations on investment in industry or housing ; discriminatory treatment in access to services by non-legal residents ; eviction of illegal residents, destruction of squatter housing
Rustication programmes	Resettle urban residents in rural areas	As above plus the following : assignment of rural residence and work place ; making ration coupons valid only in authorized place of residence
Accommodationist programmes		
Slum and squatter settlement improvement	Accommodate to existing patterns of urban in-migration and growth by improving urban habitat, especially housing and related services	Legitimization of tenure in squatter settlements ; upgrading of services and utilities ; provision of prepared sites and basic services prior to occupancy
Dormitory towns and satellite city programmes	Deconcentrate growth within metropolitan area by developing settlements in periphery	Infrastructure investment in metropolitan periphery ; development of public transport and commuting facilities ; housing project developments in peripheral settlements ; industrial relocation grants and subsidies ; zoning and other controls on further development in core

TABLE 4.4 : (*Contd.*)

Type of programme	Migration/mobility objectives	Instrument commonly employed
Promotion of growth in intermediate-size cities and regional centres	Channel migration to, and stimulate retention of, population in intermediate-size cities as alternative to metropolis	Infrastructure investments ; incentives (grants, loans, subsidies) to employing organizations, particularly industrial firms ; indirect incentives to individual migrants (housing developments, education, and medical service provision) ; direct incentives to individuals (job training and relocation grants)
Rural development		
Land colonization schemes	Resettle rural residents from overpopulated areas to new or underutilized agricultural areas	Infrastructure investments for land clearing, roads, dams, etc. transfer payments for land acquisition ; transfer of title ; investment in productive facilities ; provision of social services or technical assistance ; credit provision ; establishment of managing agency
Integrated rural development	Retain rural population and develop rural service centres	Land reform ; provision of credit and extension services ; physical infrastructure investments, including roads ; development of marketing network ; vocational training and education ; expansion of off-farm employment ; creation of rural market towns

Source : Findley (1977) ; Demko and Fuchs (1981) ; Bahrin (1981) ; Lacquian (1981) ; Simmons (1981) ; Hansen (1981) ; and Findley (1981).

substantial difficulties involved in such evaluations. The goals of population distribution policies are often imprecise and the programmes embedded in more general spatial development programmes, complicating the choice of criteria for determining effectiveness. The appropriate time periods are unclear. As with all policy evaluation, an *ex post facto* approach creates difficulties in controlling variables other than policy instruments and in isolating policy-induced effects from autonomous trends.

Nevertheless, some impression of the effectiveness of existing policies can be obtained from recent general reviews of population distribution policies in Asia (Oberai 1981 ; Day and Demko 1981 ; Laquian, Aquino and Postrado 1981). These reviews synthesize a considerable number of country or programme-focused studies, the great majority of which are descriptive assessments. The major conclusion of these reviews is that population distribution policies in the Asian region have had but limited effectiveness, a conclusion similar to that found in global reviews (Findley 1977 ; Gosling and Lim 1979 ; Demko and Fuchs 1981) and in reviews of policies implemented in other regions (Fuchs and Demko 1979 ; Abumere 1981 ; Mabogunje 1981).

Although the overall effectiveness of population distribution policies is limited, the experience varies considerably depending on the type of programme involved :

Rustication Programmes

Programmes intended to transfer urban population to rural areas have been adopted by several of the Communist nations in Asia. Judged only demographically, they have been successful in accomplishing major population transfers in brief periods. In China it is estimated that between 10 million and 15 million urban secondary school graduates were resettled in rural areas in the period 1969-73 alone (Chang 1975). In Vietnam the population of Ho Chi Minh City, some 3 million in 1975, was reduced 7,00,000 in 1977 and further annual reductions of 5,00,000 were planned, to meet an eventual city size target of 1 million (Lane and Kolb 1980).

While the demographic effects of these policies have been striking, the social and economic costs, including the effects upon rural destinations, remain unknown. It would appear that urban problems of un-and under-employment were simply shifted to rural areas (Simmons 1979 ; 1981). Also, to be effective such programmes require stringent laws and administrative measures, massive propaganda and authoritarian political regimes. Because of this, and their restriction of free movement and choice of residence and employment, considered civil rights in many countries, rustication programmes are unlikely to be adopted outside of the centrally planned economies and hold little interest for nations with other economic and political systems.

Closed-city Programmes

These represent attempts to limit primate city growth through legal and administrative measures directly or indirectly affecting inmigration. Major examples in Asia include : the attempts to limit inmigration to Jakarta through registration, residence permits, and related controls on the informal sector ; 'to' Manila through discriminatory education fees and spatial industrial controls ; and to Seoul through various taxes and land use planning (Simmons 1979). The experience suggests that although such programmes may slow in-migration, they may also create undesirable side effects. Administrative measures are often evaded through bribery. Limits on growth within the administratively defined 'closed city' are often accompanied by displacement of growth to adjoining areas with increases in commuting from and migration to these peripheral areas of the metropolis. Even if successful, the programmes may result in undesirable demographic (e.g., aging) or economic consequences for the 'closed' city.

Despite these problems, which might have been anticipated had planners and policymakers been aware earlier of the outcome of similar programmes in the USSR and Eastern Europe (Fuchs and Demko 1979 ; Houston 1979), these programmes should not be entirely dismissed. For example, elements of Korea's attempts to control the growth of Seoul—restrictions on new construction, residence tax, the removal of state-run firms and public agencies, and various measures to limit the growth of higher-education facilities in Seoul—warrant closer examination, and with modification may be suitable for transference elsewhere (Kim and Donaldson 1979 ; Song and Choe 1981).

Urban Accommodationist Programmes

In contrast to policies designed to reverse or constrain metropolitan growth, accommodationist policies attempt to adjust to inmigration and growth by improving the urban habitat. Included under this rubric is a range of programmes concerned with housing (including slum and squatter settlements), transport and communication, sanitation, government and fiscal management, and also attempts to decongest the metropolitan areas by developing dormitory towns and satellite cities (Laquian 1981 ; Linn 1979). Such programmes, currently in favour with international donor agencies and many national governments, are intended to promote urban efficiency, but raise issues of equity : do overall national welfare and development needs justify the large investments involved, which generally reach only a small proportion of the population ? In addition, the possible effect of such programme in encouraging further migration to the city, which would only aggravate the problems they were intended to relieve, requires further study.

The need is for ameliorative programmes that will increase urban efficiency without attracting excessive inmigration. Especially promising from this respect are tax and pricing policies that would pass along costs to users and reflect the agglomeration diseconomies of larger metropolises (Linn 1979). However, resistance from the politically powerful elite resident in the metropolis may prove an insurmountable obstacle in many nations. Another approach that holds promise is improved physical and land use planning for larger cities. In many Asian cities current planning rests on assumptions and models derived from Western experience which may be irrelevant. There is thus an obvious need for further study of the internal structure, including the spatial functional and activity patterns, of Asian cities as a prelude to improved planning (McGee 1981).

ATTEMPTS TO PROMOTE GROWTH IN SMALLER AND INTERMEDIATE CITIES

Urban programmes of 'concentrated decentralization' (Rodwin 1970) have been a component of many Asian development plans. In Asia, as elsewhere, such programmes initially focused on growth centre policies that involved the establishment of large-scale manufacturing enterprises producing import substitutes generally employing capital-intensive methods. The early optimism regarding such urban dispersal programmes gave way to widespread pessimism as the programmes failed to achieve expected growth in the centre and desired spread effects (Hansen 1981).[3] It is likely that, in part, the shortcomings may have resulted from failures to implement policies that remain only political pronouncements, as apparently occurred in much of South-east Asia (Salih *et al.* 1978 : 29). Also policy application was often sporadic with inadequate time allowed for the expected results to be achieved.

Since there are no clear alternatives to accommodating the enormous increases that will take place in Asian urban populations, other than permitting growth of metropolitan populations to reach unmanageable proportions, interest has revived in programmes intended to stimulate the growth of intermediate cities. However, it is now recognized that to be effective, intermediate city policies require integration with more comprehensive and balanced sets of urban and rural programmes (Lo and Salih 1978a) and that strategies must involve more than physical infrastructure and industrialization (Rondinelli 1983). Nevertheless, serious questions remain as to the precise objectives to be sought, the selection of appropriate centres, the measures to be employed, and the timing in initiating such programmes (Richardson 1977).

Rural Development Programmes

Disenchantment with the outcome of industrially based growth-centre approaches, growing concern about the need to give greater attention to the rural sector, and the desire to reduce rural-to-urban migration have led to an increased emphasis in rural development programmes. So-called integrated rural development programmes ordinarily provide for co-ordinated provision of infrastructure and services in rural regions. Although located in primarily rural regions, many such programmes also are intended to affect urban areas by slowing rural-urban migration and by injecting urban functions into rural areas through the development of an appropriate regional hierarchy of marketing and service centres (Rondinelli 1979). Examples in Asia include the Lampang project of Thailand, the

Bicol project in the Philippines, and various rural development projects in the hill areas of Nepal. The relative recency of such projects has precluded definitive evaluation of their effects upon migration, which appear to vary depending on the precise programme components and the nature of the rural region and its population. Although the programmes appear to have reduced permanent out-migration in many cases, "... a more significant though little-publicized effect ... is the opening up of other migration and commutation alternatives : to other rural areas ; to towns in the vicinity ; and to other cities economically linked to the rural market towns" (Findley 1981 : 166). As with intermediate cities, existing approaches to planning the development of rural service centres (Rondinelli 1979 ; Rondinelli and Evans 1983), would seem inadequate in guiding policymakers as to appropriate choices of centres, programmes, projects and their timing.

Not only does the experience of redistribution programmes in affecting migration or distribution greatly vary by types of programmes, as noted above, but also programmes of a similar type greatly vary in outcome from nation to nation (Lacquian *et al.* 1981). Such national differences suggest that details of programme design, administrative structure and capabilities, political commitment, and so on, play a major role in determining the outcomes. Even allowing for such variation, however, it is apparent from transnational comparisons that the majority of such programmes have been considered only fair to poor when evaluated on the basis of effectiveness, efficiency or costs. In view of the considerable expenditures on such programmes in Asia and the Pacific, and elsewhere in the developing world, this generally negative assessment raises the question of what might be done to make them more effective and cost-efficient.

FUTURE PROSPECTS

Unless some fundamental change occurs in the nature and implementation of population distribution policies, including urban policies, it is unlikely that future outcomes will be more successful than those experienced to date. What are the implications of past experience for research needs, desirable policy changes and institutional implementation? Some speculations are offered here concerning the scope and institutional framework of population distribution policies, the need for intervention, selection of appropriate instruments, policy evaluation and the political and institutional framework.

Assessing the Need for Intervention

Since the vast majority of Asian and Pacific nations have already adopted policies designed to modify the rates and patterns of internal migration and population redistribution, further assessment of the need for intervention may seem unnecessary. Nevertheless, there is a continuing debate as to the need for such policies, which serves to weaken the political commitment to and effective implementation of the policies within the individual nations and to diminish the financial commitment of donor agencies. More accurate assessment of whether intervention is required in particular nations could conceivably remove or reduce intervention where it seems unnecessary and strengthen the political and economic commitments to intervention in those cases where it appears warranted.

General evaluations of the arguments against and for intervention have appeared elsewhere (Mera 1981 ; Stohr 1981) and need not be elaborated here. Essentially, the case against intervention rests upon the automatic equilibrating mechanisms postulated under neo-classical economies, leading over time to equilibrium in the terms of trade and to spatial equilibrium in factor prices. Empirical support consists of various cross-national and longitudinal studies demonstrating convergence over time, as development proceeds, in personal and regional income disparities (Kuznets 1955 ; Williamson 1965 ; Alonso 1980) accompanied by reduced urban primacy (El-Shakhs 1972).

The arguments for intervention essentially are that the assumptions of neoclassical economic theory and its automatic equilibrating mechanisms do not apply to most developing countries and, even if they did, few nations could afford to wait for the indefinite time required for equilibrium. Intervention is, therefore, required to bring population distribution into better accord with resource and ecological patterns and to relieve political and social pressures (Stohr 1981). An additional set of arguments in favour of intervention is advanced by those scholars who, on the basis of their own empirical studies, have challenged the validity of the Kuznets and Williamson hypothesis of personal and regional income convergence (Gilbert and Goodman 1976 ; Wright 1978 ; Krebs 1982) or have raised formidable arguments against very large cities (Gilbert 1976). Also, Richardson has recently argued that population distribution policies and urban development strategies are more justified in developing countries as opposed to developed countries, since in the former patterns are less ossified, have generated conditions conflicting with national development goals, and are needed to offset implicit spatial impacts of macro-and sectoral policies (Richardson 1978 ; 1982).

Further, general assessment of the need for intervention will, no doubt, proceed through elaboration of theoretical and comparative studies. From the standpoint of determining the need for intervention in a given country, however, more fruitful approaches may be suggested. Particularly useful would be studies examining the demographic and socio-economic effects on both origin and destination areas of existing and projected mobility patterns. Comprehensive research that examined demographic consequences (e.g., on fertility age and sex structure) and the social and economic costs and benefits to individuals, households, the communities in source and destination areas—covering the full range of mobility, not only migration—would add enormously to an accurate assessment of intervention needs (Goldstein 1981 ; Simmons 1981 ; Hugo 1981b).[4] It would also serve to remedy a major weakness of the neoclassical approach, which equates migration with mobility and, in treating migration as simply a movement of a production factor, commonly ignores social costs borne by the migrants, as well as by sending and receiving areas.[5]

Goal Definition

There are grounds for concluding that the mode of specifying objectives or goals is currently a major weakness in the population distribution policy process, and that the establishment of unrealistic or inappropriate goals may be a fundamental cause of the limited success achieved by nations attempting to manipulate their population distribution (Richardson 1981). Some of these difficulties and suggestions for their amelioration follow.

First, it should be evident that in formulating population distribution goals, universal prescriptions cannot be employed. Such goals must be formulated on the basis of unique national conditions, including the size of a country, its environmental and resource base, level of development, the structure and spatial patterns of the economy, the dynamics and spatial patterns of the population, the structure of its society, the form of government and cultural and historical factors. That governments with widely differing national contexts have often adopted very similar goals, such as deceleration of rural-urban migration, dispersed urbanization and the limiting of metropolitan growth, raises the suspicion that, rather than devising nation-specific goals, planners often resort to 'borrowing' goals, or adopting those currently fashionable. The widespread emphasis on reducing rural out-migration would seem particularly inappropriate, at least on economic grounds, in view of considerable evidence of gains accruing to migrants through the development process (Stark 1982) and the evidence that individual nations may in fact be underurbanized (Pernia 1977). An urgent and obvious remedy is for planners to formulate goal based on the distinctive national context.

Second, there are no operational algorithms or other technical means of determining 'optimal' distributions of populations. Existing approaches require data beyond the capacity of developing nations to acquire, or result in indeterminacy (Conroy 1978 ; Willekens 1979 ; Tan 1980). Richardson's (1975) call for cost-benefit assessment of alternative settlement patterns has not yet resulted in operational examples. Planners must, therefore, accept the fact that for the foreseeable future, the definition of goals will remain largely judgmental. Thus, it is especially important to ensure that goals are the result of informed judgment and not arbitrary or wishful thinking.

Third, as Alonso (1972) pointed out a decade ago, spatial population goals, or spatial economic goals, are lower-order and not ultimate goals ; they must be formulated as subsets of national economic or social goals, which typically include economic efficiency and growth, social equity and welfare, environmental quality and other standards of well-being. How these national goals are to be translated into spatial population goals, however, is not very clear. Although the concepts of an optimal city size and an optimal rank-size hierarchy have been widely discredited and the enormous literature on national city-size distributions has contributed little in the way of findings useful for either academic understanding or policy (Carroll 1982), substantial and important controversy still centres on the question of the relative efficiency of various sizes of urban places. Useful here would be further empirical research in individual developing countries, perhaps applying Lo and Salih's (1978b) hypotheses that efficiency will vary with urban function as well as size (comparative sectoral efficiency), and incorporating social costs and benefits if possible. Since labour absorption will become an acute issue in the future for many Asian and Pacific nations, studies on the labour absorptive capacity within these nations in the rural and urban sectors, regionally and by settlement size would be a highly useful prologue to goal formulation. Similarly, further study of labour markets and clarification of the spatial relationships among labour supply, job opportunities, income and mobility is an urgent need (McGee 1981 ; Standing 1981).

Fourth, planners have formulated goals largely on the basis of assumed national and social benefits. Personal preferences, of course, may differ from the presumed social preferences, leading to migration and location choices that differ from planning formulated goals. Survey research on individual residential and locational preferences (e.g., Fuguitt and Zuiches 1973) would be desirable if we believe that planners' goals should to some degree reflect the wishes of the people. Similarly, survey research on intended migration and mobility, which has been demonstrated as useful for prediction in developed nations (Speare, Goldstein and Frey 1976), might assist in formulating more realistic goals. Moreover, survey research on the actual economic and social costs and benefits accruing to various types of migrants to different types of destinations (e.g., metropolis versus secondary cities) would prove useful in translating national efficiency and welfare goals into urban and population distribution goals.

Fifth, in the absence of comprehensive mobility data, planners have concentrated almost exclusively on permanent migration as a subject of policy, excluding short-term and repetitive movements. This is a serious oversight since in many Asian and Pacific nations, non-permanent forms of movement already form a large and growing part of total mobility patterns and have important economic consequences (Goldstein 1978). Goals are, therefore, less comprehensive than they should be, omitting aspects of mobility (e.g., seasonal labour movements and commuting) that may be highly relevant to the spatial development process. As a result, the interdependence among the various elements of mobility, and the degree to which one substitutes for another, are overlooked. This oversight argues for survey research to determine the relative incidence, patterns, economic significance and linkages of the full spectrum of mobility processes in individual nations.

Sixth, the time periods involved in the mobility response to economic and other stimuli is unknown in most nations. This creates difficulties in formulating population distribution goals, their articulation with economic goals and development projects, and their incorporation into short-, intermediate- and long-term development plans. Research to clarify the time periods of mobility response is thus also an obvious need from the standpoint of goal formulations.

IMPROVEMENT OF INSTRUMENT SELECTION

The various types of urban and population redistribution programmes constitute a package of individual measures or instruments that, taken as a group, are intended to achieve the programme objectives. In selecting individual instruments, policymakers can choose from a broad range of economic, social and administrative measures (Table 4.5). Such measures can be characterized as either positive, in the sense of offering incentives for inmigration to a particular locale, or negative, if they serve as disincentives or obstacles. They may be aimed directly at the individual or household, but more commonly operate indirectly through employing organizations. Most commonly employed are the economic measures aimed at the employing organizations, which in effect attempt to spatially manipulate labour demand. Measures aimed at individuals or households which attempt to spatially affect labour supply are less commonly employed. The emphasis on labour demand, it should be noted, would seem quite appropriate in view of the clear evidence that job opportunities are the primary determinants affecting migration

TABLE 4.5 : **Major population distribution measures**

Directed toward	Policy Emphasis: Incentives	Policy Emphasis: Disincentives
Employing organizations	Direct government investment and location of state-owned enterprises, offices, and facilities; subsidies (grants, loans, tax rebates) to private sector on location basis; spatial industrial infrastructure policies (transport, utilities, etc.); government, procurement, and location policies; transport rate adjustments	Relocation of government enterprises, offices, facilities; restrictions, permits, licenses, tax surcharges for private-sector locations of industry and offices; land use or indirect controls on location
Individuals and families	Social infrastructure investments (housing, education, medical services, etc.); mobility grants, allowances, and loans; employment or other information agencies; job training and human resource development programmes; various rural development measures—land reform, credit, extension services, public works	Administrative and legal measures (residence and work permits); discriminatory treatment of legal and illegal residences in access to services; slum clearance and squatter resettlement measures; residential zoning limitations

Source : Fuchs (1981b).

and the labour supply (Yap 1975 ; Todaro 1980).

To date there has been no comprehensive and systematic evaluation of the effectiveness of the individual instruments, whether aimed at labour demand or supply and such knowledge as we have come largely from studies in developed countries, raising the question of transferability.

Richardson (1982) has noted many of the apparent deficiencies in implementing policy instruments to spatially affect labour demand in LDC's. There has been a heavy reliance on public infrastructure provision and subsidies but such measures suffer from primate city bias, standards too high relative to capital availability, and allocation too far in advance of need. Grants, loans and tax incentive programmes are more often used to promote industry *per se* rather than its redistribution, have a capital-intensive bias, and may offer too little assistance to overcome attractions of the primate city. Transport subsidies have little impact because transport costs form a small portion of total costs, even for most industries in isolated locations. Most promising he suggested are programmes combining locational incentives, including labour subsidies, in the periphery, with a tax on industrial development in the core region, possibly on a self-financing basis. What appears to be needed from a research perspective, in addition to evaluation of existing programmes, is further research on the behaviour of small-scale industrial and business entrepreneurs who loom large in the economic structure of small and intermediate cities in most LDC's, to determine what incentives they would find most appealing.

In regard to the measures taken to affect labour supply, other than job creation, there would seem to be considerable scope for improving the effectiveness of these policy instruments based on our current knowledge of migration determinants (Fuchs and Demko 1981). Some specific problem areas and suggestions follow :

There appears to be too heavy reliance upon social infrastructure, for example, housing, schools, and medical and other services. With the possible exception of education, it is doubtful that these amenities are important as migration determinants. Moreover, the measures may be so widely applied over a country as to negate any desired spatial demographic effects.

The known effect of distance as a deterrent to migration, however, is largely neglected in current instruments and programmes, except for limited use of relocation grants. A possibly more useful approach would be to address redistribution programmes geographically—toward likely migrants from source areas near the proposed destination area, thereby taking advantage of the distance bias.

Migrant selectivity, which plays a major role in the composition of regional flows, also needs to be taken into account to a greater extent then at present. Programmes and measures should be addressed toward sub-groups most likely to move or stay, as called for by particular programmes and objectives.

Significance of information in including and directing migration is similarly overlooked in most redistribution programmes and measures. In the absence of formal information programmes, migration streams become biased towards the metropolis, which receives the most media attention and is likely to have the largest pool of migrants informally transmitting information to friends and relatives in source areas. Incomplete or misleading information as to the actual state of job opportunities and amenities over national space appears to be a primary factor leading to maldistribution of migration in LDCs (Jones 1979). The use of information measures—through schools, government offices, and the media—is thus potentially a powerful tool for affecting both decisions to move and potential destinations and in bringing about a more rational distribution of population. It would have the advantage of being relatively low in cost and the potential of showing results much more quickly than many economic measures.

In addition to a more rigorous evaluation of existing programmes and instruments and further research on the spatial behaviour of entrepreneurs, various types of additional research would seem desirable to bring about the selection of more appropriate policy instruments.

First, further elaboration of spatial development theory is needed. As Richardson (1973 : 140) noted some years ago, Friedmann's (1966) centre-periphery model is still the 'best formulated construct in a haphazard and fuzzy literature.' Promising directions for improvement of the basic concept have been pointed out by Richardson himself; Friedmann (1973) ; and

Lo, Salih, and Douglass (1981) with their macro-spatial model. Logan (1981) has also called attention to the need to treat national spatial development as an open system subject to trade, investment and other international forces. Work along these directions may confirm the need in spatial development and population planning for intervention in regard to such international factors and thus would add a dimension missing from current programmes.

A second need is for improved methodologies linking spatial, economic and demographic growth models. The current cleavage between these models handicaps both economic and demographic planning. Attempts to link regional input-output and multiregional demographic models seem promising (Gordon and Ledent 1980 ; 1981).

A third need is to reformulate macro-level models of migration to include more variables that are potentially subject to policy intervention. Macro-models of migration, generally based on cross-sectional analysis of census data by territorial units, have provided a broad overview of the spatial determinants of migration, useful to policymakers (Yap 1975 ; Todaro 1980). However, they also have serious limitations which derive largely from limitation of their data sources : the focus on migration and exclusion of other forms of mobility, the limited range of explanatory variables employed, and the serious and biased underestimation of migration that results from use of census temporal and territorial definitions of migration. Despite these limitations, there is still scope for substantially improving their utility for policy purposes by reformulating the variables employed. Such models commonly attempt to explain rates of migration between areas on the basis or wage or income levels, unemployment rates, degree of urbanization and distance between origin and destination. The wage or income and employment variables may be subject to policy intervention, but distance and degree of urbanization are not potential policy variables. Distance should be decomposed into the constructs that it represents in reality and that could conceivably be subjected to policy intervention : travel time and costs, and measures of information or contact. Similarly, degree of urbanization should be reformulated as composite or individual measures of the various urban services and amenities. The migration elasticities derived from such macro-level models, reformulated to accord with policy-relevant variables, should thus be much more useful in the selection and design of appropriate instruments and programmes.

To date, macro-level migration models have been more widely employed than macro-level models for policy purposes. In the future, micro-level studies focusing on the behaviour of individuals and households may prove more useful (De Jong and Gardner 1981). They afford the opportunity for covering the full scope of mobility behaviour and for a deeper and broader understanding of migration determinants than is possible for micro-level studies which must generally infer determinants from place attributes. Whether micro-level studies will achieve their potential for contributing to policy and planning will depend on the degree to which they are structured and designed to answer policy questions. The majority of such studies to date unfortunately have not adopted this perspective and have thus added little of direct use to policymakers (Simmons, Diaz-Briquets, and Laquian 1977).

Policy Monitoring and Evaluation

The limited success of urban and other population distribution policies can be traced in part to the neglect of policy monitoring and evaluation. As indicated earlier, formidable difficulties face those evaluating the effects of such policies (see also Chan 1980). As a result, most studies are descriptive and evaluate by measuring achievement versus targets. This approach is inadequate because targets are often unrealistic and, in any event, it fails to account for what would have happened in the absence of policy implementation or to identify the actual effects of policy.

Various forms of trend or time series analysis are generally employed in public policy evaluation (Nachmias 1979). The difficulty remains of separating autonomous from policy-induced effects. Here quasi-experimental research designs may prove useful. One possibility is to derive a general national regression equation with population growth or net migration rates by areal units or the dependent variable, using a set of socio-economic variables as the independent variables. The equation can then be employed to predict either autonomous population growth or net migration. Policy effects can be estimated by comparing actual and predicted change for regions subject to policy programmes or measures (Folmer 1980).

Another possibility is to use control areas as a basis for trend comparison : areas designated for integrated rural development compared with those not so designated, 'growth' centres compared with intermediate cities not subject to policies, and so on. Selection of appropriate control areas to ensure comparability in socio-economic and demographic structure is critical and can be accomplished through factor analysis, cluster analysis and shift-share techniques (Isserman and Merrifield 1982).

In addition to measuring the overall effects of policies planners will wish, if data sources permit, to study the effects of individual policy instruments in order to modify existing policies or improve future ones. Useful techniques for conducting such research include simultaneous equations and path analysis (Berentsen 1978). Special surveys may also be used for evaluation and, when combined with a battery of the macro-level evaluation methods, may result in quite comprehensive evaluations (see, e.g., Moore and Rhodes 1973 ; 1977).

THE INSTITUTIONAL AND POLITICAL FRAMEWORK

Institutional and political factors may perhaps be the most important source of failures in existing urban and other population distribution policies. The limited success of existing policies reflects in part a failure in many nations to achieve adequate institutional co-ordination (Pryor 1981). Such policies cannot be properly formulated or implemented in isolation but must be integrated with national, economic development plans, including sectoral as well as area plans. Conceivably, various institutional mechanisms can be developed to achieve the required policy integration and no single model need be adopted or would be appropriate for all nations. What is required, however, is that spatial population planning be more closely linked to spatial economic planning and that both become more closely tied to macro-economic and sectoral planning.

The explicit urban and related population distribution policies and programmes discussed earlier account for only a portion of the spatial effects of government policies. Macro-and sectoral economic policies may also have strong, if unintended and indirect, effects on population distribution as well as on economic activity (Richardson 1977). Major areas of concern include various policies regarding foreign trade and exchange, domestic and international investment, taxation, interest rates, agricultural price supports or ceilings, wage regulations, social welfare programmes, government procurement policies and government institutional structures (Renaud, 1979). From examination of selected Asian development plans there is reason to believe that the spatial effects of various sectoral policies may contradict the goals of explicit population distribution policies (Fuchs 1981a). There is evidence in the case of Korea that agricultural pricing policies may have been the most important determinant of rural outmigration rates (Rehnberg 1977).

The lack of a methodology precisely to measure the net effects of various sectoral policies remains an obstacle to a full understanding of their spatial effects and the degree to which they outweigh explicit spatial policies.[6] Further research on the subject is a priority research need (Richardson 1977 ; G. Jones 1978). The challenge for policy purposes, as Richardson has noted, is not only to recognize and identify sectoral policy effects but also to find alternative means of achieving non-spatial objectives with policy instruments that avoid undesirable spatial consequences. This must become a high priority for Asian and Pacific nations in order for them to avoid further wasteful expenditures on contradictory policies.

Also a challenge are the bureaucratic problems of translating promulgation of policy to implementation and the political problem of bringing consistency to policy application in place of the common pattern of on-off policies. This latter problem is a result of the differences between the time horizon needed for policy achievements and the short time horizon of most governments and politicians. Here there may be no solution but to work for policies of wide consensus that will appeal to successive political administrations.

SUMMARY AND CONCLUSIONS

The first part of the paper presented a general overview of population distribution policies in the Asian and Pacific region. The major findings may be summarized as follows :

All but two of thirty Asian and Pacific nations are dissatisfied with their current distribution of population ; half perceive their distribution as 'extremely unacceptable' and in need of 'radical intervention'.

The most common goal is to decelerate or reserve rural-to-urban migration. Altering rural population configurations is a more common objective than altering urban configurations.

Spatial population distribution programmes are commonly part of more general spatial development programmes. Often the population objective is secondary to economic objectives.

The instruments employed encompass economic, administrative and social measures. The emphasis generally is on economic measures intended to affect spatial demand for labour.

The results of programmes vary widely from nation to nation and from programme to programme, but in general they have demonstrated limited effectiveness or unintended adverse consequences. Rustication programmes may have slowed metropolitan growth, but at the expense of a rise in corruption and other undesirable effects. Accommodationist policies raise questions of social justice and their effect on stimulating further metropolitan growth. Intermediate-city programmes have generally failed to affect growth and spread as expected. Rural development schemes have increased mobility options for rural residents but have not stopped rural out migration.

The second part of the paper considers apparent policy deficiencies and suggests needed research and other possible areas of policy improvement. The major arguments are as follows :

The need for intervention should be reassessed in the individual nations. Studies of the demographic and socio-economic consequences of current and projected mobility patterns upon individuals, households and source and destination areas would be especially useful in such a reassessment.

Policy goals often appear to have been borrowed rather than independently designed to fit national characteristics and contexts and should be re-examined in many cases. The widespread concern with slowing rural to urban migration seems misplaced. Further study of labour markets and the spatial relationships among labour supply, job opportunities, income and mobility is urgently needed on a country-by-country basis, as is survey research to establish individual location preferences, mobility interactions and benefits and costs accruing to various types of migrants to different destinations.

There appears to be considerable scope for improving programme instruments. Current locational incentives for industry are often inappropriate or unsuited for LDCs. Research on the spatial decision-making behaviour of industrialists and other entrepreneurs in LDCs could lead to improvements in such economic instruments intended to affect spatial patterns of labour demand. Instruments for affecting spatial patterns of labour supply on the other hand, must be closely matched to the determinants of migration and mobility behaviour. Distance deterrence and migrant selectivity are currently neglected. Information measures are particularly promising because of the possibility of relatively lower costs and shorter time periods than those demonstrated by commonly employed economic measures. Both macro and micro-level studies, structured to include policy-relevant variables, have a role to play in improving programme design and instrument selection.

The past neglect of policy monitoring and evaluation should be remedied, and by more rigorous approaches employed than commonly found at present. Quasi-experimental designs may be useful.

Perhaps the most important weakness of past urban-related population distribution policies is the political-institutional framework. A broad political consensus must be built to overcome the all too common on-again off-again implementation of policies. Also spatial economic and population planning must become more closely related to national macro-economic and sectoral planning. In many Asian nations it would seem that the implicit spatial effects of macro and sectoral policies often contradict and outweigh explicit spatial policies.

As Schumpter (1950) noted many years ago that we 'plan too much and think too little'. The experience with urban and other population distribution policies in Asia as elsewhere would seem to confirm his view. As matters stand, our knowledge in most nations is insufficient to accurately guide policymakers and there is a risk that governments are intervening in processes inadequately understood to achieve ends irrationally defined.

FOOTNOTES

1. In a July 1978 survey of 116 governments in the less developed regions, 53 considered their current levels of fertility satisfactory, but only six were content with the spatial distribution of their population (UN, 1979). Whether this perception is accurate is another matter.
2. The survey response should be treated with some caution for they do not always correspond to policy objectives as found in other statements such as national development plans.
3. Despite the general disappointment with such policies, there have been notable successes, as in Korea (Mera 1976).
4. An example of our limited knowledge in these areas is the continuing controversy regarding the impact of migration on fertility (Findley and Orr 1978 ; Pernia 1981).
5. This criticism is elaborated for neoclassical studies of international migration in OECO, 1979 : 30.
6. One useful approach, although it would yield only a partial answer, would be to examine and assess spatial patterns of tax receipts and receipts and public expenditures, in other words, the redistribution effect of public sector finances (Bennet 1979).

REFERENCES

Abumere, S.I., 'Population Distribution Policies and Measures in Africa South of the Sahara : a Review.' *Population and Development Review* 7 (1981) : 421-433.

Alonso, William, 'Problems, purposes and implicit policies for a national strategy of Urbanization', in *Population Distribution and Policy*, U.S. Commission in Population Growth and the American Future, Research Reports, Vol. 5 (Washington : U.S. Government Printing Office, 1972).

Alonso, William, 'Five bell shapes in development', Papers of the Regional Science Association, 45 (1980) : 5-16.

Bahrin, Tunku Shamsul, 'Review and evaluation of attempts to direct migrants to frontier areas through land colonization Schemes.' In *Population Distribution Policies in Development Planning*, eds. George J. Demko and Roland J. Fuchs (New York : UN, Population Division, 1981), pp. 131-143.

Bennet, R.J., *The Geography of Public Finance*, London : Methuen, 1979.

Berentsen, William H., 'Austrian regional development policy : the impact of policy on the achievement of planning goals.' *Economic Geography* (1978), 115-134.

Carroll, G.R., 'Na ional city-size distributions : What do we know after 67 years of research ?' *Progress in Human Geography*, 16 (1982) : 1-43.

Chan, Paul, 'Evaluation of migration-related policies in Peninsular Malaysia.' Paper presented at the 1980 Development Studies Centre Conference, Population Mobility and Development, October 8-10, 1980.

Chang, Parris H., 'China's rustication movement.' *Current History*, 69 (1975) : 85-89.

Conroy, Michael E., 'The role of large cities in Third World development : a putative reapprisal' (Paper prepared for Regional Science Association meetings, Chicago, 1978).

Day, Frederic, and George J. Demko, 'Population distribution problems and policies in Asia.' *Population Geography*, 3 (1981) : 16-40.

De Jong, Gordon F., and Robert W. Gardner, eds., *Migration Decision Making : Multidisciplinary Approaches to Microlevel Studies in Developed and Developing Countries* (New York : Pergamon, 1981).

Demko, George J., and Roland J. Fuchs, eds. *Population Distribution Policies in Development Planning* (New York : UN, Population Division, 1981)

El-Shakhs, Salah, 'Development, primacy and systems of cities.' *Journal of Developing Areas* 7 (1972) : 11-36.

Economic and Social Commission for Asia and the Pacific (ESCAP), *Asian-Pacific Population Programme News*, 1981, Special issue on the 1980-81 census round.

Findley, Sally, *Planning for Internal Migration : A Review of Issues and Policies in Developing Countries*. U.S. Bureau of the Census ISP-RD-4 (Washington, D.C. : U.S. Government Printing Office, 1977).

Findley, Sally, 'Rural development programmes : planned versus actual migration outcomes' in *Population Distribution Policies in Development, Planning* J. Fuchs eds., George J. Demko and Roland (New York : UN, Population Division, 1981), pp. 144-66.

Findley, Sally and A.C. Orr, *Patterns of Rural-Urban Fertility Differentials in Developing Countries : A Suggested Framework* (Washington ; Office of Urban Development, AID, 1978).

Folmes, H., Measurement of the effects of regional policy instruments.' *Environment and Planning* A 12 (1980) ; 1191-1202.

Friedmann, John, *Regional Development Policy : A Case Study of Venezuela* (Cambridge, Massachusetts : Massachusetts Institute of Technology, 1966).

Friedmann, John, *Urbanization, Planning and National Development* (Beverly Hills, California : Sage, 1973).

Fuchs, Roland J., 'Conflicts between explicit and implicit population distribution policies in Asian development plans.' *Population Geography* 3 (1981a) : 41-56.

Fuchs, Roland, 'Use of survey findings for developing population redistribution policies', Paper prepared for. ESCAP Technical Working Group on Migration and Urbanization, Bangkok, December 1-5, 1981b.

Fuchs, Roland J. and George J. Demko, 'Population distribution policies in developed and Socialist and Western nations.' *Population and Development Planning* 5 (1979) : 439-67.

Fuchs, Roland J. and George J. Demko, 'Population distribution measures and the redistribution mechanism', in *Population Distribution Policies in Development Planning*, eds. George J. Demko and Roland J. Fuchs (New York : UN, Population Division, 1981), 70-84.

Fuchs Roland J. and J.M. Street, 'Land constraints and development planning in Taiwan', *Journal of Developing Areas* 14 (1980) : 313-26.

Fuguitt Glenn V. and J.J. Zuiches, 'Residential preferences and population distribution : results of a national survey', in *Where Will All the People Go* ? Report of the subcommittee on Rural Development of the Senate Committee on Agriculture and Forest, 93rd Congress, First Session, 23 October 1973 (Washington, D.C. : U.S. Government Printing Office, 1973).

Gilbert. A.G., 'The arguments for very large cities reconsidered', *Urban Studies*, 13 (1976) : 27-34.

Gilbert, A.G. and D.E. Goodman, 'Regional income disparities and economic development : a critique', in *Development Planning and Spatial Structure*, ed. A.G. Gilbert (London : Wiley, 1976).

Goldstein, Sidney, *Circulation in the Context of Total Mobility in Southeast Asia* (Papers of the East-West Population Institute, No. 53, East-West Center, Honolulu, 1978).

Goldstein, Sidney, 'Research priorities and data needs for establishing and evaluating population distribution policies', in *Population Distribution Policies in Development Planning*, eds., George J. Demko and Roland J. Fuchs (New York : UN, Population Division, 1981), pp. 183-202.

Gordon, Peter and Jacques Ledent, 'Modelling the dynamics of a system of metropolitan areas', *Environment and Planning* A 12 (1980) : 125-33.

Gordon, Peter and Jacques Ledent, 'Towards an interregional demoeconomic model', *Journal of Regional Science* 21 (1981) : 79-87.

Gosling, L., A. Peter and L.Y.C. Lim, eds. *Population Redistribution : Patterns, Policies and Prospects* (New York : UNFPA, 1979).

Hansen, Niles, 'Review and evaluation of attempts to direct migrants to smaller and intermediate-sized cities', in *Population Distribution Policies in Development Planning*, eds. George J. Demko and Roland J. Fuchs (New York : UN, Population Division, 1981), pp. 113-30.

Houston, C.J., 'Administrative control of migration to Moscow', *Canadian Geographer* 23 (1979) : 32-44.

Hugo, Graeme J., 'Some observations on the role of migration and population mobility on wealth transfers between individuals and their families in Third World societies with special reference to Indonesia', Paper prepared for IUSSP and EWPI Seminar on Interrelationships Between Demographic Factors and Income Distribution : Individuals and Families in Income Distribution, Honolulu, April 6-9, 1981a.

Hugo, Graeme J., 'Methods for evaluation of the impact of migration on individuals, households, and communities', Paper presented at ESCAP Technical Working Group on Migration and Urbanization, Bangkok, December 1-5, 1981b.

Isserman, A. and J. Merrifield, 'The Use of Control Groups in Evaluating Regional Economic Policy', *Regional Science and Urban Economics* 12 (1982) : pp. 43-58.

Jones, Gavin, *Social Science Research on Population and Development in South-East Asia* (Mexico City : El Colegio de Mexico, International Review Group, 1978).

Jones, R.C., 'Myth maps and migration in Venezuela', *Economic Geography,* 54 (1978) : 75-91.

Kim, S.U. and P.J. Donaldson, 'Dealing with Seoul's population growth : government plans and their implementation', *Asian Survey* 19 (1979) : 660-73.

Krebs, G., 'Regional inequalities during the process of national development', *Geoforum* 13 (1982) : 71-81.

Kuznets, Simon, 'Economic growth and income inequality', *American Economic Review* 45 (1955) : 1-28.

Lane, Michael H. and Barry Kolb, 'Locational components of urban and regional public policy in postwar Vietnam : the case of Ho-Chi-Minh City (Saigon)', *Geo Journal* 4 (1980) : 13-18.

Laquian, Aprodicio A., 'Review and evaluation of urban accommodationist policies in population redistribution', in *Population Distribution Policies in Development Planning,* eds., George J. Demko and Roland J. Fuchs (New York : UN, Population Division, 1981), pp. 101-12.

Laquian, Aprodicio A., R.M. Aquino, and L.T. Postrado. 'Effectiveness of Population redistribution policies : cases from South and Southeast Asia', Paper presented at IUSSP General Conference, Manila, December 9-16, 1981.

Linn, Johannes, *Policies for Efficient and Equitable Growth of Cities in Developing Countries* (Staff Working Paper, No. 342, World Bank, Washington, D.C., 1979).

Lo, Fu-chen, and Kamal Salih, eds. *Growth Pole Strategy and Regional Development Policy : Asian Experiences and Alternative Approaches* (New York : Pergamon Press, 1978a).

Lo, Fu-chen and Kamal Salih, 'Growth poles and regional policy in open dualistic economies : Western theory and Asian reality' in *Growth Pole Strategy and Regional Development Policy : Asian Experiences and Alternative Approaches,* eds. Fu-chen Lo and Kamal Salih (New York : Pergamon Press, 1978b), pp. 243-69.

Lo, Fu-chen, Kamal Salih, and Michael Douglass, 'Rural-urban transformation in Asia', in *Rural-Urban Relations and Regional Development,* ed. Fu-chen Lo (Singapore : Maruzen Asia, 1981), pp. 7-43.

Logan, Mal I., 'The shift from informal to formal : some costs and benefits of large scale industrialization in Asian countries', Paper presented at Pacific Science Association 4th Inter-Congress, Singapore, 1981.

Mabogunje, Akin L., 'Effectiveness of population redistribution policies : the African experience', *Solicited Papers,* International Population Conference, Manila, Vol. 2, pp. 527-42 (Liege; International Union for the Scientific Study of Population, 1981).

McGee, T.G. Labour mobility in fragmented labour markets, rural-urban linkages and regional development in Asia', in *Rural-Urban Relations and Regional Development,* ed. Fu-chen Lo (Singapore : Maruzen Asia, 1981), pp. 245-63.

Mera, Koichi, *Population Distribution Policies in the Republic of Korea* (Staff Working Paper, Washington, D.C. 1976), World Bank.

Mera, Koichi, 'Population distribution policies : the need for caution', in *Population Distribution Policies in Development Planning*, eds. George J. Demko and Roland J. Fuchs (New York : UN, Population Division, 1981), pp. 33-41.

Merrifield, J.D., 'Evaluating economic development programs : a new approach', Paper presented at the Annual Meetings of the Association of American Geographers, Los Angeles, 1981.

Moore, Barry and John Rhodes, 'Evaluating the effects of British regional Policy', *Economic Journal* 83 (March 1973) : 82-110.

Moore, Barry and John Rhodes, Evaluating the economic effects of regional Policy', In OECD Industry Committee's Working Party on Regional Development Policies, *Report on Methods of Measuring the Effects of Regional Policies* (Washington, D.C. : OECD Publications Center, 1977), pp. 11-82.

Nachmias, David, *Public Policy Evaluation : Approaches and Methods* (New York : St. Martins, 1979).

National Economic and Social Development Board, *The Fourth National Economic and Social Development Plan (1977-81)*, Bangkok, 1977.

Oberai, A.S., 'State policies and internal migration in Asia', *International Labour Review* 120 (1981) : 231-244.

Organization for Economic Co-operation and Development, *Migration, Growth and Development*, Paris, 1979.

Pernia, E.M., *Urbanization, Population Growth, and Economic Development in the Philippines* (Westport, Connecticut ; Greenwood, 1977).

Pernia, E.M., 'On the relationship between migration and fertility', *Philippine Review of Economics and Business*, 18 (1981) : 192-202.

Pryor, Robin J., 'Population redistribution and development planning in Southeast Asia', Paper presented at IGU New Zealand Regional Conference, Massey University, 1974.

Pryor, Robin J., 'Population redistribution : policy formulation and implementation', in *Population Distribution Policies in Development Planning*, eds., George J. Demko and Roland J. Fuchs (New York : UN, Population Division, 1981), pp. 169-82.

Pryor, Robin J. ed., *Migration and Development in South-East Asia : A Demographic Perspective*, Kuala Lumpur : Oxford.

Rehnberg, R.D., 'Agricultural price policy and rural to urban migration : the recent South Korean experience', *Journal to Developing Areas*, 11 (1977) : 509-18.

Renaud, B., *National Urbanization Policies in Developing Countries*, World Bank (Staff Working Paper No. 347), Washington, D.C., 1979.

Richardson, Harry W., *Regional Growth Theory* (London : MacMillan & Co., 1973).

Richardson, Harry W., 'The costs and benefits of alternative settlement patterns', in *The Population Debate : Dimensions and Perspectives*, Papers of the World Population Conference, Bucharest, 1974, Vol. 2 (New York : UN, 1975), pp. 131-42.

Richardson, Harry W., *City Size and National Spatial Strategies*, Staff Working

Paper, No. 252, Washington, D.C. : World Bank.

Richardson, Harry W., 'Growth centers, rural development and national urban policy : a defense', *International Regional Science Review* 3 (1978) : 133-52.

Richardson, Harry W., 'Defining urban population distribution goals in development planning', in *Population Distribution Policies in Development Planning*, eds. George J. Demko and Roland J. Fuchs (New York : UN, Population Division, 1981), pp. 7-18.

Richardson, Harry W., 'Population distribution policies', Paper presented at the Expert Group Meeting on *Migration, Population Distribution and Development* (New York : United Nations, 1982).

Rodwin, L., *Nations and Cities* (Boston : Houghton-Mufflin, 1970).

Rondinelli, D.A., 'Applied policy analysis for integrated regional development planning in the Philippines', *Third World Planning Review* 1 (1979) : 150-78.

Rondinelli, D.A., *Secondary Cities in Developing Countries : Policies for Diffusing Urbanization* (Beverly Hills : Sage, 1983).

Rondinelli, D.A. and H. Evans, 'Integrated regional development planning : linking urban centers and rural areas in Bolivia', *World Development* 11 (1983) : 31-54.

Salih, Kamal, Phisit Pakkasem, B. Prantilla, and Sugijanto Soegijoko, 'Decentralization policy, growth pole approach, and resource frontier development : a synthesis of the response in four Southeast Asian Countries', in *Growth Pole Strategy and Regional Development Policy : Asian Experiences and Alternative Approaches*, ed. Fu-chen Lo and Kamal Salih (New York : Pergamon Press, 1978), pp. 79-120.

Schumpeter, J.A., *Capitalism, Socialism and Democracy*.

Simmons, Alan B., 'Slowing metropolitan city growth in Asia : policies, programs and results', *Population and Development Review* 5 (1979) : 87-104.

Simmons, Alan B., 'A review and evaluation of attempts to constrain migration to selected urban centers and regions', in *Population Distribution Policies in Development Planning*, eds. George J. Demko and Roland J. Fuchs (New York : UN, Population Division, 1981), pp. 87-100.

Simmons, Alan B., S. Diaz-Briquets, and A.A. Laquian, *Social Change and Internal Migration : A Review of Research Findings from Africa, Asia, and Latin America* (Ottawa : IDRC, 1977).

Song, B.N. and C.S. Choe, 'Review of urban trends and policies', *Urban and Regional Report*, No. 81-2, (Washington : World Bank, 1981).

Speare, Aldan. S. Goldstein, and W.H. Frey, *Residential Mobility, Migration and Metropolitan Change* (Cambridge, Massachusetts : Ballinger, 1976).

Standing, Guy, 'Issues in analysing inter-relationships between migration and employment', Paper prepared for ESCAP Technical Working Group on Migration and Urbanization, Bangkok, 1-5, December, 1981.

Stark, O., 'Research on Rural-to-Urban Migration in LDC's : The Confusion Frontier and Why We Should Pause to Rethink Afresh', *World Development*, 10 (1982), pp. 63-70.

Stöhr, Walter, 'Evaluation of some arguments against government intervention to influence territorial population distribution', in *Population Distribution Policies in Development Planning*, eds. George J. Demko and Roland J. Fuchs (New York : UN, Population Division, 1981), pp. 42-49.

Tan, K.C., 'Solution strategies for national settlement system planning', *Geographical Analysis*, 12 (1980) : 68-79.

Todaro, Michael P., 'Internal migration in developing countries : a survey', in *Population and Economic Change in Developing Countries*, ed. R.A. Easterlin (Chicago : University of Chicago Press, 1980), pp. 361-390.

United Nations, Population Division, *World Population Trends and Policies : 1977 Monitoring Report*, Vol. II, *Population Policies*, 1979.

United Nations, Population Division, *World Population Trends and Monitoring Report*, Vol. II, *Population Policies*, 1980.

United Nations, Population Division and United Nations Fund for Population Activities (UNPD and UNFPA), *Population Policy Compendium*, 1979 and 1981.

Willekens, Frans, 'Optimal migration policies : an analytical approach : *Regional Science and Urban Economics* 9 (1979) : 345-67.

Williamson, Jeffrey G., 'Regional inequality and the process of national development : a description of patterns', *Economic Development and Cultural Change* 13, Part II (1965) : 1-84.

Wright, C.L., 'Income inequality and economic growth : examining the evidence', *Journal of Developing Areas* 13 (1978) : 49-66.

Yap, Lorene Y.L., *Internal Migration in Less Developed Countries: A Survey of the Literature* (World Bank, Staff Working Paper, No. 215), Washington, D.C., 1975.

FIVE

R.L. DWIVED

A COMPARATIVE STUDY OF EVOLUTION AND DEMOGRAPHIC CHARACTER OF KAVAL TOWNS OF UTTAR PRADESH, INDIA

THE KAVAL towns (Kanpur, Agra, Varanasi, Allahabad and Lucknow), the five largest cities of Uttar Pradesh (U.P.), although accounting for only about 25 per cent of the urban population of the state, dominate the urban landscape of this part of the Ganga Valley.[1] Significantly, these cities are all located on the river Ganga and its two tributaries, the Yamuna and the Gomati. This may be so because in the pre-railway age, the bank of a navigable river was probably the most favourable site where urban settlement could originate and grow. The high bluffs, meander loops and confluence forks of rivers were, therefore, selected as sites with obvious advantages. Cultural factors too played their role in the origin of towns. Strategically located forts on large rivers attracted urban communities to grow in their immediate vicinity. In the modern period, industrialization and new transportation routes, particularly the railways, have been factors in the evolution of urban centres. Towns situated in the same geographical area may thus owe their existence to a variety of natural and cultural factors. The objective of this paper is to make a comparative study of the origin, growth pattern and demographic character of the five most important cities of the Ganga plain of Uttar Pradesh.

ORIGIN AND EVOLUTION

Of the five cities, Varanasi (25° 18'N—83° 1'E), formerly known as Banaras and renowned for its great antiquity and religious sanctity, is the oldest city.[2] This largest city of eastern U.P. lies in the Middle Ganga Valley and is located on the bank of the river Ganga between its two small tributary streams, the Varuna in the north and the Assi in the south, hence the name Varanasi. The city no doubt had its origin in the ancient period. According to some authorities, the ancient city probably lay near Sarnath, a few miles to the north of the present city, and associated with the life of Buddha. Since the city was a great centre of Hindu philosophy and religion, the earliest settlement must have originated on the Rajghat plateau near the fort overlooking the Ganga.[3] From there the settlement first expanded southward along the four mile long ridge-like bank of the river and later towards the west and south-west. Before the invasion of India by the Muslims, Varanasi was not only a great religious centre but also an important commercial mart of the surrounding region. The holy city lay quite close to the bank of the Ganga which, along with *ghats* and nearby temples, acted as a great centripetal force in the early evolution of the townscape. After the cession of the Province of Banaras to the East India Company, the city came under the possession of the British in 1794. Consequently, the civil lines to the north of the Varuna and the cantonment to its south were laid out in the north-west. During the British period the city grew rapidly as a result of the railway communication, flourishing trade, educational facilities and provision of public utility services. The Banaras Hindu University established in the early twentieth century, and the Diesel Locomotive Works of the post-independence period set in operation the centrifugal forces, and the city expanded towards the south and south-west. Varanasi has thus witnessed a contentric growth with an inner zone clinging to the semicircular river bank, and the middle and outer zones encircling it. The city with magnificent *ghats* on the Ganga, the Banaras Hindu University, and nearby Sarnath, is one of the great tourist centres of India.

Allahabad (25° 26' N—81° 50' E) like Varanasi, has long been a famous place of Hindu pilgrimage, since it is located on the sacred confluence of the Ganga, the Yamuna and the mythical invisible Saraswati rivers. In all probability the ancient town of Prayag (ancient name of Allahabad) originated near the sangam or triveni (confluence of three rivers), which provided a naturally defensive site and a parennial water supply to the early settlement. At the time of the visit of the Chinese traveller, Hieun Tsang in the seventh century the town, with a brahmanical temple in its centre, stood near the confluence where the annual mela (fair) and every twelfth year the *Kumbh mela* were held in the month the *Magh* (January). Thus the physical factor of the confluence and cultural factor of religion were involved in the genesis of the ancient settlement.[4] The old city was however eroded away by the rivers before the Muslims arrived here. During the 16th century the Moghul emperor Akbar was impressed with the strategic importance of the confluence and commanded a fort to be built near it. As Smailes has observed, 'in innumerable cases, a nascent urban community has gathered under the shadow of a castle, for which, in turn, a naturally defensible site has been chosen.'[5] A small town, rechristened as Allahabad by the Moghul emperor Akbar, sprang up to the west of the fort, probably between the historic Grand Trunk road and the bank of the Yamuna, and which served as a transport link with the country lying to the north-west. The town assumed great importance as it was made the head-quarter of a *suba* (province) of the Moghul empire. To protect the town from floods, *buxi* and *beni* bunds so conspicuous in the present landscape, were raised along the Ganga during the reign of Akbar. Besides the city proper, scattered settlements grew up—Daryaganj in the east and Katra in the north. But during the 18th century the growth of the city was considerably slowed down, as it was a period of unsettled conditions. Allahabad was, therefore, not a large town when the British occupied it in 1801. It was made a military station and large cantonments were developed near the fort and the settlement of Katra. After the uprising of 1857 Allahabad became the capital of the north-western provinces and the present civil lines for the English people and the cantonment laying to its west were laid out. The city began to expand owing to the concentration of capital functions. The establishment of the high court and the university, the bridging of the two rivers and opening of the railways led to further growth. In the early thirties of the present century the capital was shifted to Lucknow. But the city has continued to grow around the three nuclei of the city proper—Katra—Colonelganj and Daryaganj. Among the KAVAL towns,

Allahabad perhaps illustrates the multiple nuclei theory of urban growth. In the post-independence period, a ribbon development has occurred along the Grand Trunk road in the west and an industrial zone has emerged at Naini, to the south of the Yamuna river.

Agra (27° 10'N—78° 3' E) owes its importance to its strategic location on the right bank of the meandering Yamuna river in south-western U.P. It is undoubtedly a city of mediaeval origin and can be traced back to the times of the Lodi sultans of Delhi, among whom one Sikander Lodi (1489-1517) is supposed to be its real founder.[6] But the settlement—a forward base of the Lodi sultans—lay east of the Yamuna.[7] The Moghul emperor Akbar built the present fort in red sandstone on the western bank in 1566, probably on the ruins of an earlier fort and thus moved the town to a strategically better site. 'The fort town of Agra formed the first line of defence against the invaders from the west, as well as an effective control point for the granary of the vast fertile plain of the Ganga in the east.'[8] The Moghul emperors ruled their empire from Agra for a century. The famous mausoleum, the Taj Mahal in white marble, was built by Shahjahan to perpetuate the memory of his wife, Mumtaz Mahal. After the death of Shahjahan in 1665 the old capital of the Moghuls was reduced in the status of a provincial town and began to stagnate. Its importance was however revived when the British occupied it in 1803 and made it the capital of the north-west provinces. At that time, the greater part of the town lay to the north of the fort near the river bank. The vast area lying to the south of the fort formed the cantonment, and to the north-west of the cantonment and the fort, the civil station extended in a linear fashion. During the present century, the city has considerably expanded to the north and west. But expansion towards the south was restricted owing to the presence of the cantonment. The spurt in commercial activity caused by the railways, and the facility of education and public utility services let loose the forces of urban development. In the post-independence period, not only has the city expanded towards the north and west, but also to the east of the river. Because of the Taj Mahal, the fort and other historic monuments, Agra holds a great attraction for tourists from within the country and abroad.

Lucknow (26° 51' N--80° 53' E) located on both sides of the

river Gomati, a tributary of the Ganga in Central U.P. lies at a distance of 72 km. to the north-east of Kanpur. What is now the capital and second largest city of the state, originated as a small settlement near the ancient mound of Lakshman tila, quite close to the river.[9] The city no doubt is of mediaeval origin, and judged by the then standards it was a large city during the Moghul period and served as the headquarters of the provincial administration. The expansion of the city however occurred during the times of the nawabs of Oudh, especially Asaf-ud-daullah, the fourth nawab, who made it the capital of his state. During the nawabi period (1775-1856), although the city expanded towards the east up to Aminabad, the main concentration was in the west with Chowk as its nucleus. Several historic buildings were built and spacious gardens were laid. The bridge over the Gomati stimulated growth to the north of the river. After the uprisings of 1857, the city under the British further expanded towards the east and south. A large area lying to the east of Ghaziuddin Hyder canal was developed as a cantonment, and the civil lines in the north-east near Hażratganj was developed for residential and administrative purposes. Lucknow was made the capital of the province of Agra and Oudh in the early thirties of the present century, and the university was established to the north of the river. As a result of the eastward growth, the old nucleus of Chowk has been superceded by the new nuclei of Aminabad and Hazratganj, the oriental and western style markets respectively of the city. Further east a large administrative zone has emerged, where several government offices, the governor's house and Council House are located. Since independence, the city has considerably expanded towards the south along the Lucknow-Kanpur road and to the north of the river, where several new localities such as Maha Nagar and Nirala Nagar have sprang up as planned residential areas.

Kanpur (26° 28' N—80° 24' E), the largest city among KAVAL towns, is the youngest of the five cities. The city, located on the right bank of the Ganga river in central U.P.. is of modern origin. About the time of the Cession of Banaras Province to the East India Company (1775) a small settlement 'Kampoo Kanpur' four miles to the south-east of 'Cawnpore (or Kanhpur) Kuhna' or old Kanpur, became the nucleus of the present city.[10] Originally it was a military centre and in 1778 British troops were stationed here for the protection of the European trading community. Consequently, a large cantonment comprising the area stretching along the bank of the Ganga from old Kanpur in the west to Jajmau in the east had to be developed. Unlike Varanasi, the indigenous city thus grew at some distance from the river along the southern boundary of the cantonment. In the early 19th century, the town assumed an irregular form with houses separated from one another by narrow and winding lanes. There were no broad streets till after the uprising of 1857, in which the town figured most prominently. The expansion of Kanpur began after 1852 when British owned industries (cotton and woollen mills and leather factories) were established in the western part of the erstwhile cantonment near the bank of the river. The railways and roads converged here from all directions and the Ganga was bridged in 1875. The town emerged, by the close of the 19th century as a big centre of business (especially of cotton and grain) and large-scale industry. Thus unlike other KAVAL towns, Kanpur originated and expanded under the impact of modern industrialization and railway communication. With the increase in population, the build-up area began to grow. Singh has estimated that during 1865-1961 the built up area of Kanpur increased by about 691 per cent, whereas that of Allahabad, Lucknow, Agra and Varanasi by 376 per cent, 263 per cent, 158 per cent and 64 per cent respectively.[11] During the post-independence period the city has extended towards the west along the Grand Trunk road, south-westward along the Kalpi road and also towards the south and east. While new educational institutions, namely, the medical college, Kanpur University and Indian Institute of Technology dominate the landscape in the west, new mills, industrial estates and ordnance factories have sprung up along the Kalpi road and the northern railway line up to Panki. As a result of the demand of more residential accommodation, large areas in the west, south and south-east have been developed as planned residential colonies. The city of Kanpur has thus grown through the swelling of a single nucleus, and the modern sprawling city is essentially industrial in character and appearance.

TRENDS OF POPULATION GROWTH

In the preceding pages, the growth of KAVAL towns has been traced. We may now examine the population characteristics of the five cities that rank among the first fifteen cities of India. Kanpur and Lucknow, whose populations crossed one million mark in 1981, belong to the category of million-cities in the country. During 1901-81, while the population of Kanpur has increased by more than 800 per cent, that of Lucknow and Agra rose only by more than 400 per cent. Varanasi and Allahabad have registered an increase of less than 400 per cent during the same period. The dynamism of the youngest city, Kanpur and modest growth of the oldest city, Varanasi, are indeed most striking and call for a careful analysis of their population growth trends. A study of population growth rates reveals that the first two decades of the present century were in general marked by negative growth (Table 5.1). The universal fall in the population of the cities was caused by the repeated visitation of epidemics and other diseases that took heavy tolls of human lives.[12] But from 1921 onward an upward trend is clearly discernible in respect of all the five cities. During 1921-81 the population of Kanpur increased at a faster rate (3.5% per annum) than those of other cities. Almost the same growth pattern obtained during the last eighty years (1901-81). However after independence the population of Agra has grown at a faster rate (3.5%) than that of Kanpur (3.0%).

The pulsatory nature of the population growth is revealed by the annual growth rates given in Table 5.2. While each decade recorded positive growth rates, the rates show fluctuations from one decade to another. Fluctuation is highest in the case of

TABLE 5.1 : **Growth rates per annum**[13]

(*as percentage*)

City	1901-21	1921-81	1951-81	1901-81
Kanpur	+0.3	3.5	3.0	2.7
Lucknow	–0.3	2.4	2.3	1.7
Agra	–0.1	2.3	3.5	1.7
Varanasi	–0.3	2.3	2.7	1.6
Allahabad	–0.4	2.3	2.2	1.6

TABLE 5.2 : **Growth rates per annum during the last six decades**

(as percentage)

City	1921-31	1931-41	1941-51	1951-61	1961-71	1971-81
Kanpur	1.2	2.3	3.8	3.2	2.8	2.8
Lucknow	1.4	1.1	2.5	2.8	2.2	2.0
Agra	2.2	2.1	2.8	3.0	2.2	2.0
Varanasi	0.3	2.5	3.0	3.2	2.2	2.7
Allahabad	1.6	3.5	2.4	2.7	1.7	2.3

Varanasi and lowest in the case of Agra. From Table 5.2 it will be seen that the growth rate of each city during 1961-71 was less than that of the previous decade. The reasons for this deceleration of growth rates are not clear.

Of the five KAVAL towns, Kanpur became a million-city in 1971 and Lucknow in 1981. This is because Kanpur is the biggest industrial and commercial centre of the state, whereas Lucknow is fast growing owing to a greater concentration of capital functions. As a rule the population of cities in U.P. has risen through the twin process of natural increase and immigration. But the population of Kanpur, being the biggest industrial centre of the state, has grown more by immigration than natural increase. In 1951 more than half of the population of 0.7 million was outborn—the highest percentage among the KAVAL towns.[14]

SPATIAL PATTERN OF POPULATION

A comparative study of the gross population densities (Table 5.3) shows that overall densities of KAVAL towns in 1971 vary from

TABLE 5.3 : **Gross population densities**

(persons per acre)

City	1951	1961	1971	1981
Kanpur	57.1	13.6	17.7	14.1
Lucknow	38.6	23.2	31.3	18.1
Agra	27.3	30.2	38.3	25.5
Varanasi	62.9	25.8	35.5	36.8
Allahabad	30.8	26.6	31.3	19.5

17.7 for Kanpur to 38.3 for Agra. In 1951 and also in 1981 Varanasi was the most thickly populated city. But in 1961 and also in 1971 Agra reported the highest gross density. Another significant fact is that in 1961 all towns except Agra, and in 1981 except Varanasi, registered an appreciable fall in density values from the corresponding figures of the previous censuses. This fall probably has been brought about by enormous increases in the areas of cities as a result of jurisdictional changes. In 1971 however all the five towns recorded an increase in their densities over the 1961 figure. Although Kanpur is the most populous city, its density is the lowest among the KAVAL towns. This may be owing to the presence of a vast rural area within the corporation limits.

Overall densities however do not give any idea of the prevailing high densities in the five cities. An analysis of the intra-urban population distribution shows that in all the five towns, more or less centrally located wards have exceedingly high densities and that density values decrease from the centre towards the periphery (Table 5.4). Brush has considered this problem in greater detail, and his results show that the density distance relationships generally conform to the Bleicher-Clark model based on negative exponential decline of density with increase of distance from the city centre.[15] He however admits that a simple linear regression analysis does not adequately describe the observed population patterns in the cities since the intra-urban population growth has occurred under a wide variety of circumstances.[16] The maximum density does not exist in the central bazars of Kanpur and Varanasi, but is found inwards between a quarter of a mile and half a mile from the centre.[17]

TABLE 5.4 : **Density-distance relationship**[19]

City	No. of statis-tical units	Persons per acre — miles from city centre 0	1	2	3	4	5
Kanpur	16	429.5	159.9	59.5	22.6	8.4	3.1
Lucknow	32	229 5	97.7	40.6	16.9	7.0	—
Varanasi	30	265.0	91.0	31.2	10.7	3.7	1.3
Allahabad	27	130.8	70.8	37.4	19.8	10.5	5.5

The spatial pattern of population in dual centred cities—Lucknow and Allahabad—differs from that of Kanpur and Varanasi, densities are generally lower and populations are more dispersed.[18]

FUNCTIONAL CHARACTER

In the occupational structure of the five cities what is most striking is the preponderance of non-workers. On the basis of three previous censuses, the situation as it obtained in 1961, 1971 and 1981 may be compared. It will be seen from Table 5.5 that in 1961 in all the KAVAL towns except Agra, the non-workers constituted over two-thirds of the total city dwellers, while this proportion in 1971 and 1981 rose to a little less than three-fourth in all the cities. The situation is rather alarming and calls for immediate plans for providing jobs to the jobless people. As expected among the non-workers, females outnumbered males, as women account for about 60 per cent of the total number of non-workers.

The statistics regarding workers have been reported under nine categories in the two censuses of 1961 and 1971.* In Table 5.6 the first four categories have been combined under 'cultivation and other primary occupations', to afford easy comparison. From a perusal of the percentage figures it may be stated that Varanasi and Kanpur employ larger percentages of total workers in industry ; Agra and Varanasi in trade ; Agra and Lucknow in transport ; and Allahabad and Lucknow in services, especially in

TABLE 5.5 : **Non-workers in KAVAL towns**

City	Percentage of non-working population to total population			Percentage of females in non-workers	
	1961	1971	1981	1971	1981
Kanpur	67.4	70.4	73.0	59.9	59.0
Lucknow	68.8	72.1	71.9	60.5	59.7
Agra	63.7	74.2	71.8	60.0	60.4
Varanasi	68.0	72.3	72.9	60.0	60.8
Allahabad	69.5	72.2	74.8	58.4	57.2

*Figures on a comparable basis have not been reported in 1981 census.

TABLE 5.6 : **Occupational structure of KAVAL towns**

(*as percentage of total workers*)

City	Cultivation and other primary occupations		Manufacturing including household industry		Trade and commerce		Transport		Construction		Services	
	1961	1971	1961	1971	1961	1971	1961	1971	1961	1971	1961	1971
Kanpur	3.9	3.9	37.8	34.4	18.0	20.9	8.0	8.6	2.5	1.2	29.8	31.0
Agra	2.1	3.2	31.2	32.5	19.8	21.8	9.5	12.6	3.3	2.9	34.1	26.8
Varanasi	2.5	4.3	40.1	37.1	19.6	21.8	8.4	9.9	2.3	1.6	27.1	25.3
Allahabad	3.7	5.7	21.3	20.5	16.6	19.6	12.8	9.2	3.3	1.7	42.3	43.3
Lucknow	2.5	3.6	23.7	20.8	17.4	18.7	11.8	13.9	3.6	1.6	40.9	41.4

Source : Based on Census of India, 1961, Paper No. 1 of 1962, pp. 260-65 and Census of India, 1971, U.P. Part II-B(i) (Series 21), 1973.

administration. Varanasi is the home of household industry whereas Kanpur is the biggest centre of large-scale industries. Lucknow being the state capital and Allahabad enjoying the status of a secondary capital, give employment to a large number of people in government and other offices. A significant change occurred during 1961-71. While the industrial base of each city except Agra has weakened, all the five towns have gained in commercial importance. The position of Allahabad as a transport centre has declined. But it remains unquestionably the biggest administrative centre, closely rivalled by Lucknow, the state capital.

The high percentage in any livelihood class may, however, not be adequate to prove the dominance of that particular function. The 'standard deviation method' as a measure of the intensity of functional specialization shows that Allahabad specializes in cultivation and services ; Lucknow in transport and services ; Varanasi in commerce and manufacturing ; and Agra in commerce and construction. These four cities are thus bifunctional. The degree of specialization in the four bifunctional cities is however not more than +1 SD. Kanpur with specialization in no function, displays functional diversity.

REFERENCES

1. The KAVAL towns in order of population in 1981 are : Kanpur 1,688,242 ; Lucknow 1,006,538 ; Varanasi 793,542 ; Agra 770, 352 and Allahabad 642,420.
2. Banaras was renamed as Varanasi on 24 May 1956.
3. R.L. Singh, *Banaras—A Study in Urban Geography*, 1955, p. 32.
4. R.L. Dwivedi, 'Origin and Growth of Allahabad', *The Indian Geographical Journal* 38 (1965) : 16.
5. A.E. Smailes (1953), *The Geography of Towns*, London, p. 44.
6. H.R. Nevill (1905), *Agra—A District Gazetteer*, Allahabad, p. 191.
7. O.H.K. Spate and E. Ahmad 'Five Cities of the Gangetic Plain', *The Geographical Review*, April 1950, pp. 260-78.
8. U. Singh, 'Changes in the Built up Areas of KAVAL Towns', *National Geographical Journal of India* 2 (1966) : 203-04.
9. M.N. Nigam, 'Evolution of Lucknow', *National Geographical Journal of India* 6 (1960) : 30-46.
10. U. Singh, 'The Origin and Growth of Kanpur', *National Geographical Journal of India* 5 (1959) : 1-11.
11. U. Singh, 'Changes in the Built up Areas of KAVAL Towns' 12 : 205,

12. R.L. Dwivedi, 'Demographic Features of Allahabad City', *Geographical Review of India* 27 (1965) : 163-88.
13. The formula used for calculating growth rates per cent per annum is : $P_n = P_o (1+\frac{r}{100})^n$, where P_n =Population after n years, P_o=Population of base year, n=number of years between P_o and P_n and r=increase or decrease per cent per annum.
14. U. Singh, 'The Origin and Growth of Kanpur', 5 : 8.
15. J.E. Brush, 'Spatial Patterns of Population in Indian Cities', *The Geographical Review* 58 (1968) : 362-91.
16. J.E. Brush, 'Application of a Model to the Analysis of Population Distribution in Indian Cities', in *Urban Geography in Developing Countries*, R.L. Singh, Varanasi, 1973, p. 31.
17. J.E. Brush, 'Spatial Patterns of Population in Indian Cities', 58 : 376.
18. Ibid., 58 : 386.
19. Ibid., p. 370 (No data for Agra).

SIX

ASHOK K. DUTT
MOHDUDUL HUQ

CHRISTALLER'S CENTRAL PLACE MODEL
Fifty Years of Test

THE CENTRAL PLACE (CP) theoretical model was formulated by Walter Christaller. His theoretical work on the size and space of CPs first appeared in Germany in 1933. Since then researchers in geography and regional science have enjoyed themselves in exploring the theoretical and pragmatic dimensions of this field. This model consisted of a series of assertions and definitions, logical consequences of which were a hierarchical ranking of urban centres according to function which associated with market areas and the transportation network.[1] Basically, Christaller attempted to find out the laws that determine the number, size and distribution of hamlets, villages, towns and cities. He believed that 'just as there are economic laws which determine the life of economy, so are there special economic-geographic laws determining the arrangement of different types of settlement.' He pointed out that the basic function of any city is to be a CP, providing retail goods and services to its surrounding tributary areas. This model was first applied to the external relationship between a city and its hinterland, the city being the focal point for different types of services, trades and financial institutions. Later, this model has been applied to the internal relations of tertiary activities, that is the actual distribution of business and financial institutions in the city.[2]

DEVELOPMENT OF CP MODEL

This model is based on some assumptions and certain constraints. First, Christaller begins with the location economists' traditional medium, the isotropic plain, which is a physiographically uniform surface. Second, a population or farming population is assumed to be evenly distributed, and disposable income per farm family is assumed to be constant and demand is evenly spread. Third, movement/transportation is held to be equally unrestricted in all directions from all points and transportation costs are seen to be directly proportional to distance travelled.[3] Besides these physical constraints, there are several behavioural assumptions ; a maximum number of demands for goods and services should be satisfied ; the incomes of the people offering the goods and services should be maximized ; distances moved by consumers to purchase those goods and services should be minimized or, in other words, goods should be purchased from the closest point or source ; the number of central places should be the minimum possible ; and there should be no intra-product differentiation.[4]

THE BASIC MODEL OF CP

Within these assumptions and constraints, Christaller postulated a number of events in the isotropic plain which are briefly described in this section. First of all a series of evenly spaced service centres would be established. Then, the distance between two service centres should be 2 hours of non-mechanized travel. Which means that 2 hours is the equivalent of about 7.2 km, that is, the distance from one service centre to another (Fig. 6.1), since a man could walk or drive by horse cart about 3.60 km in one hour. Theoretically, the trading area of the service centre or town would be a circle of 3.60 km in radius. Again, given the uniform density of population, a hexagonal market area would in fact require the least average distance of movement to the centres for consumers, minimize the number of centres, and avoid areas of market area overlap. Finally the service centres would be small and provide only limited services. Therefore, CPs of this kind would be numerous. The denser the population, the greater will be the service function of these first order centres. In southern Germany, Christaller identified seven orders of settlement, each

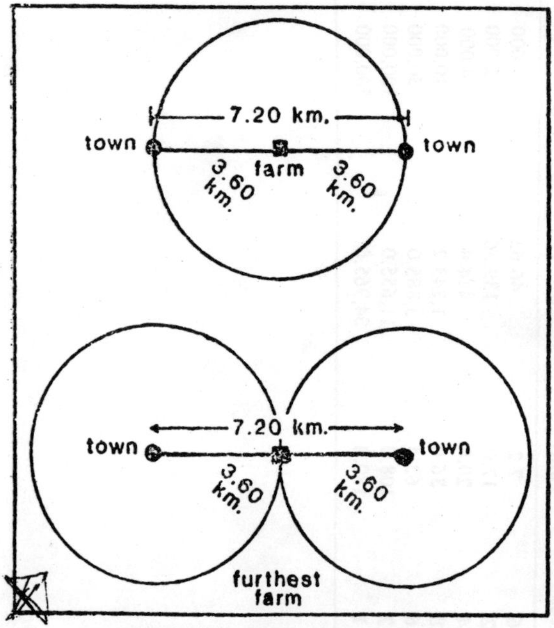

FIG. 6.1 : *Spacing of lowest level (Marktort or Market town) central place according to Christaller's central place model*

Source : Based on Hurst (1972), p. 199.

with a precise quantitative relationship to others (Table 6.1).

Two concepts are fundamental to Christaller's grouping of service centres, that is, range and threshold. According to classical economic theory, an assumption is made that, as the cost of the good rises, so the quantity demanded declines. In spatial terms this means that the quantity demanded declines as distances from the supplier increases. Sooner or later a point will be reached where the demand for the good is zero, since farmers or people are not willing to pay the transportation costs necessary to reach the supplier. In this case the supplier market area will have a circular perimeter, beyond which the demand for his good is zero. The radius of this market area is termed the range of the good. It will not be possible for the supplier to stay in business unless he is partronized by a certain number of customers ; below

TABLE 6.1 : Levels of central places : Their numbers and characteristics according to Christaller

Nature of central place	German name	Number of places	Distance apart (km)	Service area (km²)	Estimated population
Market Town	Marktort	486	7.2	46.62	1,000
Township Centre	Amstort	162	12.0	139.86	2,000
County Seat	Kreisstadt	54	20.8	414.4	4,000
District City	Bezinkstadt	18	36.0	1,243.2	10,000
Small State Capital	Gaustadt	6	62.4	3,885.0	30,000
Provincial City	Provinzstadt	2	108.0	11,655.0	100,000
Regional Capital	Landstadt	1	180.8	34,965.0	500,000

Source : Based on Murphy and Hurst.

this critical level, the supplier will not be able to meet his own costs and business will fail. This minimum number of consumers necessary to support the supply of a particular good is termed the good's threshold.[5]

The Christaller model includes various levels or orders of central places. He calculated the population of these places, their distance apart, and the sizes and populations of their tributary areas in accordance with a hexagonal system (Fig. 6.2). The market town is the smallest unit in the CP system with a population of 1,000 and it is assumed to lie 7.2 km from it's nearest market town. This distance was, according to the model, calculated so that from any part of its service area the market town could be reached in approximately an hour's walk. The costs of reaching a CP increased outward in all directions upto an economic limit set by transport costs and travel time ; beyond that point it costs less to travel to another centre of equal or higher rank. Christaller estimated that in Germany the market had a tributary area of 46.67 km² and with a population of approximately one thousand. There are seven levels of CP in Christaller's model (Table 6.1). Some of the places performing

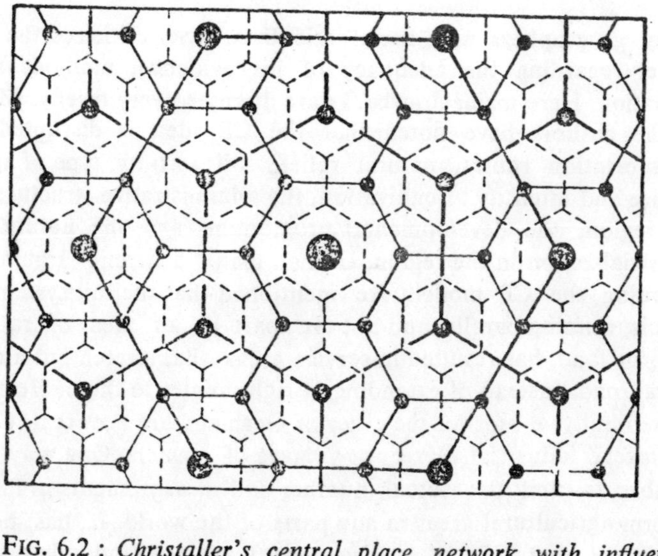

FIG. 6.2 : *Christaller's central place network with influence areas delineated for four levels*

the township centre function were spaced at intervals of about 12 km, while the county seats averaged a spacing of about 20.8 km, and so on through the several orders of urban centres. Each CP performed the functions characteristic of its rank and all those of lower order centres as well. The distance between adjacent similar CP's amounted to the distance between those of the preceding category times 3. According to Christaller, the numbers of representatives of the seven levels of CPs follow a theoretical framework from the largest to the smallest centres which are as follows : 1 : 2 : 6 : 18 : 54 : 162 : 486.[6]

Christaller was mainly concerned with administration, culture, health, social services, organization of economic and social life, finance, trade, service industries, the labour market and traffic.[7] He decided that telephone services explain best the connection between a CP and its hinterland region which was measured by formula $TZ\text{-}EZ\,(Tg/Eg)$ where TZ is the number of telephones in the place, EZ the number of inhabitants of the place, and Tg and Eg the number of telephones and inhabitants respectively in the surrounding region.[8]

GENERAL CRITICISMS OF THE CENTRAL PLACE MODEL

Many geographers and social scientists have criticized the CP model regarding the adequacy of his evidence, methods and validation. Here major drawbacks are discussed very briefly. Most of the authors have noted that the CP idea is disrupted by transportation routes, natural relief, soil fertility, type of agriculture and intensity of cultivation, the administrative structure of the region, the developmental programme and the nature of industrialization in the region. Ullman made a strong argument regarding the CP model. He mentioned that the rail system in Illinois, running parallel and not far apart in an area of rather uniform land, has resulted in service areas that stretch out along the railroads instead of extending at right angles to them. He also pointed out that most of the cities of Germany are not real CPs, but instead industrial centres or villages of farmers. One may not be able to find the actual number and measurements in fairly uniform agricultural areas in any parts of the world. It has been established that CPs of the same rank are not randomly distributed. According to Ullman, in the USA there are strong forces that lead to the establishment of CPs. Goods brought into an area are handled through brokers, wholesalers and retail systems in central cities. It seems that these are supporting a CP system of settlements, and are not actually spaced according to Christaller's model. Ullman concluded that the growth of industry and trade during the last hundred years has often not explained the actual relationship of the cities to the countryside. There is much evidence that the rapid growth of some large cities is not based on their being industrial centres, but rather as commercial and service centres for the hinterland. Finally, Ullman suggested that the CP model is a theoretical norm and planning for new areas. The number of telephones as a measure of centrality has been questioned by different authors who have tried to apply this method empirically. According to Christaller, the telephone was a reliable index to measure centrality since these were used for business purposes in southern Germany. At the present time it is not valid in southern Germany. In western countries such as the USA telephones are as common in homes as in business establishments and such an index would be of little use to measure centrality. But Ullman mentioned in his article, that out of town telephone calls could be applied in the USA to measure centrality.[9,10]

Christaller assumed that a given area of land will support a town to provide for its social and commercial needs. He envisioned a hierarchy of towns, each providing a higher grade of service for an area more extensive than that served by towns of the next lower level in the hierarchy. According to Christaller, a hexagonal area of influence is produced for each town. The formulation of his theory has helped geographers and social scientists to grade towns in a hierarchy and grade the distance between towns of the same and different levels which he worked out in theory and applied in southern Germany.[11] A major limitation of his work is that he did not give any specific criteria for the meaningful functional importance or status of towns. The CP model has major weaknesses regarding its application such as : the model deals only with urhan systems which are composed of CPs providing service and market activities ; it does not include any important urban functions or activities such as manufacturing ; it does not take into account any economic aspect of urbanization such as conurbation formation or the scale of economies ; and the model is algebraic, and, therefore, it seems that it is deterministic in nature rather than probabilistic, because it does not allow for any deviations from precisely stated relations and locations of CPs There is no evidence to verify this model in the real world.[12] Finally, there is a great need to introduce the time dimension to central place studies which is a major limitation in the Central Place Model.

EMPIRICAL EXAMPLES FROM THE OLD AND NEW WORLD

Brush and Bracey's analysis of rural service centres in Wisconsin and southern England is a classic example of application of the CP concept in two highly developed countries immediately after World War II when the carts and horses were replaced and automobiles had started to replace the established railroad system. They found that : higher-order centres occurred at 33.60 km apart serving 331.52-334.11 km^2 and the lower-order at 13-16 km serving 83-124 km^2 ; higher-order centres formed clusters with few or no centres of lower order around them ; the mean population of the higher-order centres in Wisconsin was only 2,515 compared to England's 13,850 and similarly, the lower-order centres of Wisconsin had only 400 people compared to England's 5,080 ; no significant difference in the agricultural pattern could be observed in spite of the fact that close agricultural settlement existed in south England for 1,500 years and in Wisconsin only for 100-25 years. Brush and Bracey concluded that the functional hierarchy was much more variable from place to place and year to year than the rigid system first described by Christaller.[13]

Dutt and Harrity developed a composite index of Newspaper Influence Area based on population and area served by each city in the state of Ohio, U.S.A. They came up with four hierarchic levels : 3 ; 3 ; 11 ; 50 (Fig. 6.3). The service areas of the three levels have been mapped, the fourth level CPs have been identified in Fig. 6.3D.

> Although the primary emphasis of this paper was the establishment of a hierarchial pattern, the research and statistical procedure resulted in a secondary finding transcending the importance of the case study. On the basis of the

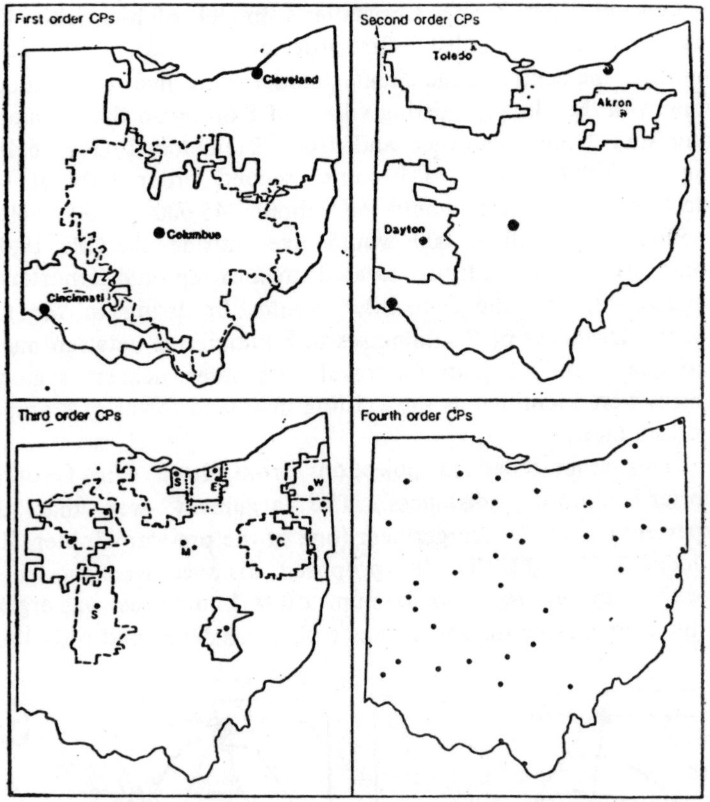

FIG. 6.3 : *Four levels of influence areas for the state of Ohio, USA, based on daily newspaper circulation*

Source : Based on Dutt and Harrity (1971), p. 128.

findings of this study, it appears as if newspaper circulation is an excellent yardstick by which to measure hierarchy.[14]

Thus it is evident that not only retail business but tertiary activity such as newspaper circulation can be used for designating levels of CPs.

In the Netherlands, before the north-east polder (part of the Zuider Zee Project) fell dry in 1957, the planners tried to apply CP model to evolve a two-level hierarchy of settlements. Dutt,

Costa, van der Wal and Lutz have examined the application of the concept. First, Walter Christaller's model of hierarchic CPs was adopted to establish a first order central place, Emmeloord, almost at the centre of the circular polder and five new villages along with the already existing village of Kuinre on the mainland, equidistant from each other and from Emmeloord (Fig. 6.4A). The six villages were to become second order CPs. It was calculated that there would be about 45,000-50,000 polder dwellers, of which 15,000 would live outside the CPs, 10,000 would live in a centrally located town of regional importance, Emmeloord, and the remainder would be living in the five villages. Distances of 7 kilometres to 8 kilometres between nuclei were considered adequate for travel. The other nearest regional centres were identified as the existing mainland towns of Kampen and Lemmer.

This scheme led to objections from those who favoured shorter commuting distances. The argument was that the experience in the Wieringermeer (one of the pre-war Zuider Zee polders) had taught that the proposed distance were too length (for the bicycle) and a maximum of 4 kilometres was argued to be a reasonable distance. Thus a plan was developed with 45-55

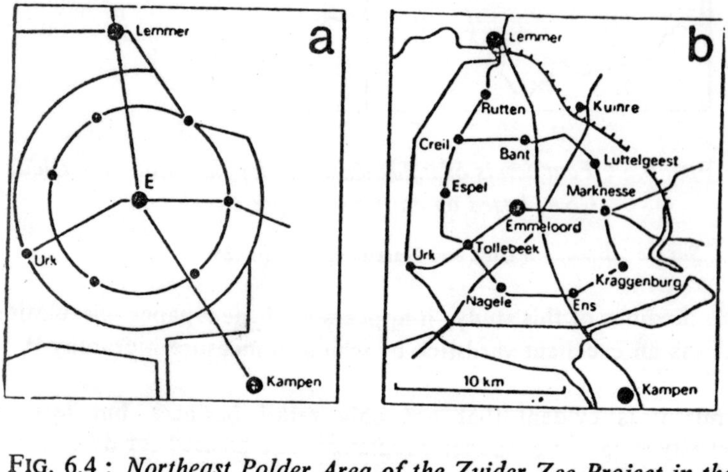

FIG. 6.4 : *Northeast Polder Area of the Zuider Zee Project in the Netherlands and two stages of planning (A, B) for developing the number and hierarchy of central places in the newly reclaimed polder lands in the 1950s*

hamlets, each of which were no more than 4 kilometres from a village and 3 kilometres from the farthest farmsite. Finally a compromise was reached and in 1948 a plan was approved which contained one central nucleus, Emmeloord, and ten villages in a ring around it (Fig. 6.4B). The villages were generally located near road intersections (canals had become less important as a mode of transportation). About half of them were located on irregularly shaped parcels, which was presumably good for the design of a physical plan. Construction began on Emmeloord in 1946, a year after the war. In 1948 village construction began with Ens, Marknesse and Kraggenburg. The second group of villages, Luttelgeest, Bant, Rutten and Creil, was built between 1950 and 1953. Finally came Nagele, Espel and Tollebeek in the years 1954-57.[15]

Constandse criticized the above plan because it did not foresee the growing affluence of the people, increased automobile ownership and the increased mechanization of farming.[16] As automobile ownership increased, the mobility of the people increased and this led to the weakening of the CP functions of the ten villages because most found it more convenient to drive to a larger CP at Emmeloord and shop there, on a weekly basis, for their basic shopping needs. Moreover, mechanization of farming meant a smaller number of people were needed to operate farms and the projected number of people in the villages was over-estimated. It is, thus, evident that in the era of automobile transportation and mechanized farming, CPs, to be viable, need to be located at greater distance compared to the era of carts, bicycles and mopeds.

Dutt has applied the CP concept in an urban context with a case study of Calcutta. Based on an index developed with consumers and catchment areas of the different markets in the city four hierarchic levels were identified and their ratio was in the order of 1 : 4 : 18 : 36 (Fig. 6.5).[17] This study represents the existence of different levels of CPs within a city with the highest level located almost at its geometrical centre.

CONCLUSION

The CP model continues to be a viable concept. Christaller's ideas gave it a boost in conceptualization. Those who tried to

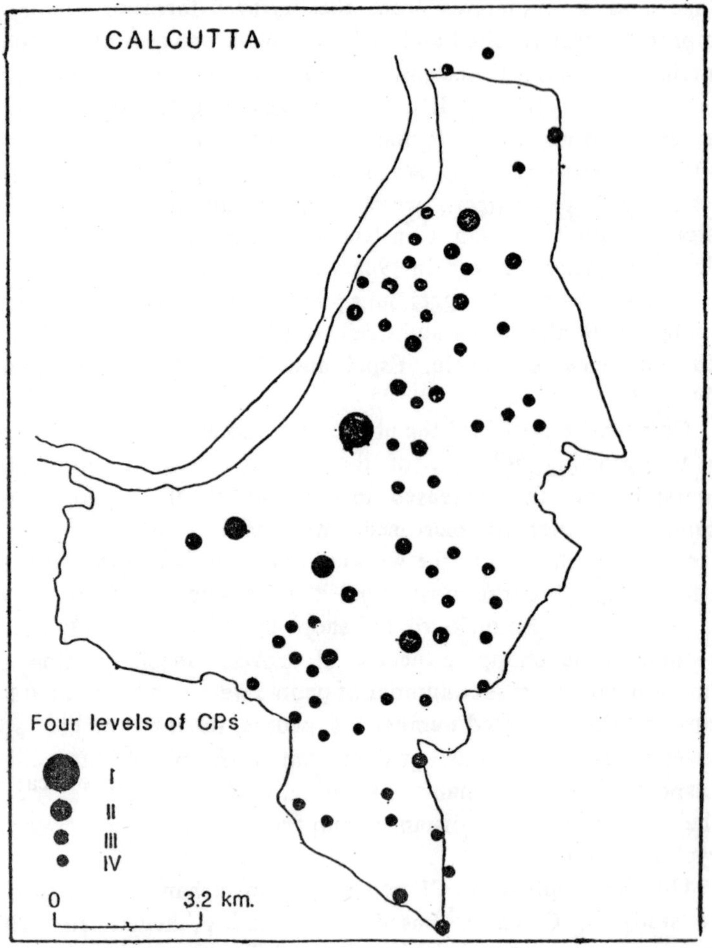

FIG. 6.5 : *Four hierarchic levels of central places in the city of Calcutta*

follow him rigidly failed to understand the main spririt of the model. The fact remains that the central places remain in hierarchy but their size, catchment area and numbers vary from city to rural area, from one region to another, from one kind of transportation technology to another, and from one time to another. The model, therefore, is not a static one, but should be considered in a dynamic context showing different kinds of regional variation. In a broad sense, Christaller's model has stood the test of time for about half a century and will continue to stir the investigative interests of geographers and regional scientists in the future.

REFERENCES

1. B.J.L. Berry and W. Garrison, 'Recent Developments of Central Place Theory', in *Papers and Proceedings of the Regional Science Association*, 4 (1958) : 107-20.
2. M.F. Hurst, *A Geography of Economic Behaviour* (Duxhury Press, 1972), p. 198.
3. J.H. Marshall, *The Location of Service Towns : An Approach to the Analysis of Central Place Systems* (Toronto : University of Toronto Press, 1969), p. 12.
4. Hurst, *Geography*, p. 198.
5. Marshall, *The Location of Service Towns*, p. 13.
6. R.E. Murphy, *The American City : An Urban Geography* (New York : McGraw Hill Book Company, 1966), p. 77.
7. Murphy, *The American City*, pp. 78-79.
8. Murphy, *ibid.*, p. 81.
9. E. Ullman, 'A Theory of Location of Cities', *American Journal of Sociology*, 46 (1951) : 853-64.
10. U.D. Lipman, 'Town and Country', *Public Administration*, 30 (1952) : 204.
11. M. Dacey, 'A Probability Model for Central Place Location', *Annals of the Association of American Geographers*, 56 (1966) : 550-68.
12. M. Dacey, 'A Probability Model', pp. 550-68.
13. John E. Brush, and Howard E. Bracey, 'Rural Service Centers in Southwestern Wisconsin and Southern England', *The Geographical Review* 45 (1955) : 559-69.
14. Ashok K. Dutt and Peter B. Harrity, 'Newspaper Circulation and City Hierarchy ; Ohio as a Case Study', *Preconference Publication Papers of the Annual Meeting of the Canadian Association of Geographers*, 1971, p. 129.
15. Ashok K. Dutt, Frank J. Costa, Coenraad Van der Wal and William Lutz, 1984, 'Evolution of Land Uses and Settlement Policies in Zuiderzee Project Planning', in *Public Planning in the Netherlands*, ed. Ashok K. Dutt and Frank J. Costa (Oxford, England : Clarendon Press of Oxford Uuiversity, 1984).
16. A.K. Constandse, 'The Ijsselmeerpolders, and old Project with New Functions', *Tydschr. Voor Economische en Sociale Geografie* 63 (1972) : 200-10.
17. Ashok K. Dutt, 'Intra-City Hierarchy of Central Places : Calcutta as a Case Study', *The Professional Geographer* 21 (January 1969) : 18-22.

SEVEN

R.P. MISRA

URBANIZATION PROCESS
A Cross Cultural Perspective

INTRODUCTION

THE HISTORY of human settlements appears to have witnessed four major changes in the pattern of growth and spatial organization. The first was the evolutionary transition from hunting and fishing to sedentary agriculture from which rural settlements emerged. The second was the transition from the 'village' to the 'city' which can be said firstly to have occurred in the river valleys of West Asia soon after 3500 B.C. and later in many other parts of the world, chiefly in Egypt, Pakistan, India, China, Mexico, Peru, etc. The advent of the Industrial Revolution and accompanying colonization of Latin America, Asia and Africa, heralded the third revolutionary transition characterized by increased agglomeration and emergence of large metropolitan centres in Western countries. The Second World War and the decolonization of the present-day Third World countries brought about the fourth major change, that is rapid urbanization of the Third World countries without accompanying industrialization. According to current projections, by 1990 more of the larger metropolitan centres will be in the developing countries of Asia and Latin America (see Fig. 7.1).

FOUR STAGES

Let us elaborate the main characteristics of each of these stages :

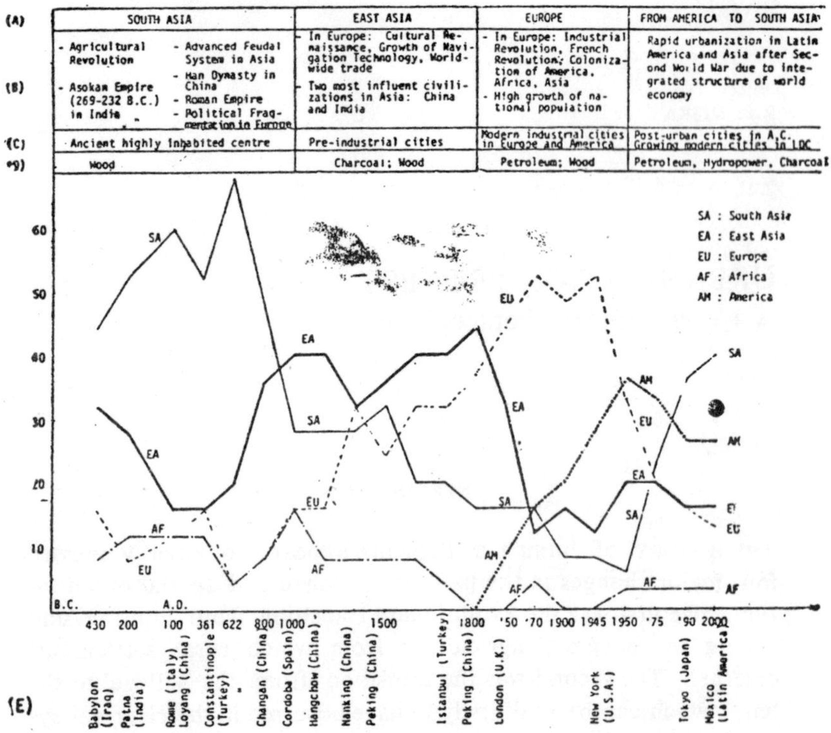

FIG. 7.1 : *Evolution of large cities in the world : General trends by region*

Notes : (A) Region with highest share (HSLC); (B) Major historical characteristics; (C) City characteristics; (D) Major energy sources; (E) Name of largest city.

Phase I (B.C. 430-1000 A.D.)

The highest share in the percentage of 25 (or 30) of the largest cities in the world (hereafter referred to HSLC) was in South Asia, and the next highest was in East Asia. In Asian history, this is the splendid era of the Asokan Empire 269-232 B.C. in India where the largest city in the world (in 200 B.C.), named Pataliputra (Patna), was located ; and also of the Han Dynasty (206 B.C.-A.D. 221) in China which had the largest city in the world (in 100 A.D.), named Loyang with an estimated population of around 6,50,000. The advanced agricultural societies in favourable tropical climate with stable feudal systems which characterized Asian civilizations at that time could be seen as an explanatory factor for this phase. In 400 B.C. the estimated population in Asia was about 80 million, mainly in peasant villages ; 12 million in West Asia ; 30 million in China ; 30 million in India ; and in Europe it was about 20 million, mainly concentrated in southern Europe, that is, Greece, Rome and Spain.[1] Close to the beginning of the Christian Era, the Roman Empire (127 B.C.-A.D. 476) emerged to dominate the West and established Rome as a city, competing with Loyang and China.

Many of the magnificent cities in ancient times, as landmarks of luxury, autocracy and exploitation of the people, fell into ruins after the downfall of the Sumerian, Indo-Gangetic, Egyptian, Greek and Roman civilizations. Only a few survived the ravages of history. In the medieval period, the cities grew mostly as capital and commercial centres. However, the Western cities could not expand further owing to the political fragmentation of Europe and hence limited support from the hinterland during this period.

Phase II (A.D. 1000-A.D. 1800)

In this pre-industrial period, East Asia assumed the leading role in HSLC. Europe, with the cultural renaissance, showed a rapid growth and took second place in HSLC, while South Asia exhibited a sharp decline and became the third in the last half of this period. This period might be called as a 'gestation period' of Western civilization. Even in this period, for all regions, the primary sector (largely agriculture) was of paramount importance. There was some secondary activity (mostly construction and handicraft works), and little tertiary activity (services to primary and secondary sectors). In Western Europe the new navigation technology expanded commercial and trade links with other regions of the world.

Phase III (A.D.1800 to Second World War)

This is the 'flourishing era' of modern European civilization marked by the colonization of non-Western countries and trade expansion : by the Industrial Revolution from a technology and production viewpoint ; and, philosophically and politically by the

French Revolution (1789). This led to amazing and rapid urbanization in Europe and Anglo America, resulting from huge migration from rural to urban settlements. While in Asia, both East and South, there was in sharp contrast a declining trend that could be explained by stagnant agriculture, high absolute growth of population and colonization by Western powers.

The process of industrial capitalism in Europe was characterized by increasing agricultural surplus, a class of free status and dynamic entrepreneurs, a process of capital accumulation by them to finance large-scale investments, the development of a practical science, and the 'market expansion and machine production' structure of the economy. Industrialization was the 'engine' for urbanization. These essential elements made Western cities very different from Asian cities. The overseas market expansion led to the colonization of non-Western societies. It is the colonization and exploitation of these countries which were the foundation of Western capitalism. 'Indeed without that exploitation, one may doubt whether the West would ever have industrialized in the first place.'[2]

Phase IV (since Second World War)

Among other factors, the uncontrolled conflicts, resulting from economic expansion. between industrialized European countries led to the Second World War. After the war one notes a secular and sharp decline in HSLC in the industrialized countries. America reached its peak in 1950. South Asia is expected to take the top position in 1990 as projected by the United Nations.[3] This period is characterized by low rates of population growth in the developed countries, and high rates in developing countries.

The present patterns and characteristics of urbanization in the world are simply not determined by industrial factors as in previous phases, but rather by multi-integrated factors : on the vertical dimension are the regional, national and international forces ; and on horizontal dimension are the economic, political and social forces.

The post-colonial urbanization in the Third World, especially in Asia, entered a new phase with the further growth of a colonial 'primate cities', mainly owing to persistent high rates of natural increase in population and large-scale 'rural-urban' migration. It is characterized by a dual structure of the economy—modern and traditional, Western and non-Western, urban and rural, formal and informal—by increasing multi-national enterprises ; and finally by the continuing dependence in many respects on the Western advanced industrial countries for technology and finance. All of these have brought about the existence of a great number of squatter camps and vast areas of slums in the periphery of metropolitan cities, and have created a chronic and severe shortage of employment, housing and public services.

Determinants of Growth

By the year 2000 about half of the population of the world will be classified as urban (in 1900 only the United Kingdom had more than 50% of its population living in urban areas). Africa will be 40 per cent urbanized, Latin America 75 per cent and Asia 39 per cent. All other continents will have 75 per cent or more of their population in urban areas. And much of this urban population is likely to be in metropolitan cities.

Table 7.1 indicates the basic trends as projected by the United Nations. One interesting trend that can be observed from Tables 7.1 and 7.2 and Figures 7.2 and 7.3 is that by the year 2000, while the more developed regions (MDC) have a higher share of population against the world total in city size less than one million, the less developed regions (LDC) show a contrasting trend with a higher share of population against world total in city size of one million or more. This trend is a 'switching phenomenon' between MDC and LDC, and the turning point is some time in the 1980s. This phenomenon can also be observed when the number of cities criterion is taken into account (see Fig. 7.3.)

Table 7.2 shows that the number of cities in the world with a population of one million or more was 71 in 1950 (23 in LDC ; 48 in MDC). It is likely to increase to 414 by the year of 2000 (264 in LDC ; 150 in MDC). Thus, during this period the number of cities of this size group in the LDCs is likely to increase 3.7 times faster than that in MDCs. During 1950-80 the rate of increase was 2.2 for the city-size class of one million or more, and was approximately 1.7 for the other two city-size classes of less than one million. In terms of population the prospects reveal a greater contrast. In 1975 there were 262 million people living in cities of one million or more in MDCs and 244 million in LDCs. By the year 2000 these cities will contain 914 million in

TABLE 7.1 : **Size distribution of cities in developed and developing countries, 1950-2000**

Population in cities (thousands)

City size (thousands)	1950	1960	1970	1975	1980	1990	2000
World total							
1 million or more	476,704	282,761	418,659	505,845	631,124	955,568	1,359,526
	(45.10)	(48.14)	(50.41)	(51.45)	(54.26)	(59.51)	(62.72)
0.5—1 million	69,062	95,949	124,336	153,846	171,701	208,403	273,641
	(17.60)	(16.33)	(14.97)	(15.65)	(14.76)	(12.98)	(12.62)
0.1—0.5 million	146,281	208,683	287,525	323,543	360,380	441,769	534,363
	(37,30)	(35.53)	(34.62)	(32.90)	(30.98)	(27.51)	(24.66)
Total	392,047	587,393	706,184	983,234	1,163,205	1,605,740	2,167,530
	(100.00)	(100.00)	(100.00)	(100.00)	(100.00)	(100.00)	(100.00)
More developed regions							
1 million or more	129,329	181,035	235,820	262,244	303,477	379,190	444,792
	(49.40)	(50.97)	(52.04)	(52.12)	(54.48)	(57.21)	(58.81)
0.5—1 million	41,630	56,839	68,715	78,752	81,083	89,456	103,784
	(15.90)	(16.00)	(15.16)	(15.65)	(14.55)	(13.50)	(13.72)
0.1—0.5 million	90,821	117,289	148,663	162,165	172,522	194,097	207,775
	(34.70)	(33.03)	(32.80)	(32.23)	(30.97)	(29.29)	(24.47)
Total	261,774	355,163	453,198	503,161	557,082	662,743	756,331
	(100)	(100)	(100)	(100)	(100)	(100)	(100)

TABLE 7.1 : (*Contd.*)

Less developed regions							
1 million or more	47,381	101,726	182,839	243,601	327,647	576,378	914,734
	(36.37)	(43.80)	(48.46)	(50.74)	(54.06)	(61.12)	(64.82)
0.5—1 million	27,432	39,110	55,621	75,094	90,618	118.947	169,857
	(21.06)	(16.84)	(14.74)	(15.64)	(14.95)	(12.61)	(12.04)
0.1—0.5 million	55,460	91,394	138,862	161,378	187,858	247,672	326,608
	(42.57)	(39.35)	(36.80)	(33.62)	(30.99)	(26.27)	(23.14)
Total	130,273	232,230	377,322	480,073	606,123	942,997	1,411,199
	(100.00)	(100.00)	(100.00)	(100.00)	(100.00)	(100.00)	(100.00)

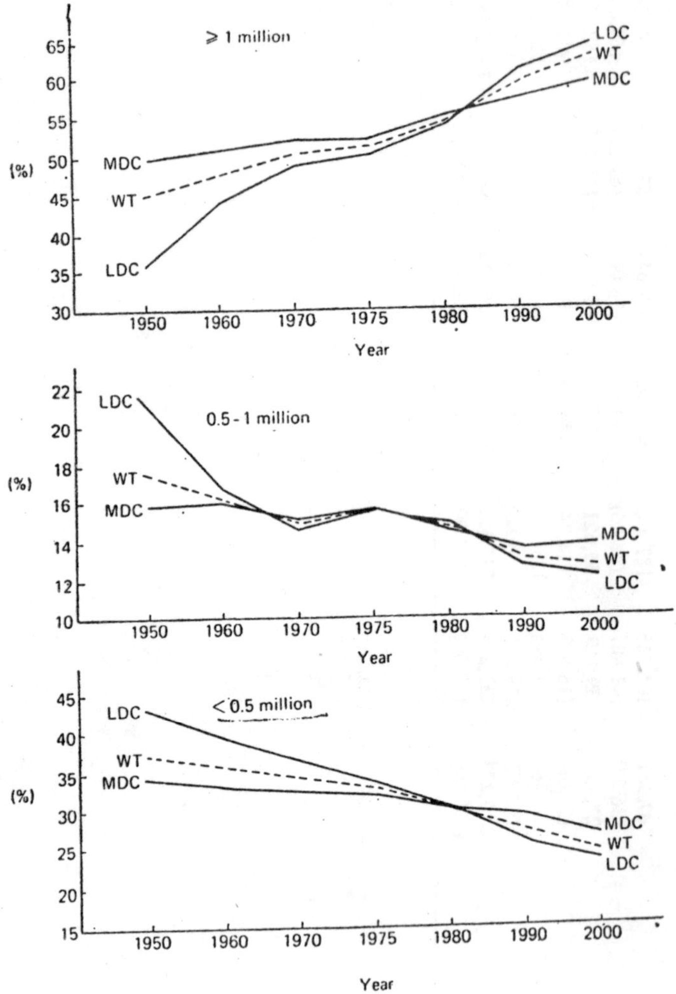

FIG. 7.2 : *Percentage share of population by city size*

Source : See Table 7.1.
Note : More developed countries (MDC)—Less developed countries (LDC)—World Total (WT).

LDCs and 444 million MDCs.

The growth rates apart, the nature of the expansion is so

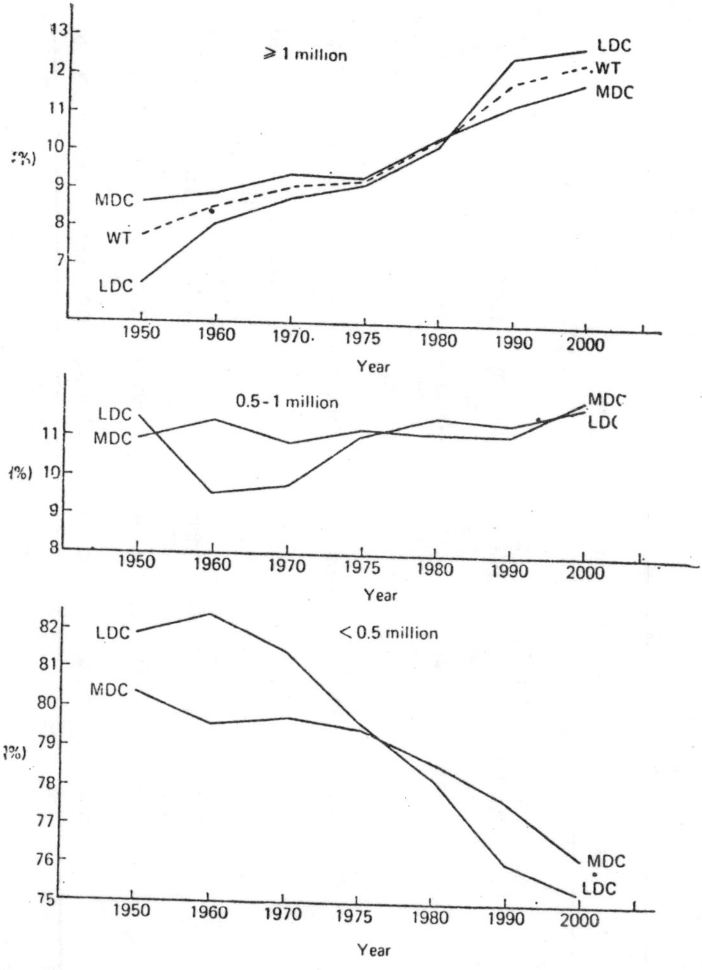

FIG. 7.3 : *Percentage share of number of cities by city size*

Source : See Table 7.2.
Note : More developed countries (MDC)—Less developed countries (LDC)—World total (WT).

different that it produces not only dissimilar but often contrasting results. The developed countries are undergoing post-industrial urbanization, at times referred to as mature urbanization. It is characterized by the transfer of population from one urban region

TABLE 7.2 : **Size distribution of cities in developed and developing countries, 1950-2000**

Number of cities

City size (thousands)	1950	1960	1970	1975	1980	1990	2000
World total							
1 million or more	71	110	157	181	225	322	414
	(7.84)	(8.56)	(9.07)	(9.20)	(10.22)	(11.95)	(12.44)
0.5—1 million	101	136	179	220	249	303	396
	(11.15)	(10.52)	(10.34)	(11.19)	(11.31)	(11.25)	(11.90)
0.1—0.5 million	734	1,039	1,395	1,566	1,728	2,069	2,519
	(81.01)	(80.86)	(80.59)	(79.61)	(78.47)	(76.80)	(75.66)
Total	906	1,285	1,731	1,967	2,202	2,694	3,329
	(100.00)	(100.00)	(100.00)	(100.00)	(100.00)	(100.00)	(100.00)
More developed regions							
1 million or more	48	64	85	91	109	133	150
	(8.62)	(8.92)	(9.35)	(9.28)	(10.25)	(11.24)	(11.82)
0.5—1 million	61	82	99	110	118	131	152
	(10.95)	(11.44)	(10.89)	(11.21)	(11.10)	(11.07)	(11.98)
0.1—0.5 million	448	571	725	780	836	919	967
	(80.43)	(79.64)	(79.76)	(79.51)	(78.65)	(77.69)	(76.20)
Total	557	717	909	981	1,063	1,183	1,269
	(100.00)	(100.00)	(100.00)	(100.00)	(100.00)	(100.00)	(100.00)

TABLE 7.2 : (*Contd.*)

Less developed regions							
1 million or more	23	46	72	90	116	189	264
	(6.59)	(8.10)	(8.76)	(9.13)	(10.18)	(12.51)	(12.82)
0.5—1 million	40	54	80	110	131	172	244
	(11.46)	(9.51)	(9.73)	(11.16)	(11.50)	(11.38)	(11.84)
0.1—0.5 million	286	468	670	786	892	1,150	1,552
	(81.95)	(82.39)	(81.51)	(79.72)	(78.31)	(76.11)	(75.34)
Total	349	568	822	986	1,139	1,511	2,060
	(100.00)	(100.00)	(100.00)	(100.00)	(100.00)	(100.00)	(100.00)

Source : United Nations, Population Division, Department of Economic and Social Affairs, *Trends and Prospects in the Population of Urban Agglomeration, 1950-2000* (ESA/P/WP50) New York, November 1975.

Note ; Numbers in parentheses are percentages against total.

to another, with a growing tendency towards suburbanization.

These shifts result from differences in the quality of life among urban centres and between the core and periphery of the same city. The migrants are urbanized ; most of them are trying to relocate themselves in order to lead a socially and culturally fuller life. In this process of change, the core urban regions are losing the native population and opportunities in the face of competition from more attractive locations and places. This tends to create several problems. The cores of the major metropolitan regions such as London, Paris and New York are unable to maintain themselves without external assistance, and tend to degenerate. Second, the ethnic character of the core regions changes rather dramatically as native-population losses are largely made up by foreign immigrants who are less skilled and have a lower income and a lesser stake in the metropolis. Third, with this change in ethnic character and with low income, alienation and deteriorating urban amenities, including housing, the core region becomes a den of crime and corruptive social work and values. All this leads to a vicious circle of deterioration leading to more deterioration.

As against this, in the less developed countries urban growth is the direct consequence of rural-urban migration. During the decade 1950-60, such migration constituted 60 per cent of urban growth. This figure is likely to come down to 42 per cent during the decade 1980-90 indicating a continuing high rate of natural increase of urban population. This again is in contrast to the situation in the more developed countries, where the overall population growth is low (less than 2%). The metropolitan cities of the developing countries contain much of the modern sector of the national economies and hence they attract large numbers of migrants.

The impact of large-scale migration to large cities is many sided. First, the density of population in relation to available facilities increases. Demand for housing shoots up and people start doubling up, renting out rooms to families, and piecing together shelters in places that were uninhabitable to begin with. The density of the central business district or the city core increases fastest, for it is here that the migrants can get casual work both in the formal and informal sectors of the urban economy. This leads to a sharp rise in land values, making it more difficult for the poor to acquire or build a house. In due course, squatter settlements extend beyond the metropolis, creating new problems of urban management.

This uncontrolled growth makes it almost impossible for the metropolitan regions of the developing countries to expand and up date urban amenities. Most of them are unable to provide water, sewerage and drainage to a large section of the urban population. The problems acquire more serious proportions when 'many migrants burden the city's economy through increased underemployment and unemployment levels, lower levels of productivity, greater demand for costly municipal facilities and services, lower per capita tax revenues, and higher city budget deficits.'[4] Even where a migrant has full-time employment, the income so generated does not remain in the city ; it spreads to his immediate and extended family living in the city or back home in the village.

The social fabric of these cities also undergoes change owing to large-scale migration. At times, migrants establish homogeneous ethnic communities and function as pressure groups to legalize squatter settlements or to acquire access to urban amenities and services. Participation in civic elections is noted to be low. Once this happens, it becomes difficult to integrate the community into social fabric of the city.

To sum up, the urbanization processes in the developed or post-industrial and developing or pre-industrial societies are marked by the following characteristics.

Developed or Post-Industrial Societies

—A very slow rate of population growth.

—Almost no rural-urban migration. The trend is rather for population to move from congested city-cores to the countryside.

—The dispersal of manufacturing from metropolitan centres to small and intermediate-sized towns in the country. The high cost of land and labour and environmental pollution are the main causes for this shift.

—The service-oriented tertiary and quaternary sectors of the economy are becoming more prominent in the metropolitan regions, replacing manufacturing and production-oriented tertiary activities.

—Outlying rural areas have become urbanized, and the distinction between urban and rural is narrowing down rather rapidly.

—Disparities in income and earning capacities have been reduced largely because the income of the poorest has risen well above the subsistence level. Poverty in the form seen in pre-industrial societies and industrializing societies does not exist in the post-industrial society.

—With further advances in production technologies, the role of people in production is decreasing. Moreover, workers are becoming increasingly alienated from work.

—There is a growing interest in a life-style which, without doing away with economic gains, would give a more satisfying meaning to life and living. The tendency is towards humanism and a peaceful world, but the framework for the conceptualization of this world is confined to the narrow limits set by an individual's own culture and experience.

Developing or Pre-Industrial Societies

—The natural increase in population is at times rather high, reaching from 3 per cent to 4 per cent per annum in both urban and rural areas.

—These countries are essentially rural ; less than a quarter of their population is urban. The capital city and a few large centres often dominate the urban scene.

—There is a high rate of migration of people from rural areas to a few urban centres, especially the metropolitan regions.

—Manufacturing industries still constitute a major activity in the metropolitan regions, although a tendency for the tertiaries to grow is quite visible in quite a few of them.

—There is a vast difference in earning capacity and quality of life between urban and rural areas and between the richer and poorer sections of the urban population. The mass of people is still below the poverty line in both urban and rural areas, although the problem is more accentuated in the latter.

—The labour force in both urban and rural areas is largely unskilled or semi-skilled, and hence earning capacity is low.

—Since the majority of people are still below the poverty line, improving the quality of life basically means increasing income. This results in people moving to urban slums in the hope of earning more and creating opportunities for a new start.

—All this leads to overcrowding and environmental deterioration in urban areas. The impact is felt more severely in the metropolitan regions, which attract large numbers of rural migrants.

—The problem in the metro-regions does not therefore boil down to inner city versus outer city. It is a problem that affects the whole region. It is essentially a problem between the metro-region and its hinterland.

PERSPECTIVES

The challenges faced by the urban centres of the developing countries are thus quite different from those faced by their counterparts in the developed parts of the world. It is a problem of overcrowding, of underemployment and unemployment, of inadequate social services and urban amenities, of poor urban finances, of *ad hoc* planning and development measures, of lack of attention by the national government, of rural poverty and agricultural stagnation, and of poorly developed human resources. In effect, it is a problem of national development with all its varied ramifications. The problem acquires more serious proportions when many of those who migrate to urban centres have really no stake in what happens to the city so long as they can earn a bit more than what they could back home in the village. Many of those who fail to make the grade in the competitive world of the city turn to urban vices, such as shoplifting, burglary, pickpocket, smuggling, and worst of all, prostitution, in order to make ends meet. Another effect of low or no income is the poor housing that dominates the metropolitan scene in all developing countries. It is estimated that more than half of the urban population of developing countries lives under poor and inadequate housing conditions.

There are about one million slum dwellers each in :

> greater Bombay and Calcutta (city corporation area), constituting about 25 and 31 per cent of the total population, respectively, of the two areas. In the year 1961, 4,00,000 persons, or 24 per cent of the total population of Madras City, were slum dwellers. In 1950, there were about 300 slum areas in the city. Since then, 180 areas have been cleared and yet, about 700 slum areas are estimated to exist even today. One-fourth of the residential areas of Lucknow are reported to consist of workers' bastis or slums.[5]

Another example concerns the Philippines. 'In 1963, Manila had an estimated 282,730 squatters, compared with 23,000 in 1946, and 98,000 in 1956. This grim picture has since deteriorated even further . . . . An official survey carried out in March 1968, revealed that there were 7,67,112 persons living as squatters . . . , while a later report . . . has it that by 1971, the number of squatters had reached almost 30 per cent of Manila's total population . . . .'[6] The situation has deteriorated further since then.

Poverty and inequality in developing countries, both in urban and rural areas, are not a natural phenomenon of an inevitable, temporary process of development as described by 'Kuznets curve'. In these countries, poverty will be perpetuated given the present 'North bias' in the international economic order, through which 'the Western countries exploit the weakness of the poor countries to weaken them further', and the national development policies that tend to bypass the poor. According to Harrison, 'Development policies have been biased against the poor, the backward, the rural, and skewed in favour of the rich, the westernized and the urban, for a very simple reason. It is politics that determines the direction of development, and the privileged groups have indefinitely more access to political power than the poor.'[7] Thus, 'urban bias' in development policies is not a product of metropolitan lobby, but rather a by-product of the 'rich biased' politico-economic order in both the national and international spheres. To explain poverty in the LDCs in terms of national conflict 'between the rural and urban classes' is a 'necessary but not sufficient' argument.

The bitter lesson that emerges from this reality is that the spatial strategies of distributing socio-economic activities or national policies on urban areas cannot be disassociated and treated independently from the broader issues of the national economic, social and political development on the one hand, and of international economic structure on the other hand. In the developing countries, the political commitments at the highest level and the appropriate adjustments of the governmental structure and modes of operation in response to the new and severe problems, in both national and international contexts, are required as the most crucial pre-requisites for handling the emergent and emerging problems.

The prospects for the future seems still worse, when one looks at what is happening to the economies of the developed countries. Despite the 'North biased' international economic order, the industrialized countries, which were doing well after the Second World War until the 'oil shock' in 1973 and 'the first few years of post-Vietnam War', are now facing a number of serious economic problems. According to the World Bank Report 1982 :

> Continuing recession and heavy unemployment in the industrialized countries are accompanied by real interest rates at unprecedently high levels, most commodity prices are at their lowest levels in three decades ; the volume of international trade has ceased to grow. Many developing countries, already struggling with large debt payments, have thus seen their problems exacerbated by the rise in interest payments, adverse trends in the terms of trade, and depressed export volume.[8]

One of the main causes for this is the continuing recession in the industrialized countries which are highly urbanized. This recession is however closely linked to the declining purchasing power of the LDCs where the majority of the poor of the world live. Whatever manufacturing potentials these countries have, is limited to the metropolitan areas. While it is true that the major front to fight poverty lies in the rural parts of the Third World countries, the urban centres offer the best and immediate environment to channelise the investment resources for socio-economic transformation. The issue is not whether urban centres are good or bad ; the issue is how to make them good and strong so that they can better serve the national development objectives.

REFERENCES

1. C. McEvedy and R. Jones, *Atlas of World Population History*, New York, 1978.
2. P. Harison, *Inside the Third World*, Penguin Books, 1982.
3. United Nations, *Patterns of Urban and Rural Development* (United Nations Population Studies, No. 68), New York, 1980.
4. S. Findley, *Planning for Internal Migration : A Review of Issues and Policies in Developing Countries* (Washington D.C. : United States Bureau of Census, 1977), p. 79.
5. Town and Country Planning Organisation, 'Towards a Settlement Policy in India—2001', New Delhi, mimeographed, 1975, p. 37.
6. D.J. Dwyer, *People and Housing in the Third World Cities* (London, Longman, 1975, p. 32.
7. P. Harison, *op. cit.*, 1982.
8. World Bank Report, 1982.

EIGHT

C.R. PATHAK

INDUSTRIAL DEVELOPMENT AND URBANIZATION PROCESS

INTRODUCTION

OUR SEARCH for linkages between the urbanization process and regional economic development in India has been on for quite sometime, although no conclusive finding as yet has been obtained. In this paper we would like to explain the linkage mechanism between the urbanization processes and industrialization patterns in India. India's efforts for rapid economic development through agglomerated industrialization have made the urbanization process a mere spatial consequence. Regional locational patterns of economic activities and specialization thereof, also bring about a restructuring of the regional systems of human settlements, and any distortion in one would have a tendency to influence the other.

The search for a spatial theory of urban-industrial growth applicable to all developing countries, and not merely to the urbanized and industrialized ones will be a difficult exercise, first because of a lack of spatial integration between these two processes in the former, and second because of the uniqueness characterizing the developing countries.

Urban-industrial growth as a cultural process is yet to result in a spatial order in developing countries, and India is no exception. However, India's urban-industrial growth processes have taken three distinct paths, wherein the emergence of the following divergent spatial patterns is quite discernible :

—entrepot port-based agglomerated industrialization with strong colonial heritage that continues to attract large-scale rural-urban migration, causing a lop-sided metropolitan urban growth :

—emergence of administrative-cum commercial regional capitals that have some national level resources based capital intensive industries, their decentralized concentrated location at favoured pockets showing symptoms of urban dispersion at a regional level ;

—penetration of market economy and the green revolution technology in the agricultural rural hinterland, leading to the emergence of rural market centres and further distorting the urban-industrial linkages.

These three processes together have manifested themselves in a relatively balanced distribution of urban settlements at the national aggregate level. But the spatial organization often breaks down at sub-national regional level, and vividly reveals the distortion in the spatial structure of urban and industrial growth. In a vast country such as India, three different growth patterns and spatial trends resemble three differen ttypes (if not more) of urban-industrial growth (or lack of it) at the regional level, each comparable to that of many smaller countries of the developing world.

While the first type has exerted, and is still exerting, a dominant influence on agglomerated urban industrial growth around the metropolises ; the second one brings modifications in the spatial trend by dispersing both the phenomena in conjunction with regional aspirations and socio-cultural consciousness ; whereas the third one is the result of a continuous de-industrialization process in the rural areas. The growing commercial exploitation of the rural market centres as a matter of fact is very much anti-urbanization. The above divergent regional trends of urbanization and industrialization in India clearly demonstrate a lack of integration, rather than interdependence between these processes, and need their in-depth analysis in time and space for a meaningful explanation of their regional variations and divergent patterns.

The relationship between urbanization and industrialization is to be understood first by treating the processes and their levels of development separately, and then by linking them in a functional form. The first step pertains to the various components that make up India's urban and industrial patterns of subnational levels and the second step pertains to the quantitative relationship between them at subnational and aggregate level. The urbanization pattern has to be explained within the framework of the spatial structure of the national economy.

The search for a causal relationship between urbanization and industrialization has a preconceived implication : that a healthy urban structure would be achieved when urbanization is integrated with industrialization. Urban growth without a corresponding economic growth would result only in a spurious relationship creating 'urban accretion' or 'pseudo-urbanization' and/ or 'parasitic' cities and towns. These cities and towns with a dominant tertiary sector, make themselves burdensome even at a low level of overall urbanization because of the inability of the space economy to sustain them.

URBANIZATION AND INDUSTRIAL PROCESSES

The proponents of urban growth theories often try to link the urbanization process with industrialization, and even endeavour to explain the former with the help of the latter process and extend such explanations to most of the countries irrespective of their distorted history of urban or industrial development. These twin processes have also been conceived as having a cumulative causal or circular relationship. Obviously, urbanization, metropolitanization and megalopolitanization are conceived as serial stages of industrial agglomeration and broadening employment opportunities, with the improvement of infrastructural facilities for spatial integration of the twin processes. Such a relationship even if it holds good at the national aggregate level, definitely would have variation at the subnational level because of varying regional factors. Besides, although the urbanization-industrialization relationship holds good at the initial stages, and continues until diseconomies start in any of them, it is the growth of tertiary and quaternary functions which explains the modifications of urban structure and form, and causes horizontal expansion of the metropolitan agglomerations. If such a relationship shows any spatial variation in the international or cross-country situation, we get concerned about it.

A hasty generalization of urbanization as a demographic counterpart of industrialization is an over-simplification of both the processes, particularly in the case of India.

Although relatively rapid industrialization has undoubtedly accelerated urbanization, particularly in and around large metropolises and new emerging industrial complexes, and both of them are getting spatially polarised, we do expect strong a spatial linkage between them. Industrial location policies were accordingly formulated for decentralized urban-industrial growth. Our expectation of a rapid urban growth analogous with industrial growth in India has to a great extent been belied. Although both the processes are simultaneous, they lack spatial integration and display wide variations in spatial pattern. Massive investments in capital goods industries in the resource rich but relatively backward regions, seemingly decentralized industrial locations but not the urbanization process. The reasons lie in the very pattern of industrialization. Incentives and subsidies have failed to counter the urban-industrial polarization process in small and medium size towns that have failed to respond positively. Under these circumstances it is desirable to identify the sub-processes of urbanization and industrialization at subnational levels, and look for patterns of spatial integration between them. According to Prakasa Rao and Tewari (1978) :

> In India as in other developing countries, the sub-processes operate independently or function with disproportionate mix. This results in a disjointed and primate system regionally. This system is unintegrated with metropolitan system at the apex, sub-standard rural-urban base, and weak intermediary links through "subsistence" urban-commercialization.

URBANIZATION TREND AND PATTERN IN INDIA

Urbanization in India has experienced a relatively slow growth rate throughout this century, with a little higher growth rate in the post-independence period. The slow rate of urbanization in colonial India could be explained by the destruction of the industrial system in the country-side, the far less than compensatory growth in the factory industry in towns, and the distortion in the economic development processes. In 1901 when the level of urbanization was as low as 10.85 per cent, small towns with a sound industrial base could account for as much us 48.8 per cent of the total urban population (Table 8.2). The process of de-industrialization in the countryside, which has started in the mid-nineteenth century with the establishment of factory industries in the cities, continued unabated until 1921. As a matter of fact, employment in non-agricultural occupations declined. For example, the secondary sector employment declined from 19.69 per cent to 9.70 per cent and in the tertiary sector from 16.83 per cent to 13.9 per cent during 1901-21.

Since 1921 India has been urbanizing at a relatively higher rate than before (Table 8.1) and could raise its urbanization level from 11.4 per cent in 1921 to 17.3 per cent in 1951, the urban population increasing by more than two-fold. The relative shares of city population increased from 25.3 per cent in 1921 to 38 per cent in 1951 with a corresponding decrease in the share of small towns from 48.8 per cent to 32 per cent during the same period. There was practically no change in the relative share of the medium size towns from the population point of view. This trend of urbanization could partly be explained by the relative increase in non-agricultural employment from 23.62 per cent in 1921 to 25.18 per cent in 1951. To meet the specific need of World War II, factory industries grew and were concentrated in the cities. The railway network also expanded from coast to coast, linking the metropolises and large towns to maintain uninterrupted supplies. The industries were disproportionately concentrated in Bombay and Calcutta which together accounted for 42 per cent of the country's registered factory employment and 67 per cent of the entire paid-up capital of manufacturing, and contributed 50 per cent of the total manufacturing output. The concentration of factory industry which started in the middle of the last century in the metropolises of today, was thus further accentuated. Of course spatial imbalance in industrialization preceded the urban imbalance : for example, cotton and jute textile manufacturing industries, employing 44.38 per cent of industrial workers and accounting for about 83.71 per cent of the total value-added by manufacturing in 1946, were located in cities whose share in the urban population of the country was

TABLE 8.1 : **Urbanization in India during the twentieth century**

Census year	Total population in million)	Percentage variation per decade	Urban population (in million)	Percentage variation per decade	Percentage urban to total population
1901	236.3	—	25.7	—	10.9
1911	252.1	+5.7	26.6	+2.4	10.6
1921	251.4	+0.3	28.6	+7.3	11.4
1931	279.0	+11.0	33.8	+18.4	12.1
1941	318.7	+14.2	44.3	+31.1	13.9
1951	361.1	+13.3	62.6	+41.2	17.3
1961	439.2	+21.5	78.8	+25.9 (34%)	18.0
1971	546.9	+24.5	108.9	+38.2	19.9
1981*	683.8	+25.0	144.0	+32.1	21.1
1981 (Excluding Assam and J.K.)	658.1	+24.4	156.2	+46.0	23.7
1980**	659.4	+19.7	143.8	+32.1	21.8
1985**	722.4	+17.9	165.6	+30.5	22.9
1990**	786.0	+16.6	189.4	+28.7	24.1
1996**	869.0	+15.6	220.7	+27.5	25.4

Source : Census of India, 1971 General Population Tables, Part IIA (1).
*Provisional Population Tables, Census of India, 1981.
**Projected in the Sixth Five-Year Plan, 1980—85.

Note : If the urban population of 1951 is adjusted to the 1961 definition of urban places, the quantum of urban population in 1951 would get reduced and the rate of urban population growth during 1951—61 will be 34 per cent.

hardly one-third.

Independent India's rate of urbanization has been faster than during the pre-independence period, but with continued concentration in large cities, and without any significant change in the level of urbanization. During 1951-61 the urban population grew at the rate of 26 per cent (34 per cent if 1961 definition of urban place is applied to the 1951 data), in 1961-71 by 38 per cent and in 1971-81 by 46 per cent. The tempo of urbanization which was the highest ever during 1961-71 cannot be considered very high when compared to the 25 per cent increase in total population, which had further increased in the seventies and is not expected to slow down in the eighties and nineties. The level of urbanization by the end of this century is expected to increase to 35 per cent. The large quantum of urban population we are going to have (220 million in 1996) should be a matter for concern, as it will be that of 1971 (Table 8.1).

In 1971 about 50 per cent of the urban population were found in the cities of 1,00,000 or more. While the medium size towns maintained their share, the small towns' share had come down to less than 20 per cent (Table 8.2). The process of urbanization in India has essentially been a process of migration to the big cities which have grown much faster than any other size-class during the last three decades. On the other hand, there have been structural stagnation in the lower class towns. If the present trend continues, the big cities and the metropolitan region are going to urbanize more, thus further accentuating the present day urban imbalance. Regionally also, it is the largest city that dominates the urban system. During the past three decades, although all the states have improved their urbanization levels, the ones having higher urbanization level have urbanized faster. The less urbanized states, in spite of their relatively high growth rate, have suffered most as they unfortunately are also the least industrialized. An analysis of the statewise urbanization

TABLE 8.2 : **Urbanization trends 1901—71 : Percentage of total urban population in each size class**

Year	Cities	Medium size towns	Small towns
1901	22.9	28.3	48.8
1911	24.2	28.6	47.2
1921	25.3	29.3	45.4
1931	27.4	30.7	41.9
1941	35.4	29.5	35.1
1951	38.0	30.0	32.0
1961	44.5	32.0	23.5
1971	48.9	31.7	19.4
1981	60.4	26.0	13.6

Source : Calculated from Census of India, 1971.

Note : *Cities* : having population 1,00,000 and above.

Med. Size Towns : Towns having population between 20,000 and 99,999.

Small Towns : Towns having population less than 20,000.

pattern clearly brings out the regional variations as well as departures from the national aggregate pattern. The metropolises (Calcutta and Bombay) themselves are infact going to experience decelerated growth, but metropolitan expansion (suburbanization) along the main arteries of transport, and urban growth in the administrative-cum commercial centres are likely to take place rapidly in response to increasing tertiary activities. On the other hand, rural-subsistence urbanization with a tertiary economic base in the backward states and undeveloped parts of the developed states, will be trailing behind.

Metropolitan urban growth in India has brought about regional primacy in the distorted city-size distribution. The level of urbanization of the more urbanized states has been kept high by metropolitan population growth, and the tempo of urbanization in these states will be determined only by the pattern of metropolitan growth. Urban imbalance is more severe in the states that have metropolises : as a matter of fact, the larger the metropolis, the greater is its shadow effect on its hinterland. This type of urban imbalance had its beginning in the colonial period and has been continuing ever since.

The spatial pattern of urbanization in India thus could be summarized as follows :

1. There is an extra-ordinary concentration of urban population in the metropolises. The metropolises themselves have of late not been growing very fast, resulting in a corridor urban pattern along the main transport arteries.

2. Since the beginning of modern industrialization in India in the 1950s, while factory industries were being located in the metropolises, the rural economy—both agriculture and village-based industries—was declining. As a matter of fact, urban-based industries have grown at the cost of their rural counterparts.

3. The urban-based industries were not functionally linked with the rural economy, hence both of them were gradually detached from each other, leading to an unfair and unhealthy competition between the two.

The result was a decline of the rural industries. The rural artisans had only two options : either to fall back on agriculture as labourers or migrate to the city ; in both the places they were found to be redundant.

Hence, India's urbanization imbalance has been shaped by

the colonial economic structure and has continued to influence the urban pattern even in the post-independence period. A real break in the quality of Indian urbanization is yet to be achieved (Pathak 1980).

INDUSTRIALIZATION PROCESS AND REGIONAL PATTERN

An attempt will be made to analyse the process of industrialization and growth of regional economy in order to seek explanations for the spatial urban processes and patterns as described earlier.

The spatial imbalance in urbanization during the colonial period had been brought about by the pattern of factory industrial location and capital accumulation in the major cities. Independent India has been trying to achieve rapid economic growth through industrialization within the inherited framework of a predominantly, centralized spatial structure, with an exploitative relationship between the metropolises and the rural hinterlands. This has resulted in regional imbalances as this process has mobilized the productive resources of specific areas of concentrated development and thus destroyed the rural hinterland industrial base. A severe hiatus between industrial-cum-commercial metropolitan and rural country side economy has thus been created.

A similar urban-industrial relationship has been extended—within the framework of a modified space economy—even to new industrial locations at aggregate level and regional level, thus replicating the metropolis-hinterland exploitative relationship. In this process the growing urban-industrial imbalance at the national level gets transferred to the regional level, resulting in a pseudo-balance at the national level. This phenomenon is often interpreted as the achievement of national economic planning and industrial dispersal policy. The very basic question of growth versus equity has remained unresolved.

India's overall industrial growth during the past 30 years has been relatively impressive, although not continuous. Industrial production increased at an average rate of 6.1 per cent per annum during 1951-78. The structure of industrial production has also undergone significant changes during this period. The output of basic and capital goods industries now accounts for about 50 per cent of industrial output, compared to only 20 per cent in 1951. The overall industrial growth has decelerated since 1966. Annual growth rate was around 7.8 per cent between 1951 and 1965 but declined to 4.5 per cent between 1965 and 1978 (Table 8.3).

TABLE 8.3 : **Industrial production increased an average rate**

Year	Index Number 1960=100	Year	Index Number
1951	54.8	1966	151.9
1952	57.2	1967	152.6
1953	58.1	1968	163.0
1954	63.4	1969	184.4
1955	72.7	1970	186.1
1956	78.4	1970=100	
1957	82.7	1971	104.2
1958	84.4	1972	110.2
1959	90.3	1973	112.0
1960	100.0	1974	114.3
1961	110.6	1975	119.7
1962	120.5	1976	132.4
1963	130.6	1977	138.3
1964	142.4	1978	147.8
1965	151.5	1979	149.5
		1980	150.6

Source : Monthly Statistics of the Population of Selected Industries, 1951—78, *Economic Surveys* 1970—78.

Table 8.4 reveals the growth performance of the Indian economy in various plan periods. Such growth at the aggregate level during the planning periods could hardly be related to the urban growth, but nevertheless some generalizations could be made. For example, in the period of rapid economic growth in the earlier half of the last three decades, urban growth was slow but when deceleration started in mid-sixties in industrial growth, urban growth was maximum. Again, in the late seventies when the economy was recovering from stagnation, urbanization rate slowed down. If urban and industrial growths are compared, two types of observations can be put forward. In the positive sense, industrial growth, with emphasis on capital goods industries, would have at least a 5-10 years gestation period, making

TABLE 8.4 : **Growth performance in the plans**

Sl. No.	Items	First Plan 1951-52 to 55-56	Second Plan 1956-57 to 60-61	Third Plan 1961-62 to 65-66	Annual Plans 1966-67 to 68-69	Fourth Plan 1969-70 to 73-74	Fifth Plan* 1974-75 to 78-79	1950-51 to 1978-79
	Annual Growth rates							
1.	National income	3.6	4.0	2.2	4.0	3.3	5.4	3.5
2.	Agricultural Production	4.1	4.0	(—) 1.4	6.2	2.9	4.2	2.7
3.	Industrial Production	7.3	6.6	9.0	2.0	4.7	5.9	6.1
4.	Per Capita Consumption	1.7	1.8	0.1	2.0	0.4	2.3	1.1
5.	Gross Fixed Investment	3.0	5.8	8.7	1.5	3.1	6.6	5.5

Source : Sixth Five Year Plan 1980—85, New Delhi : Planning Commission, Government of India.

Notes : Col. 9 gives the trend rate calculated from a semilong regression. All other columns give compound growth rates between the base year before the plan and the last year of the plan.

* For the time period originally envisaged for the Fifth Plan.

urbanization a consequential phenomenon. There may not even be any significant positive link between industrial and urban growths at the aggregate level, but only spatial linkage at the subnational level. In the latter case, an analysis with district level data is called for. Cross classification of Indian districts according to levels of industrial and urban development and their correlation analysis would have demonstrated the pattern of spatial linkage. This exercise will be taken up as a separate study.

An earlier study on spatial variation of urban and industrial growth for 1961 (Pathak 1975) has revealed a close relationship between urbanization and industrialization in the case of 89 industrialized districts out of a total 329 districts. Urban growth if any, in the non-industrialized districts could only be explained by growth in the tertiary sector.

Raza and Kundu (1979) have brought out the characteristic features of the regional structure of Indian industrialization and have related them to the urbanization process. They have identified two patterns of industrialization : agglomerated and dispersed. The former is based mainly on the major port towns of colonial India, and the persistence of agglomerated industrialization therein suggests that the inherited industrial structure still constitutes the hard core of the industrial base of the country, and that the development impulses generated in independent India have only modified the enclave characteristics of its space economy. The pattern of dispersed industrialization is however more evenly distributed and the associated process shows interregional disparity of a much lower order.

The pattern of industrialization in India could be summarized as follows :

1. The dominant pattern of industrialization in India is the result of deconcentration of the old clusters of traditional industries, over which technological clusters have brought about territorial expansion and diversification in industrial structure. This has resulted in the formation of broad industrial regions such as Ahmedabad-Baroda, Bombay-Pune, Calcutta-Damodar Valley, Madras-Chinglepet, etc.

2. A new region has emerged, mainly because of technological clusters, in north-west India comprising the states of Punjab, Haryana, western Uttar Pradesh and northern Rajasthan with high level productivity. Such industrial growth is being supported

by equally high rural agricultural and tertiary sector economic growth.

3. The coal-steel belt of India, traditionally known as Chotanagpur plateau, has grown recently, attracting a number of resource based industries of national importance. These spatial and technological clusters, having multi-nuclei location with every possibility of developing into industrial complexes, have yet to integrate themselves into the regional economy.

4. There are at least three dozen isolated industrial districts spread out in the industrially backward or developed states as pockets with one or two large industries, but without being integrated into the regional economy. These exogenous industrial city districts have grown with heavy public investments under social and political compulsions to achieve industrial dispersion. These isolated industrial districts are functionally linked with each other and with the metropolises rather than with their immediate hinterlands. In the case of resources-based industries, even though there is backward linkage in the region for some natural resources inputs ; for forward linkage they look to metropolitan-based industries, and the national market for output marketing. Unlike the major industrial regions they are least diversified and do not themselves form any industrial region. These capital-intensive technological clusters and/or resource-based industries, with products having nation-wide demand, could neither generate large-scale employment for nor could get themselves integrated into the regional economy, and hence could not solve the problems of industrial backwardness in the regions. The decentralization policy of industrial location, particularly in case of large public sector projects, together with the development of a rapid interregional transit system, have strengthened the functional linkage with the metropolitan mercantile economies rather than developing sub-regional industrial economies.

District Level Analysis of Levels of Urban and Industrial Development 1961-71

In the absence of 1981 data we shall construct districtwise urban, secondary and tertiary activity indices for the two time periods 1961 and 1971, to examine their relationship at the national aggregate level.* The following variables have been used to construct the respective indices :

Urban Activity Index—(U)

1. Share of urban population per thousand of total population.
2. Density of urban population per km² of total area.
3. Urban density of population in thousand per km² of urban area.
4. Average size of population of towns in thousand.

Secondary Sector Activities (Industrial—I)

1. Density of labour in secondary sector per km².
2. Share of labour in secondary sector per thousand of total labour.
3. Share of labour in large factories employing more than 100 workers per thousand of labour in secondary sector.
4. Density of labour in large factories employing more than 100 workers per thousand km².
5. Average size of labour of large factories in thousand.
6. Density of large factories per thousand km².

Tertiary Sector Activities (Commercial—T)

1. Density of labour in tertiary sector per km².
2. Share of labour in tertiary sector per thousand of total labour.
3. Density of labour in trade and allied sector per km².
4. Share of labour in trade and allied sector per thousand of total labour.

The variables have been logarithmically transformed before deriving composite indices in respective categories. The composite indices have been derived in the form of

$$X=a_1x_1+a_2x_2+a_3x_3+\ldots a_nx_n$$

where X is the desired index and $x_1, x_2, \ldots x_n$ are the transformed variables and $a_1, a_2, a_3 \ldots a_n$ are the logical weights.

Thus derived Urban Activity Indices (U) for the year 1961

* Interstate and interdistrict analysis has been done but not reported here and will be used in a separate study.

and 1971 have been regressed on the Secondary Sector Indices (I) and Tertiary Sector Indices (T) for both periods taking 334 districts of India :

Regression equations taking 334 districts of India for 1961 and 1971 are as follows :

$$U\,61 = 0.042\,I + 0.916\,T - 0.0753, \quad R^2 = 0.67 \qquad \ldots(1)$$
$$(4.77) \quad (13.30) \quad (1.49)$$

$$U\,71 = 0.042\,I + 0.985\,T - 0.0036, \quad R^2 = 0.69 \qquad \ldots(2)$$
$$(1.54) \quad (14.80) \quad (0.08)$$

Multiple Correlation Matrix between Urbanization Index (U) and Secondary (I) and Tertiary Sector (T) indices for 1961 and 1971 are :

	1961				1971		
	I	*T*	*U*		*I*	*T*	*U*
I	1.000	0.749	0.703	*I*	1.000	0.807	0.697
T		1.000	0.805	*T*		1.000	0.829
U			1.000	*U*			1.000

At aggregate level urbanization in 1961 has been significantly correlated ($R^2=0.67$) with both secondary and tertiary activities ; on the other hand, in 1971 urbanization index is positively and significantly correlated with tertiary sector whereas positive correlation between secondary sector and urbanization is significant only at 10 per cent level. In other words, rapid urbanization during 1961-71 could be explained mainly by tertiary activity growth. Besides, the analysis indicates a non-linear relationship between urbanization and other economic variables. In other words, urbanization cannot fully be explained in terms of industrial and commercial activities.*

CONCLUSION

Urban growth in India had originated during the colonial period in the port/administrative centres with a view to exploiting the

* The data base is derived from an unpublished Ph.D. thesis of ISI by R.N. De.

country's resources to feed the economy of the Empire. The pattern of urban growth is far from indigenous. It was exogenously superimposed over the traditional village centred economy and has not only distorted the indigenous economy but also has resulted in an exploitative mercantile relationship between the countryside and the administrative-cum-capital towns and cities. While urban growth in the developed economies was supported by industrial and technological innovations, in the developing countries it destroyed the industrial economy and growth impulses in the countryside. Both processes being cumulative in nature, they gradually revealed the difference in the pattern of urban growth.

The pattern of relationship that was established during the colonial period between the industrial-cum-mercantile urban centres and their rural hinterland of primacy produce, continued without much change even after independence. The unfair trade relationship that grew between agriculture and rural industries on the one hand, and the urban-based modern industries on the other, had penetrated into the remote rural areas.

The urbanization process in a country such as India is a very complex phenomenon. India is a unique country and at the same time a path-finding representative country in the developing world in the process of socio-economic change, particularly in the field of urban and regional change. India, with a huge rural population base and low level of urbanization has been industrializing at a relatively impressive rate, but has experienced a relatively lower rate of urbanization which is expected to further decelerate. Modernization of its productive technology and increasing industrial and agricultural production will not materially alter the processes and pattern of urbanization.

Industrialization in India since 1950 has been quite impressive, but only in terms of growth rates. Industrial employment and production have multiplied in absolute terms but without any significant change in the occupational structure, and only a slight one in its relative contribution to the economy. Industrial diversification, aimed at import substitution in capital goods and consumer durables, have mainly benefited major cities.

As a matter of fact, rural India is not effectively participating in the process of urbanization. Urbanization in the rural areas has been mainly restricted to the growth of market towns with weak industrial bases. As a result, commercial (tertiary) activities have increased and are gradually converting the rural hinterlands into dependent markets of the urban economy. The growth potential and market demand of rural economy would therefore determine the pattern of urban growth. In the agriculturally prosperous areas, market towns could sustain their growth only by commercial activities. The vast areas of subsistence agricultural economy, on the other hand, would support only a few market towns, their size being determined by the extent of the service area rather than by economic growth.

The urbanization processes and patterns in India could hardly be explained by the process and pattern of industrialization, at least at the national aggregate level. Tertiary activities are rather more important in explaining the impulses of urban growth. The urbanization processes, which have been mainly the result of rural-urban migration are primarily tertiary activity oriented. At the subnational level, there could be a significant positive relationship between urbanization and industrialization, but that too is limited to the industrial cities. Commercial urbanization is the most dominant pattern elsewhere.

REFERENCES

Deshpande, C.D. and L.S. Bhat, 'India', in *Essays on World Urbanization*, 1980, pp. 358-376.

De, Rabindranath, 'Economic Regionalization of India 1950-61 and 1970-71 : A Study in Quantitative Methods' (1981) (Unpublished Ph.D. dissertation submitted in Indian Statistical Institute, Calcutta).

McGee, Terry G., 'The Urbanization Process and Industrialization in the Marketing Economies of Asia', in *Regional Development Dialogue*, Vol. 1, No. 2, UNCRD, Nagoya, (1980), pp. 135-159.

Pathak, C.R., 'Spatial Variation in Urban and Industrial Growth in India', *Indian Journal of Regional Science* 7 (1975), 1-10.

Pathak, C.R., 'Urban Problems and Politics in India', in *Indian Journal of Regional Science* 12 (1980), 71-80.

Prakasha Rao, V.L.S. and Tewari, V.K., *Urbanization in India : Spatial Dimensions* (Bangalore, Institute for Social and Economic Change, 1978).

Raza, Moonis and Kundu, Amitabh, *Agglomerated and Dispersed Patterns of Industrialization in India—A Spatial Analysis*, Nagoya, UNCRD, 1979.

NINE

N.B.K. REDDY AND M. UTTARA PRABHU

THE COMPONENTS AND CHARACTERISTICS OF CITIES AND TOWNS OF TAMIL NADU

INTRODUCTION

A STUDY of urban functions suffers from the constraint of limited range. Knowledge of urban geography is aided by developments in multivariate analysis, and the manipulative abilities of modern computers have extended the study of urban characteristics to include simultaneous consideration of much wider ranges of variables, including the social characteristics of the residents and the physical properties of the urban structures. The result of this study indicate that there are in fact but a few basic ways in which cities vary. The large numbers of characteristics that may be collected for each town show high degrees of intercorrelation, indicating that whole groups of variables give exactly the same information about town to town variation. Each group of characteristics, in effect, behaves like the multiple symptoms of a more fundamental 'syndrome' or 'dimension' of variation. By employing multivariate analysis to study the patterns of interrelationship among the characteristics, the basic dimensions may be isolated. The type of multivariate analysis used most commonly is factor analysis.

The complexity of different urban centres have been analysed here with the help of principal component analysis. A wider application is found in the settlement studies by Ahmad (1965), King (1966), Abiodum (1907), Berry (1967), Swaminathan (1974, 1976, 1978, 1979), Swaminathan et al. (1979), Jayamani (1979).

For the purpose of analysis, all the 241 urban agglomerations and towns of 1971 have been included since 1981 data are not yet available. In this study, urban agglomerations are considered, along with individual cities, as equal elements of urban settlement space. The main aim is to bring out the underlying dimensions of the urban settlements of Tamil Nadu.

THE LIST OF VARIABLES USED IN THE ANALYSIS

1. Population of the city or town
2. Density of Population per km^2
3. Area in km^2
4. Density of Houses per km^2
5. Houses per 1,000 population
6. Households per 100 population
7. Percentage literacy
8. Female literacy
9. Administration
10. Civic administration
11. Revenue expenditure
12. Total number of workers
13. Workers engaged in household industry
14. Workers engaged in manufacturing
15. Workers engaged in trade and commerce
16. Workers engaged in transport and communication
17. Workers engaged in other services
18. Percentage of primary workers
19. Percentage of secondary workers
20. Percentage of tertiary workers
21. Ratio between secondary and tertiary workers
22. Ratio between primary and non-primary workers
23. Sex ratio of workers
24. Sex ratio of population
25. Educational facilities
26. Medical facilities
27. Source of protected water supply
28. Capacity of protected water supply
29. Sewerage
30. Latrines
31. Banks

32. Road length per km^2 in towns
33. Nodality of transport lines.

The data is reduced to a symmetric 33 × 33 correlation matrix. Most of the correlations in this matrix are low, a phenomenon also noticed by Moser and Scott in their study of British towns. According to them, the low correlations are a sign of the diversity of the British towns. The same can be said about the Tamil Nadu towns. Since many of these 33 variables may not be significant it is necessary to eliminate the redundancies in the set of variables by developing an orthogonal basis for the space in which 241 towns were distributed. The correlation matrix therefore is subjected to a principal axis solution which has yielded eigen vectors which are rotated to a normal vari-max position. These five components are also identified to have some recognisable significance.

In Tables 9.1 and 9.2, the proportion of the total inter-town variance accounted for by each of the five components is summarized. The five rotated components together explain 72.2 per cent of the total variance of the 33 variables. Thus, the variation between the towns as contained in 33 variables can be attributed to five components, while the remaining components may be treated as minor in terms of specific or random variance.

Table 9.4 shows the percentage of the variance of each primary variable accounted for by all the five components. These values are the total sum of the squares of the rotated factor loadings of the particular variable with all the five components or communalities.

TABLE 9.1 : **Cities and towns of Tamil Nadu : Percentage of total variance of components**

Component	Eigen value	Percentage variance	Cumulative variance (in percentage)
I	9.972	30.2	30.2
II	6.090	18.5	48.7
III	3.821	11.6	60.3
IV	2.058	6.2	66.5
V	1.890	5.7	72.2

TABLE 9.2 : **Factor loading matrix for 33 variables (unrotated)**

Variable	I	II	III	IV	V
1	—.859	.478	.102	.035	.017
2	—,448	—.341	—.506	—.610	.020
3	—.580	.592	.250	—.039	.018
4	—.399	—.327	—.526	—.613	.151
5	.346	.267	.104	.041	.722
6	.249	.214	.050	—.49	.706
7	—.499	—.555	.017	.043	.381
8	—.432	—.564	.140	.074	.457
9	—.810	—.039	—.058	—.067	—.039
10	—.603	—.120	—.308	—.196	—.083
11	—.836	.456	.110	.034	.016
12	.358	.693	—295	—.205	.047
13	.200	.353	—.655	.241	—.052
14	—.258	.026	—.454	.139	.140
15	—.403	—.633	—.006	.118	—.108
16	—.372	—.295	—.015	—.024	—.103
17	—.374	—.560	.309	.122	.134
18	.521	.423	.575	—.381	—.028
19	—.067	.266	—.871	.338	.071
20	—.545	—.745	.146	.131	—.032
21	.251	.474	—.681	.257	.037
22	.363	.435	.446	—.468	—.098
23	.360	.612	—.152	—.222	.160
24	.154	—.148	.162	.145	.301
25	—.854	.403	.082	.038	.020
26	—.849	.364	.100	—.029	.058
27	—.735	,413	.123	.077	.049
28	—.726	.446	.166	.053	.026
29	—.078	.147	—.148	—.090	—.050
30	—.742	.483	.170	.057	.028
31	—.894	.377	.042	—.001	.011
32	—.264	—.269	—.369	—.535	.124
33	—.721	—.035	—.116	.009	—.124
Eigen values	9.972	6.090	3.821	2.058	1.890
Cumulative percentage of Eigen values	30.2	48.7	60.3	66.5	72.2

The degree of representation of each of the initial variable is characterised by the value (h^2) of communality which shows

TABLE 9.3 : Varimax rotated factor matrix for 33 variables

Variables	Components					h²
	I	II	III	IV	V	
1	993					99.6
2				930		92.6
3	807					72.9
4				949		93.4
5					851	80.0
6					825	71.3
7		427				26.2
8		350				19.1
9	586	337				51.9
10	317			375		32.7
11	969					95.1
12		– 649	399			654
13			873			80.1
14						12.2
15		684				58.4
16		410				22.9
17		567				41.5
18		—743	—447			81.5
19			872			77.5
20		841				83.7
21			844			78.7
22		—763	—397			75.3
23		—726	312			67.5
24						2.3
25	920					85.6
26	881					78.7
27	841					71.1
28	902					81.9
29						3.0
30	940					88.8
31	936					90.5
32				799		64.5
33	492	349				40.2

Note : Decimals are omitted.
Factor loading greater than 0.30 is considered.
Factor loading greater than 0.70 is considered for factor representation of the variable.

percentage of the variance of each primary variable accounted for extracted components.

It is clear from Table 9.3 that some of the variables are better represented than others by the five dimensions. The variables that are least accounted for by the five components are : female literacy, workers in manufacturing, sewerage and sex ratio of workers.

On the other hand, about 70 per cent of the total number of variables are summarized quite well by the five components. This implies that the variations in most of the 33 characteristics are effectively captured by the five components.

The results of the calculations are given in two matrices, that is : (i) the matrix of paired correlations of all variables, and (ii) the matrix of factor loading. The factors are interpreted according to the values of factor loadings. Although the matrix of correlations is an intermediate result, its analysis is of an independent interest (Goryacheva 1976, p. 288).

A brief description of the data set is contained in the matrix of factor loadings, which is the result of processing the correlation matrix, according to the principal axis solution. Factor loadings are the coefficients of strength of relationship between the factor and the corresponding variable. The signs of factor loading characterize directions of relationship.

As a result of calculations, five factors (components) have been obtained which together explain 72.2 per cent of the total variance of 33 variables. The share of the first two components in the total variance accounts for 48.7 per cent.

It is interesting to note that the five components obtained show a high degree of representation of the primary variables. There are four variables that are poorly accounted for. Twelve variables represent more than 80 per cent of communality.

AN ANALYSIS OF THE FIVE COMPONENTS

As explained earlier, about 70 per cent of the 33 variables form into five principal components each comprising interrelated variables. Some of the variables have recurred in more than one principal component. In each component, significant variables are located on the basis of their loading. The variables that have a loading of about 0.7 or more are considered as significant. Here it may be noted that the decimal point is omitted when the loadings are tabulated against the respective variables. The significant variables of each component thus isolated are supposed to participate not only in differentiating the urban settlements on the basis of their characteristics but also function in spatial organization and region-forming processes.

A detailed account of each of the five components is given hereunder (Tables 9.4, 9.5, 9.6, 9.7 and 9.8).

Component I : The Component of Urban Amenities and Demographic Size

The first component with 30.2 per cent of variance accounts for the largest proportion of total variance. Out of the 12 variables in this component, 9 of them have significant loadings. This component represents the basic or the major dimension of the characteristics of the urban settlements of Tamil Nadu. Hence all the variables both significant and non-significant have positive loadings. The variables such as population, revenue expenditure, latrines, banks, educational facilities, capacity of protected water supply, etc. carry very high loadings of more than 0.9.

Broadly speaking, component I includes essentially urban amenities and facilities besides population size. Hence, it may be appropriately designated as the 'Component of Urban Amenities'.

TABLE 9.4 : **Component I**

Sl. No.	Variable	Loading*
1.	Population	993
2.	Revenue expenditure	969
3	Latrines	940
4.	Banks	936
5.	Educational facilities	920
6.	Capacity of protected water supply	902
7.	Medical facilities	881
8.	Source of protected water supply	841
9.	Area	807
10.	Administration	586
11.	Nodality	492
12.	Civic administration	317

*Decimals are omitted.

Component II : The Tertiary Component

This component includes 12 variables in all. However, only four of the variables are significant and show the second largest variance of 18.6 per cent. Out of them only one variable, namely, tertiary workers has a highest positive loading, and the remaining three variables (ratio of primary to non-primary workers, primary workers and sex ratio of workers) have significant negative loadings. It would be appropriate to treat the variable of workers in trade and commerce also as significant since its loading is 0.68.

Obviously, component II has a strong orientation towards tertiary workers and it represents the second dimension of the characteristics of the urban settlements of Tamil Nadu.

The variables representing the primary working force also are significant, differentiating negative variables in the sense that they indicate the strong or weak rural setting of the urban settlements. Component II may be aptly described as "Tertiary Component".

TABLE 9.5 : **Component II**

Sl. No.	Description	Loading
1.	Tertiary workers	841
2.	Ratio of primary/non-primary workers	—763
3.	Primary workers	—743
4.	Sex ratio of workers	—726
5.	Workers in trade and commerce	—684
6.	Total workers	—649
7.	Workers in other service	567
8.	Literacy percentage	427
9.	Workers in transport and commerce	410
10.	Female literacy	350
11.	Nodality	349
12.	Administration	337

Component III : The Secondary Component

With a variance of 11.6 per cent, this component represents the third important dimension of the characteristics of the urban settlements of Tamil Nadu. Among the seven variables it is comprised of, three of the variables, namely, workers in household industry, secondary workers and ratio of secondary to tertiary workers are significant. Incidentally, all of them have positive loadings. This component may be termed as the 'Secondary component'. This component represents the significance of industry, especially the household industry, as a factor in differentiating the urban settlements of Tamil Nadu.

TABLE 9.6 : **Component III**

Sl. No.	Variables	Loading
1.	Workers in household industry	873
2.	Secondary workers	872
3.	Ratio of secondary/tertiary workers	844
4.	Primary workers	—447
5.	Total workers	399
6.	Ratio of primary/non-primary workers	—397
7.	Sex ratio of workers	312

Component IV : The Compactness Component

This component comprises only four variables out of which three variables, namely, density of houses, density of population and road length have high loadings and hence are significant. Of these three variables, the first two have very high loadings. This component represents the fourth important dimension of the characteristics of the urban settlements of Tamil Nadu. This component may be designated as a 'Density Component' or 'Compactness Component'. It tends to differentiate the urban settlements on the basis of densities of population, dwellings and road length. It throws light on the pressure of population, housing problems and circulation in the urban settlements of Tamil Nadu.

TABLE 9.7 : **Component IV**

Sl. No.	Variables	Loading
1.	Density of houses per km²	949
2.	Density of population	930
3.	Road length	799
4.	Civic administration	375

Component V : The Components Houses and Households

This component which comprises only two variables is the fifth and also the last dimension of the characteristics of the urban settlements of Tamil Nadu. Both the variables, namely, houses per 1000 population and households per 1000 population, are significant. In fact this component is very much akin to component IV in its character. This component may be described as the component of 'Houses and Households'.

TABLE 9.8 : **Component V**

Sl. No.	Variables	Loading
1.	Houses per 1000 population	851
2.	Households per 1000 population	825

In conclusion it may be stated that five principal components could be discerned representing 23 variables out of a total of 33 variables selected for the purpose of the present analysis. In order of importance based on the dimensions of the characteristics of the urban settlements of Tamil Nadu the following are listed, (i) component of amenities and demographic size, (ii) component of tertiary activity, (iii) component of secondary activity, (iv) component of density or compactness, and (v) component of houses and households.

Broadly speaking, the tertiary activity and urban amenities have been shown to exert a profound differentiating influence on the urban settlements of Tamil Nadu. The primary activity which carried negative loadings also differentiates the urban settlements on the basis of the strength of rural setting. The secondary activity is no doubt significant in the process of urbanization, but its role is only third in importance. In most of the urban settlements the density of population and houses as well as houses and households per 1000 population are high, since most of the urban settlements are very crowded. Hence, as a differentiating factor they play a relatively less important role when compared with the urban amenities, tertiary and secondary activities. At the risk of repetition it may be emphasized that the population size of the urban settlements which is the first important variable in component I, itself has rightly executed its primacy as an important differentiating factor. This inference is in agreement with the census classification of urban settlements on the basis of population size.

GENERAL EVALUATION OF THE FACTOR STRUCTURE

The principal component analysis of the 33 variables thus extracted 5 major dimensions of variation with reference to the 241 urban settlements of Tamil Nadu. A general evaluation of the factor structure rests on a proper appreciation of the significance of the following points :

1. The factors or components have the characteristic of being independent of each other.

2. The factors are additive, that is, each component only tells part of the story, since it explains only part of the total variance. In the words of Moser and Scott the components are 'weighted average of the primary variables, the latter contributing in very different degrees to the different components.' In the case of the first and fourth components, the weights are greater and therefore the association stronger in the case of those variables that are highly correlated with them. The other components though accounting for decreasing proportion of the total variance, play a role which is both complementary and supplementary. The analysis was done on a fairly large and hetrogeneous set of variables. Yet the five factors account for about 70 per cent of the total variance. In fact, the number of factors reflects the great diversity of the cities and towns of Tamil Nadu.

COMPONENT SCORES

The factor scores of components I and II capture the spatial incidence of the components and characteristics of the towns of Tamil Nadu.

The positive scores on component I which is also the 'Component of Urban Amenities' exhibit the following pattern (Table 9.9) : The highest score is obtained by a class IV town, that is O' Valley (13,785) which is located in the hilly district of the Nilgiris. This high score on component I is perhaps owing to the high municipal expenditure on socio-economic infrastructural development in a small town such as this. The expenditure

TABLE 9.9 : **Component I**

Positive Scores		Negative Scores	
Town	Scores	Town	Scores
O'Valley	14.84	Madras	110.74
Highways	14.55	Coimbatore	39.00
Samsikapuram	14.14	Madurai	30.61
Vennandur	12.93	Tiruchirapalli	23.28
Ithalar	12.75	Vellore	16.28
Thirumullaivasal	11.62	Salem	14.24
Proddaturpet	11.24	Tirunelveli	14.22
Morattupalayam	10.94	Thanjavur	13.66
Gudalur	10.69	Nagercoil	13.25
Devarshola	10.67	Virudhunagar	13.09
Valparai	10.27	Dindigul	12.54

is relatively very high when compared with the expenditure on the primate city of Madras and on other larger cities such as Coimbatore, Madurai, Tiruchirapalli, Salem, Tirunelveli. The great aggregations of people in these cities have obviously caused extraordinarily low level of per head civic amenities in these cities. Yet the degree of modernization in these larger cities is high. Their degree of accessibility, functional status and magnitude of aggregation are perhaps responsible for this.

It can also be noticed that most of the plantation towns in the Nilgiris and the foothill towns are comparatively new and have relatively high scores on component I. They include Highways, Samsikapuram, Ithalar, Gudalur. Devanshola, Valparai. The small towns in general have relatively higher socio-economic infrastructure development per head. All the class I cities of 1971 have negative scores. In general, high densities of population are negatively correlated to lower levels of socio-economic infrastructure.

The negative component scores on component I are more widely distributed throughout the state. The most striking feature is that 13 cities have very low scores. The coastal region, particularly the Cauvery delta region, which has a high agricultural potential shows clustering of large settlements with negative scores and the small settlements with positive score.

The clustering of towns with positive scores on component II or 'Tertiary Component' has a wide distribution throughout the state. No clearly defined pattern of towns emerged with negative scores. The range of scores, both positive and negative, varies between 46.04 and —16.16. The variables explained by component II are mostly workers employed in tertiary, primary, trade and commerce. The higher positive scores are associated with the large and medium towns and the lowest negative scores are associated with the smaller towns.

TABLE 9.10 : **Component II**

Positive Scores		Negative Scores	
Town	Scores	Town	Scores
Madras	46.04	O'Valley	16.16
Vennandur	15.83	Proddaturpet	11.17
Valparai	14.54	Sivaganga	10.56
Coimbatore	14.48	Ramanathapuram	510.13
Samsikapuram	14.12	Arkonam	9.26
Devarshola	12.19	Thisayanvilai	8.54
Naduvattam	11.54	Pudukkottai	8.33
Courtallam	10.95		
Pillanallur	10.81		
Mallasamudram	10.74		
Gudalur—N	10.48		
Pallapatti	10.02		

Thus it may be observed that locational importance and accessibility of the large towns have only reduced the per capita availability of basic socio-economic facilities, owing to the high magnitude of population. The study revealed that a differential developmental policy is required for providing amenities and infrastructural facilities for different sizes of towns.

REFERENCES

Berry, B.J.L. and A. Mohammad, 'Spatial Incidence of Urban Induced Modernity in Andhra Pradesh', in *Indian Urbanisation and Planning*, A.G. Noble and A.K. Dutt (ed.), Tata McGraw Hill, New Delhi, pp. 93-113.

Carter, H., "Structure and Scale in the City System in M. Chisholm and B. Rodgers (ed.), *Studies in Human Geography*, London, Heinemann, 1973 pp. 172-202.

Census of India, *A District Census Handbook –Village and Town Directory, Madras City District*, Series 19, Tamil Nadu, pt. 10.

Census of India 1971, *A District, Census Handbook—Village and Town Directory Chingelput District*, Series 19, Tamil Nadu, pt. 10.

Census of India 1971, *A District Census Handbook—Village and Town Directory North Arcot District*, Series 19, Tamil Nadu, pt. 10.

Census of India 1971, *A District Census Handbook—Village and Town Dirctory South Arcot District*, Series 19, Tamil Nadu, pt. 10.

Census of India 1971, *A District Census Handbook—Village and Town Directory Dharampuri District*, Series 19, Tamil Nadu, pt. 10.

Census of India 1971, *District Census Handbook—Village and Town Directory Salem District*, Series 19, Tamil Nadu, pt. 10.

Census of India 1971, *A District Census Handbook—Village and Town Directory Coimbatore District*, Series 19, Tamil Nadu, pt. 10.

Census of India 1971, *A District Census Handbook—Village and Town Directory Nilgiris District*, Series 19, Tamil Nadu, pt. 10.

Census of India 1971, *A District Census Handbook—Village and Town Directory Madurai District*, Series 19, Tamil Nadu, pt. 10.

Census of India 1971, *A District Census Handbook—Village and Town Directory Tiruchirapalli District*, Series 19, Tamil Nadu, pt. 10,

Census of India 1971, *A District Census Handbook—Village and Town Directory Thanjavur District*, Series 19, Tamil Nadu, pt. 10.

Census of India 1971, *A District Census Handbook—Village and Town Directory Ramanathapuram District*, Series 19, Tamil Nadu, pt. 10.

Census of India 1971, *A District Census Handbook—Village and Town Directory Tirunelveli District*, Series 19, Tamil Nadu, pt. 10.

Census of India 1971, *A District Census Handbook—Village and Town Directory Kanniyakumari District*, Series 19, Tamil Nadu, pt. 10.

P. Jayamani, 'Aspects of Potential Urban Nodes and Urban Rural Linkages in Ramanathapuram District' (Ph.D. Thesis, University of Madras), 1979, pp. 116-56.

R.J. Johnston, *Multivariate Statistical Analysis in Geography*, London : Longman, 1978), pp. 127-82.

L.J. King, *Statistical Analysis in Geography*, Prentice Hall, 1969, pp. 166-93.

A. Mitra, S. Mukherjee and R. Bose, *Indian Cities Their Industrial Structure in Migration and Capital Investment*, New Delhi (Abhinav Publications, 1980), pp. 8-10, and 20-30.

K. Prabha, *Towns A Structural Analysis : A Case Study of Punjab*, Delhi (Inter-India Publications, 1979).

B.T. Robson, *Urban Growth : An Approach* (London ; Methuen & Co., 1973).

E. Swaminathan, 'Coimbatore Metropolitan Area', *The Indian Geographical Journal* 59 (1974) : 78-86.

E. Swaminathan, 'Market Centres and Consumer Preference within Coimbatore Metropolitan Area', (Ph.D. Thesis, Madras University, 1976), pp. 77-100.

E. Swaminathan, 'The Occupational Structure of Small Towns in Coimbatore District, Tamil Nadu, A Factorial Approach', *Indian Geographical Journal*

52 (1977) : 68-78.

E. Swaminathan, 'A Quantitative approach to the study of Occupational Structure of Small Towns : A Comparative analysis between Coimbatore and Thanjavur Districts', *Journal of Madurai Kamaraj University* (1978) : 37-46.

E. Swaminathan, 'A Quantitative Approach to the Study of the Spatial Distribution and Functional Hierarchy of Shopping Centres in Coimbatore City', *National Geographical Journal of India* 25 (1979) : 231-53.

E. Swaminathan, A. Rajmohan and V. Nagarajan, 'Small Towns of Coimbatore and Thanjavur Districts : A Comparative Study of Pattern, Characteristics and Hierarchy, in (ed.) R.L. Singh and R.P.B. Singh, *Place of Small Towns in India* (Varanasi : National Geographical Society, 1979), pp. 127-42.

P.J. Taylore, *Quantitative Methods in Geography* (Boston : Houghton Mifflin Co., 1977).

M. Yeates, *An Introduction to Quantitative Analysis in Human Geography* (New York : McGraw Hill, 1974).

TEN

AMAR SINGH

URBAN ACCRETION VERSUS URBANIZATION
A Case Study of the National Capital Region, India

INTRODUCTION

THE National Capital Region of India is an interstate region covering an area of about 30181.1 km² (1975) and having a population of 14.18 million (1971). It includes besides the Union Territory of Delhi, the districts of Gurgaon, Rohtak, Sonepat, and the tehsils of Panipat and Rewari, of Karnal and Mahendergarh districts, respectively (Haryana State); the districts of Meerut, Bulandshahar and Ghaziabad[1] (Uttar Pradesh); and the tehsils of Alwar, Behror, Kishangarh, Mandawar and Tijara of district Alwar in Rajasthan (Fig 10.1).

The genesis of the concept of the National Capital Region lies in the Master Plan of Delhi (1962), wherein it has been recommended that not only the Delhi Metropolitan Area but also a larger region comprising several districts adjoining Delhi should be planned as a cemposite unit on a micro-level. A comprehensive regional plan was drawn and the draft was approved on 17 September 1973. The objective was to deflect the population coming to Delhi in search of employment. It was presumed that the development of smaller towns and cities, forming various rings round Delhi city and the encircling rural areas would help to reduce the rate of migration to Delhi to zero level by the end of the century. Unfortunately, the National Capital Region plan did not work well. As a result it was found that Delhi and its population continued to grow unabated and ring towns did not

FIG. 10.1 : *National capital region : administrative set-up 1975*

develop in accordance with the plan.

The horizontal movement of the people within the National Capital Region has resulted in the increase of urban population and the growth of cities. It has been noted that though the industrial capacity of the cities has not expanded much and that there is a lack of jobs for new-comers, big cities such as Delhi, Meerut, Ghaziabad, Rohtak and Gurgaon continue to attract large number of immigrants from rural areas who constitute a floating population of economic refugees. It is, therefore, tempting to examine whether such urban growth constitutes urbanization or is merely urban accretion. Urbanization, in the real sense, means an expansion of non-agricultural, specifically, secondary activities and constitutes a step forward in economic development. Urban accretion, on the other hand, is merely the result of the social, political and economic attraction of the cities.

THE HYPOTHESES

The issue can be resolved in terms of the following criteria :

1. If the flow of immigrants is not absorbed by and large into an equivalent expansion of the secondary and tertiary sectors in the cities, it will result in urban accretion and not urbanization.[2]

2. If urban growth is dysfunctional, that is, more tertiary than secondary with sharply defined and restricted areas of urban influence, it again would constitute urban accretion and not urbanization.

3. Urbanization reflects opportunities for developing new central places and manufacturing locations. If urban growth is not accompanied by such a reorganization of the space economy, it would be urban accretion rather than urbanization.

TEST OF HYPOTHESES AND RESULTS

The first hypothesis has been tested in terms of the relevant data of the cities of Delhi, Meerut, Ghaziabad and Rohtak. The percentage rates of immigrant workers and non-workers as well as that of primary, secondary and tertiary workers to the total workers have been taken into account to study the process of urbanization and urban accretion in the cities of the National Capital Region.

For analysing the dysfunctional nature of urban growth, data relating to the occupational structure of 27 towns and cities has been used. But the quantitative testing of the second hypothesis is confined only to the data of Hapur city. The same method has been adopted for other towns of the National Capital Region.[3] But owing to constraints of space the results have not been presented here.

The third hypothesis has been tested on the basis of size, spacing and hierarchic order of 62 towns and cities in the National Capital Region. The range of size index has also been computed.[4]

Results

According to the census of 1961, 81.62 per cent of the total population were enumerated as immigrants in the Union Territory of Delhi. In 1971 this share was reduced to 49.88 per cent. Similarly, of the total immigrant population in 1961, 57.56 per cent were non-workers and 42.44 per cent were workers. In 1971 the share of migrant non-workers was reduced to 53.84 per cent but that of migrant workers marginally increased by 2.73 per cent. The share of tertiary sector in the work force was 29.60 per cent in 1961 and 31.24 per cent in 1971. The secondary sector's share in these years was 11.03 per cent and 13.04 per cent, respectively. Though both the secondary and tertiary sectors marginally grew during the decade 1961—71 (by 2.01 per cent and 1.64 per cent respectively), this had no impact upon the size of immigrant non-workers. These growth indices are, therefore, indicators of urban accretion rather than urbanization.

After Delhi, Meerut is the biggest urban agglomeration in the National Capital Region. In 1971, 37.2 per cent of the total population of Meerut urban agglomeration were immigrants. The immigrant population in 1971 consisted of 38.74 per cent workers and 61.26 per cent non-workers. The tertiary sector's share of total work force had increased from 67.66 per cent in 1961 to 68.47 per cent in 1971. But during the same period the share of the secondary sector declined from 28.68 per cent to 27.77 per cent. The almost stagnant tertiary sector and a declining secondary sector could have no role to play so far as the absorption of migrant non-workers is concerned. Meerut too, like Delhi, was growing by urban accretion.

In the third urban agglomeration of the National Capital Region, Ghaziabad, 50.03 per cent of the total population were immigrants in 1971 ; 61.62 per cent of whom were non-workers and 38.38 per cent workers. The tertiary sector declined in terms of its work force from 62.40 per cent of the total workers in 1961 to 56.68 per cent in 1971. The secondary sector's share of the work force increased from 35.09 per cent in 1961 to 39.24 per cent in 1971, with an increase of 3.05 per cent during the decade. The decline in the number of tertiary workers coupled with the higher percentage share of migrant non-workers in 1971, negated the benefits of increase in the share of the secondary sector.

Ghaziabad, therefore, had no urbanization in the real sense during the decade 1961—71.

Rohtak city, the fourth urban centre in the National Capital Region, had 48.68 per cent migrants in its total population in the year 1971. Of the total migrant population, 64.39 per cent were non-workers and 35.61 per cent workers in that year. This city too experienced a decline in the share of workers in the secondary sector from 24.40 per cent in 1961 to 19.39 per cent in 1971. The secondary sector, therefore, did not expand and could not absorb the majority of migrants. Rohtak, like other cities of the region, witnessed urban accretion during the decade 1961—71.

Taking Ghaziabad as an example, the hypotheses have further been tested in terms of the quantitative formulations given here :

$$\left(\frac{d(EU)}{dp}\right)=nE$$

if $n>1$ Urbanization

if $n<1$ Urban accretion

where, EU=urban wages

p=total rural migrant population

E=economy (attraction wages including other facilities)

In Ghaziabad in the year 1971

EU=6.80 (average daily wages per worker)

p=26,635 (total rural migrants)

E=18.67 (attraction wages computed by the author and denoted as economy of the worker)

$$\left(\frac{d(EU)}{dp}\right)=nE$$

$$\text{Rs. } 6.80=n\times 18.67$$

$$n=\frac{6.80}{18.67}$$

$$n=0.3642.$$

Since the value of n is <1, Ghaziabad comes out as the city of urban accretion. Similarly, the value of 'n' of all other towns/ cities have been computed and found to be <1 in all cases. The towns of the National Capital Region, therefore, underwent urban accretion and not urbanization during the decade 1961—71.

The test of the second criterion shows that in every town except Ballabgarh, there was a preponderance of tertiary activities in the years 1961 and 1971. The share of workers in the tertiary sector varied between 38 per cent to 72 per cent in 1961, and 39 per cent to 76 per cent in 1971 ; whereas the share of secondary sector varied between 22 per cent to 46 per cent in 1961, and 16 per cent to 45 per cent in 1971. The results reveal that a low percentage share of tertiary workers in the total work force was related to the small size of urban population (e.g. Jhajjar and Firozpur Jhirka), whereas the higher percentage share was related to the big size of urban population (e.g. Delhi and Meerut). It is interesting to note that the secondary sector does not reflect such a relationship. On the whole, both cities and towns have experienced an expansion of the tertiary sector rather than the secondary sector during the decade 1961—71 with the exception of Faridabad, Modi Nagar and Ballabgarh. Hence the majority of cities and towns have become centres of greater consumption rather than production. This again is an indication of urban accretion.

The quantitative test given below of the second criterion also reveals the same characteristics.

Quantitatively, the character and area of urban influence of the city/town may be stated as,

$$\frac{duf}{dx} = \text{the actual measure} \qquad \ldots(1)$$

where, duf = the dysfunctional character of the city

$dx = \sqrt{\text{area of city/town/tehsil/district}}$

or dx = small distance of influence with character of urban accretion

$$duf = n < 1 \qquad \ldots(2)$$

Applying the above formula to the data of Hapur city for the year 1971, the following characteristics have been noted-where,

duf = 11,109 tertiary workers/18,123 total workers or 61.30 per cent.

Area of the city = 5.83 km^2.

$\therefore$ = 2.42

According to equation 1,

$$= \frac{61.30}{100 \times 2.42} = 0.2533$$

The value of *n*, which is <1, reveals that Hapur city has a dysfunctional character with a restricted area of urban influence and facilities around it. The formula has also been applied to compute the value of '*n*' of other cities and towns of the National Capital Region. The results indicate that the index '*n*' governs the dysfunctional character of the city in the following two respects :

1. The value of '*n*' is influenced by the quality of urban living and hence greatly varies (0.03 for Delhi to 0.75 for Sardhana).

2. The dysfunctional character decreases with the increase of the value '*n*'. Besides, it may be said that as the value of '*n*' decreases, the dysfunctional character of city or area increases and urban facilities for a living decrease.

A test of the third criterion reveals that the spacing of nodes and sub-nodes of urban settlements has no proper organization of space through a network of hierarchy. Table 10.1 brings out the distribution of towns classwise :

TABLE 10.1 : **Classwise distribution of towns**

Class of Towns	Size of Population (1971)	Total No. of Towns	No. of Towns in Haryana Sub-Region	No. of Towns in U.P. Sub-Region	No. of Towns in Raja-sthan Sub-Region	Union Terri. tory of Delhi
1.	100000 and above	8	1	5	1	1
2.	50000>—<99999	7	4	3	—	—
3.	20000>—<49999	10	3	7	—	—
4.	10000>—<19999	18	7	10	1	—
5.	5000>—< 9999	15	6	9	—	—
6.	Less than 5000	3	2	1	—	—

Source : Census of India, 1971, *General Population Tables*, part II-A(i).

It is clear from Table 10.1 that the distribution of cities and towns in various sub-regions of the National Capital Region is unequal. As regards the distribution of population among the urban centres of the National Capital Region, the most outstanding characteristic is the excessive hypertrophy of Delhi metropolis, which claims 65.89 per cent to the total urban population of the region, leaving 34.11 per cent to the remaining 60 towns. The class 1 towns alone account for 78.90 per cent of the total urban population, whereas the class 6 towns have only 2.33 per cent. The analysis further reveals the eccentric nature of the distribution of urban population among cities and towns. The distribution of cities and towns shows that spacing of towns is closer in Uttar Pradesh sub-region than other sub-regions of the region. Figure 10.2 showing the distribution of classes and hierarchy of towns proves this fact. The study also illustrates that there is no

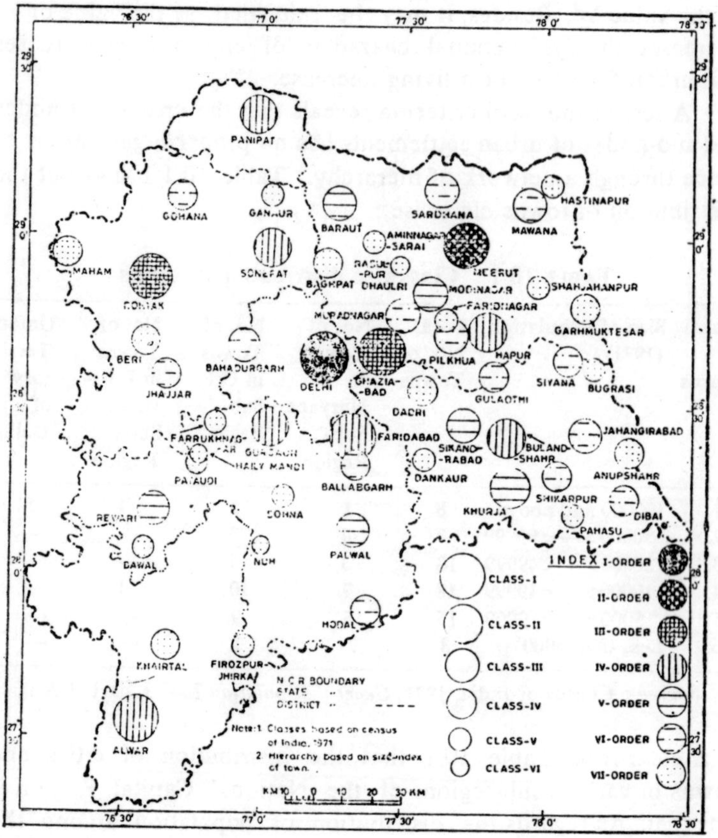

FIG. 10.2 : *National capital region : Distribution and spacing of towns (1971)*

proper hierarchic order.[4] For example, there only one town of the first order and only two of the second and third orders. Thus there is a wide gap between the number of first order and second order towns and fifth and sixth order towns in the region.

The spatial organization of urban settlements reveals that high order towns (e.g. Delhi, Meerut and Ghaziabad) are closely spaced. For proper spatial organization, such high order centres should be more widely spaced than the lower order ones. Besides, the town-village ratio is also not satisfactory. The National Capital Region has 61 towns and 6,753 villages (1971) which means that there is one town for as many as 110.70 villages. If the towns with a population of less than 10,000 are excluded, as they do not possess the necessary infrastructures to support villages (18 towns) then the number of villages per town is 157. It is, therefore, clear that the National Capital Region lacks a proper organization of space through a network of hierarchy of nodes and sub-nodes.

In the light of the above analysis of urban growth in the National Capital Region it can be stated by way of conclusion that during the decade 1961—71, the cities of the region grew by more urban accretion rather than by urbanization in the real sense of the term.

FOOTNOTES

1. The District of Ghaziabad was carved out of the districts of Bulandshahr and Meerut in 1976.
2. C. Chattopadhyay, and M. Raza, 'Regional Development : Analytical Frame Work and Indicators', paper presented at the First Indian Geography Congress, January 1, 1973, New Delhi.
3. A. Singh, 'Nature of Economic Development in the National Capital Region', University Grants Commission, Major Research Project, 1981.
4. *Ibid.*

SECTION THREE

GEOGRAPHY AND SOCIAL PROBLEMS

ELEVEN

MOONIS RAZA AND Y.P. AGGARWAL

INEQUITIES IN THE LEVELS OF LITERACY IN INDIA
The Regional Dimension

INEQUITIES IN EDUCATIONAL DEVELOPMENT : AN INTRODUCTION

ALPHABETIZATION provides the basal stratum for the multilevel educational pyramid ; and may, therefore, be viewed as an important input in and is vitally linked with the totality of the development process. The spatial patterns of alphabetization and of the regional variations therein acquire special significance in this context as strands in the fabric of imbalances as well as inequities in national development. The differential spread of literacy can, therefore, be properly analysed as an element in the system of interdependencies having bi-directional linkages with the development process. In consonance with the marked disparities in socio-economic conditions across regions in India, the level of and response to educational opportunities also varies in space ; and as a consequence, certain regions acquire relative advantage over others in terms of human resource development and human capital formation.

A high degree of temporal stability is an important feature of the regional disparities in the educational development of the third world. It is largely the composite effect of the distortions embedded in the system during colonial times, on the one hand, and of the infirmities and limitations of the current development strategies, on the other. (Raza Moonis, et. al., 1983). In view of the persistence of inequities of a high order, these countries are faced with the problem not only of raising the general level of literacy but also of giving a high priority to the objective of interregional equity in their developmental policies. Thus, 'growth with equity' is now considered to be one of the major goals of social planning in many of the developing countries. It may, however, be noted that there is no contradiction between the demands for 'equity' and 'growth'. Equity without growth is a stagnant cesspool wherein only misery, ignorance, obscurantism and superstition can be equitably distributed. Growth without equity leads to the accentuation of the structural disequilibrium that constrains growth itself. The social concern for the two can be handled together, sustaining and sustained by each other.

An attempt has been made in this paper to examine the different aspects of inequities that have acted as impediments in the path of the educational development of the third world countries in general, and that of India in particular. With a view to unravelling these complexities, the issues relating to the identification, measurement and explanation of inequities in the territorial spread of literacy in India have been examined with the help of districtwise data.

Since the measurement of inequalities in the level of literacy rates constitutes a crucial step in our methodology, these have been discussed at some length in the second section. Some of the commonly used indices of inequality have been reviewed, and a new class of measures has been proposed. With a view to delineating the underlying regularities in the complex mass of data and to taking our study deeper into the basal stratum of the Indian polity, the district level data on different indicators of inequities in literacy rates have been analysed in the third section. The fourth section focuses attention on the nature of the relationship between the levels of and inequities in the literacy rates, on the one hand, and some selected indicators of development, on the other. The findings of our analysis, on the basis of district level data, have been put together in a system of interdependencies and a model of the vicious circle of underdevelopment has been proposed in the fifth section. By way of conclusions, some of the major results of the present study have been stated in the sixth section.

Five Aspects of Inequities

The inequities and distortions embedded in the socio-economic structure in many of the newly independent countries are deeply rooted in their historical processes and particularly in their colonial heritage. Since the dawn of civilization, education and development have been intrinsically linked through bidirectional causality. Inequalities between the educational levels of different groups have been both the cause and the effect of the differentials between their levels of socio-economic development. This is especially true of literacy, which provides the essential precondition for educational development. [While in the developed countries of today, the industrial revolution had created conditions wherein the need for a literate workforce had slowly and steadily led to universalization of literacy, the accompanying process of the development of underdevelopment in the colonial empires necessitated the persistence of inequalities in the levels of literacy and of illiteracy in the case of the mass of the people.]

The requirements of colonial rule thwarted their economic development, acted as an impediment in the path of technological advances in agriculture, did not permit the emergence of a selfreliant industrial sector, and consequently acted as a serious constraint on the intersectoral transfer within the workforce from the primary to the secondary sector—at best permitting the development of underdevelopment. While the collapse of the handicraft-based secondary sector led to the overburdening of the technologically stagnant primary activities, the needs of colonial administration were met by a bloated tertiary sector, which was divorced from and contemptuous of work. The socio-economic conditions that had led to the narrowing of the gap between work and education in the developed countries of today in the wake of the Industrial Revolution, were thus not allowed to emerge in the countries of the Third World.

Let us recapitulate the socio-economic context within which the present day disparities in the educational system have emerged on the Indian scene. The following five aspects of intra- and interregional disparities would be of considerable interest in this connection :

— between scheduled castes and others
— between scheduled tribes and others

— between males and females
— between rural and urban settlements
— between developed and less developed regions.

Let us probe a little into the genesis of these disparities.

First, the disparities between scheduled castes and others emerged as a consequence of the historical separation of work from knowledge and from power (Raza, 1983). An enquiry into the distribution of the workforce, based on the 1971 census data in India has revealed that the participation rate in the workforce in the case of scheduled castes and scheduled tribes is quite high as compared to non-scheduled population ; it is 38.5 per cent for scheduled tribes, 36 per cent for scheduled castes and 31.6 per cent for the non-scheduled population. The problem with the scheduled caste population is, therefore, not that of the low participation in the workforce but is that most of them are engaged in the primary sector of the economy as agricultural labourers. The data shows that nearly 80 per cent of the scheduled caste workers are engaged in the agricultural sector and nearly 52 per cent of them are agricultural labourers. The corresponding share for the non-scheduled population is 65 per cent and 20 per cent, respectively. It is a well-known fact that most of the agricultural workers are illiterate and live under miserable conditions and continue to be below the poverty line. Another interesting aspect of the distribution of the workforce is that the share of scheduled caste workers in the organized manufacturing sector is only 3.5 per cent as compared to nearly 7 per cent for the non-scheduled workers. In the organized sector because of their low levels of education, they are usually engaged as semi-skilled workers.

The persistence of disparities in literacy between scheduled castes and others is also evident from the 1971 census data. A comparative study of the literacy rates was carried out by considering those districts wherein the share of rural scheduled castes to total rural population is 5 per cent or more. Out of a total of 356 districts in the country, 286 districts satisfied this criterion and were labelled as predominantly scheduled caste districts. It is interesting to observe that the scheduled caste literacy rate was less than 12 per cent in as many as 131 districts as compared to the fact that there was no district that had less than 12 per cent literacy for the non-scheduled population (Raza and Aggarwal 1983).

Second, the disparities between the scheduled tribes and others are rooted in the continued isolation of some tribal communities within cul-de-sacs and blind alleys in the hilly, forested or arid terrains. The low levels of educational development, regional strains and stresses that still persist are the consequence of the different levels of development between the tribal and the adjacent non-tribal communities ; the strains can be eliminated only by minimizing such disparities. The 1971 census results show that the overall literacy rate among the scheduled tribes was only 11.30 per cent as compared to 29.45 per cent for India as a whole. Among the scheduled tribes, the rural female literacy rate for 1971 was only 4.85 per cent as compared to 37.09 per cent of urban males.

Third, the disparities between male and female educational development are rooted in the persistence of some attributes of the spurious modernization of the colonial era. This expresses itself quite sharply in the denial of educational opportunities to women. The relative deprivation of women in the field of education was particularly significant because it underlined all other attributes of deprivations. The scheduled castes were deprived, no doubt, but the scheduled castes women were more deprived than their menfolk. The doors of educational institutions were closed to women as an integral part of the backward looking colonial policy of buttressing the obscurantist male chauvinistic tendencies in the polity. The persistence of a low level of female literacy is evident from the fact that it increased from 0.60 per cent in 1901 to only 7.3 per cent in 1941. The rural population was deprived no doubt, but rural women were more deprived than their menfolk. The rural scheduled caste female literacy, even after 25 years of independence, was less than 1 per cent in as many as 73 districts and less than 2 per cent in 116 districts out of a total of 286 districts with rural scheduled caste population of more than 5 per cent of the total rural population (Raza & Aggarwal 1983). A movement for women's education in the Indian social context is, therefore, very much more than a movement for women's education. It strikes at the very roots of the passive system of social inequities intertwining the Indian polity, continuously sucking out its life blood and rendering it anaemic and weak. Women's education is an instrument of liberation of not only of women but of Indian society as a whole.

Fourth, the persistence of disparities in rural-urban continuum can be traced to the spatial organization of underdevelopment. The colonial process, which not only affected the qualitative but the quantitative attributes of educational development as well, was embedded in the space economy. It is only by correcting the spatial distortions introduced and or strengthened during the colonial era that the limitations and infirmities of the present day development programmes can be properly understood. The colonial process of spatial articulation, which was strongly embedded in the space economy, effected both the qualitative and quantitative attributes of urbanization. In the educational sector their ability to play this dominant role was considerably influenced, first, by creating educational opportunities that formed the basis for the adoption of the modern means of production, and second, by creating a variety of job opportunities in the productive sectors of the economy located therein. A reference may be made in this context to the fact that in 1931 the literacy rate in the largest 34 cities was 34.8 for males and 14.9 for females as compared to 13.3 and 2.5, respectively, for India as a whole. The literacy in rural areas would thus be very much less than these figures.

Fifth, the persistence of regional imbalances and inter-regional disparities are reflected in a constricted industrial base and prolification of the tertiary sector, which by and large is non-productive. The relationship between the vertical shifts in the workforce, on the one hand, and of its horizontal mobility, on the other, is of crucial significance in the development process. The exogenous imposition of modern industry on a traditional economy was almost entirely concentrated in a few large cities—generally in port conurbations which acquired the character of islands in a vast ocean of technological backwardness in the countryside. The port towns, unlike their counterparts of the ancient and mediaeval periods, served as the focal points of a suction mechanism and became the centres for the export-import oriented commodity flows. These processes have led to the emergence of regional disparities in educational development as well. The 1981 literacy rate varied from 6.94 per cent in the case of east Kameng in Arunachal Pradesh to 81.35 per cent in Kottayam district of Kerala.

It is significant to note that all these factors do not operate separately. They are intrinsically intertwined in the thick rope of the socio-economic opression that ties down the social system. As we move down from high caste, male and urban in relatively developed regions to scheduled caste as well as tribes, female and rural in relatively less developed regions, the inequities become more and more sharp.

The results of the 1971 census show that the literacy rate varies from 78.22 per cent for non-scheduled caste, urban male population in Kottayam district of Kerala to 0.09 per cent for scheduled, rural female population in the Jaisalmer district of Rajasthan. It has also been observed that urban scheduled caste literacy in 1971 was as low as 0.29 per cent in the Tikamgarh district of Madhya Pradesh. The pyramid of the inequities at the aggregate level is shown in Figure 11.1.

Inequities in the Levels of Literacy : The Regional Dimension

If follows from the above that interregional disparities in literacy may be examined either in terms of variations in the general level or in the magnitude of a selected measure of intraregional inequities. The analysis of the variations across regions in the literacy rate of the rural females or of the rural males, for example, would no doubt focus attention on interregional disparities in literacy rate of rural females or of the rural males, as the case may be. It would not, however, reveal anything about the nature and magnitude of intraregional inequities as between male and female literacy in the rural areas.

With a view to examining these multifaceted and complex phenomena, it is desirable to define the binomial elements in terms of which the interregional variations in intraregional disparities may be analysed. It may be of considerable interest, for example, to probe into the nature and magnitude of interregional variations in the level of intraregional disparity between male and female literacy rates. The first stage in such analysis would thus relate to the measurement of the gap between male and female literacy rates for each region. This would then be followed by an examination of interregional variations in the measured values.

It is significant to note that a rise in the level of literacy rate does not necessarily result in the narrowing down of intra- and interregional disparities therein in the different stages of development. The variations in the magnitude of inequities between

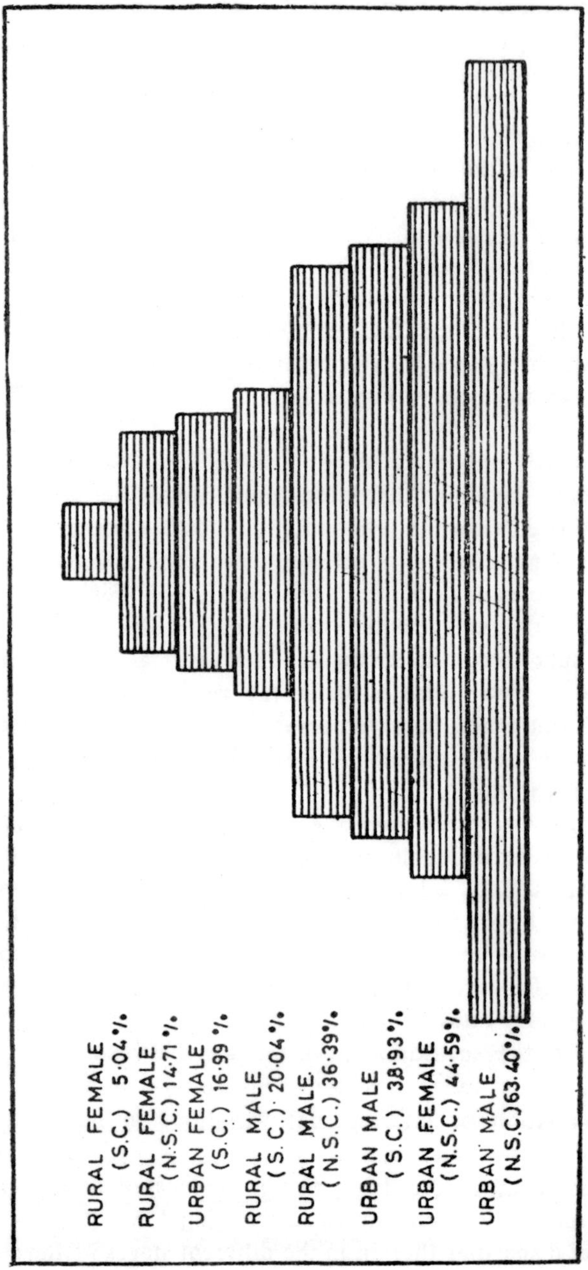

FIG. 11.1 : *Hierarchy of inequities in literacy, India, 1971*

binomial elements of a region may follow the path of 'divergence-stability-convergence' in relation to the different stages of rise in literacy rate. One should be cautious in generalizing this dynamic relationship and should not draw normative generalizations from historical behaviour. It may also be noted that the exact shape of the curve would vary from situation to situation and each of these would call for a different set of policy measures in educational planning and resource allocation. Whether literacy is looked upon as an input in the development process or as its consequence, the crucial step in the analysis of its spread across national space, therefore, does not consist of a futile search for an ideal growth curve, but rather of the identification and explanation of the inter- and intraregional differentials in its levels.

It is, therefore, surprising that these aspects of identification, measurement and explanation of the inequities in the sphere of education have not been adequately probed either by social scientists or by educational planners. Concerned scholars have been almost entirely preoccupied with the socio-economic correlates of disparities in a wonderland without space, and its regional dimension has not been duly recognized. This is particularly unfortunate because the problems of inequities are deeply rooted in the regional sub-system of interdependencies and can, therefore, be satisfactorily analysed and remedial action for their minimization can be effectively undertaken only within a regional frame.

It is only during the recent past that a few halting steps have been taken in this direction. A systematic attempt to measure the nature and patterns of regional disparities in education was made in a series of studies sponsored by the International Institute of Educational Planning (Gabriel and Ngoc, 1980). The relationship between intra- as well as interregional disparities in education and the process of socio-economic development was illustrated with the help of specific case studies that made use of a number of innovative tools in the diagnosis of the existing malaise. A comparative study of male and female illiteracy in the world has shown that the magnitude of disparities between the sexes in regard to literacy is very high in some cases and that this attribute would continue to persist. Reference may be made to the fact that in a number of countries—Chad, Ethiopia, Mali, Niger, Somalia and Upper Volta in Africa as well as Afghanistan, Saudi Arabia and Yemen in Asia—literacy rate was close to the zero mark as late as in 1971 (UNESCO, 1980).

Ryba (1979) examined the territorial aspects of educational inequality on a world scale by considering educational provisions in relation to educational achievements. He showed that territorial inequalities may be noticed at every scale of analysis and that these have tended to persist over time. Williamson's (1977) study on the patterns of educational inequality in West Germany validated the hypothesis that educational inequalities are deeply rooted in the social milleu and continue to constrain development even in the otherwise developed countries of Europe.

A number of studies on the nature and the patterns of inequities in the context of Indian literacy have been undertaken by scholars during recent years. Based on the results of the 1961 census, Gosal (1964) has brought out the essential characteristics of interstate variations in different components of literacy. A comparative study on the nature and patterns of inequities in literacy rates between 1961 and 1971 has been undertaken by Kundu and Rao (1982). The study is based on the state as the unit of analysis. Interstate differences in the rural-urban literacy rates have been examined for the year 1981 (D'Souza 1982). The study has probed into the nature of association between rural-urban disparities with urbanization, on the one hand, and with industrialization, on the other. Ahmad (1982) has examined inter-district inequities in literacy rates of the tribal population of Bihar for the year 1971. These studies have noted converging trends and the narrowing down of disparities in literacy rates.

Usha Rao (1982) has studied the intrastate disparities in literacy rates for the scheduled caste population in Karnataka. A probing study into variations in the literacy rates of Gujarat and some of the economic correlates of the same has shown that the proportion of agricultural workers in the rural workforce exerts a significantly strong influence on literacy rates in the selected villages (Shah 1981). It may thus be noted that the studies undertaken in the Indian context have noted converging trend and narrowing down of disparities in literacy rates of different regions as well as different strata of the society.

One of the serious limitations of the studies on India lies in the fact that the disparities have been examined either with states as the unit of analysis or within the districts or similar subregions of only one state. It may be noted that the state as a unit of analysis for examining different aspects of regional disparities would be quite unsuitable because of their heterogeneity both in terms of elements of diversity and disparity. It may not give any meaningful result if we club Chota Nagpur plateau with the North Bihar plains or the Gujarat plains with Saurashtra. Quite a few of the studies undertaken by economists assume away these diversities that render their conclusions to be of doubtful validity. The same can be said with equal force about the magnitude of disparities within the boundaries of a state. How does one, at any level of rationality, club together Jaisalmer and Jaipur, Burdwan and Siliguri, Bangalore and Gulbarga, Hyderabad and Mahboobnagar or Meerut and Ghazipur ? The only alternative is, therefore, to consider the district as the unit of analysis. The district, as an administrative unit of the Indian polity, has a particular vitality that is the function of its sociological homogeneity, dialectal uniformity and historical continuity. It is at this level that the macro processes get rooted into the micro worlds providing the unity in diversity of the Indian ethos. It is at this level that the emphasis shifts from the system of vertical linkages of sectoral planning to the system of horizontal linkages of regional planning. This is particularly true of the sphere of educational development, wherein specificity and universality are intrinsically linked together.

It must, however, be recognized that these pioneering studies have brought out the need for and showed the path towards detailed emperical investigations at a disaggregated level. As a response to this need an attempt is being made here first, to examine the pattern of inequities in the spread of literacy and second, to examine the interrelationships between the economic base, levels of literacy and different types of inequities in literacy rates on the basis of a cross-section of districtwise data.

Data base and framework of analysis

A comprehensive set of data pertaining to literacy and some related aspects of socio-economic development are collected every 10 years on a census basis in India. The present study makes use of this valuable source. The nature and patterns of inequities in literacy rates have been examined for a district level cross-section for 1981 (Census of India, 1981).

The first stage in the analysis related to the selection of

binomial elements, in terms of which disparity had to be measured. In the light of data availability, we have identified the following six groups for the measurement of disparities in literacy rates.

—Male-female
—Rural (male-female)
—Urban (male-female)
—Rural-urban
—Male (rural-urban)
—Female (rural-urban)

The second stage of analysis related to the choice of a suitable measure of disparity. The review of literature revealed that both absolute and relative measures have been used to measure disparities (Sen 1973). However, the major difficulty with all these measures is that the disparity between alphas and non-alphas is not the same as between non-alphas and alphas. To overcome this difficulty, Sopher has proposed an alternative measure that gives the same value in the two cases mentioned above (Sopher 1974). Sopher's index has been further modified so that it satisfies all the axioms that are normally used for the evaluation of any inequality measure. These axioms are additive and have multiplicative monotonicity, redistribution and repetitive transfers (Kundu and Rao 1982). In the present exercise we have proposed an alternative formulation of Sopher's disparity index and the new index that satisfies all the above stated axioms has been defined as

$$w = \log (L_2/L_1) + \log [(100+I_1)/(100+I_2)]\ L_2\ 3/8\ L_1 > 0,$$

where L_1 and L_2 are the literacy rates and I_1 and I_2 are the corresponding illiteracy rates for the two elements between which the disparity index is to be calculated.

The third stage of analysis related to the selection of suitable indicators so as to examine the nature of the relationships between the literacy rate, the disparity indices and the socio-economic characteristics. Keeping in view the framework of the present study and the availability of data, the following attributes of the economic base of the district have been selected :

X_1 =Per cent total literacy
X_2 =Disparity index of male-female literacy

X_3 =Disparity index of rural (male-female) literacy
X_4 =Disparity index of urban (male-female) literacy
X_5 =Disparity index of rural-urban literacy
X_6 =Disparity index of male (rural-urban) literacy
X_7 =Disparity index of female (rural-urban) literacy
X_8 =Percentage agricultural labourers
X_9 =Percentage cultivators
X_{10}=Percentage other workers
X_{11}=Percentage rural literacy
X_{12}=Percentage urban literacy
X_{13}=Percentage urban population

It may be noted that the definitions of the variables X_8, X_9, X_{10}, X_{11}, X_{12} and X_{13} are the same as in the Census of India, 1981.

The relationships of disparity indices with literacy rates, on the one hand, and with economic base indicators, on the other, have been examined through regression analysis. By using the rate of urbanization as an explanatory variable, it became possible to draw some interesting conclusions about the difference between the urban and rural areas with regard to the levels of literacy as well as to the inequities in the incidence of literacy.

While interpreting the findings of the study, a few limitations of its data base and methodology should be borne in mind. First, unless otherwise stated this study is based on literacy rates for all age groups. Second, the districtwise data in the case of the 1981 Census was not available for two states, that is, Assam and Jammu and Kashmir. The findings, therefore, relate to the remaining 388 districts of India. Third, in spite of its many positive features, the proposed measure of disparity (w) suffers from two major limitations : it does not take into account the base level of literacy, and the solution arrived at is not a unique solution. As a consequence, the index may in some circumstances take the same value for even highly disparate levels. It is hoped that in spite of these limitations this study would contribute to the sharpening of our understanding with respect to an important aspect of social reality in India : the inequities and disparities in literacy rates.

SPATIAL PATTERNS OF DISPARITIES IN LITERACY

The results of the Fourth All India Educational Survey have revealed that in 1978 about 93 per cent of the rural population has primary educational facilities either in their own habitation or within a distance of 1 km (NCERT, 1980). It is thus clear that the accessibility to educational facilities, even in rural areas, may not now be considered to be a major constraint in achieving the target of cent per cent enrolment. With a view to covering the out-of-school population, the Directorate of Adult Education has also been operating a number of schemes for more than a decade. In spite of these efforts to eradicate illiteracy, we find that the situation continues to be highly unsatisfactory.

It was in this context that the First Five-Year Plan recognized the equitable distribution of educational opportunities as one of the important objectives of educational planning in India (Govt. of India, 1951). The successful implementation of these promises would have at the same time resulted in reducing regional disparities in literacy to the zero level. The reality, however, is quite different. The Sixth Five-Year Plan has specifically recognized the continued persistence of intra- and interregional disparities at all levels of the educational pyramid (Govt. of India, 1980). UNESCO (1978) projections have shown that even by the year 2000 A.D. India would not be able to achieve 100 per cent literacy. Projecting the trends for the period 1901-81, it has been estimated that the literacy rate is likely to increase to about 55 per cent by 2000 A.D. (UNESCO, 1978). These projections further reveal that the overall gap in literacy rate of males and females would fall only marginally, and consequently the intraregional disparities may even continue to persist up to the year 2000 A.D. The policy of growth with equity acquires significance in this context for the countries of the third world. Such a policy is an imperative not only on ethical, moral or philosophical grounds, but more importantly on pragmatic grounds: just as there would be little equity without growth, there would be little growth without equity. This is the only strategy to come out of the existing morass of underdevelopment. To achieve these objectives, the identification, measurement and explanation of intra- and interregional disparities becomes a necessary prerequisite.

A plan of action, for example, for the achievement of cent per cent literacy cannot be based on an analysis at the aggregated level but on a highly disaggregated system of diagnosis and prescription. A district with a high concentration of tribal or rural population would, for example, calls for a different set of incentives, of school mapping techniques, of teacher deployment and hostel arrangements than a district with a high concentration of non-scheduled or urban population. The following analysis on the basis of the district level data has been done with such an end in view. It may be noted that the analysis in this section has been undertaken on the basis of the incidence of different types of disparities as measured through the indices outlined in Section 2.

Before examining the interrelationships among the six indices of intradistrict disparities in literacy rates it would be advisable to consider the extent of their variations across different districts. Table 11.1 gives the mean, standard deviation and coefficient of variation for each of these six indices.

TABLE 11.1 : **Coefficient of variation for disparity indices : 1981**

Sl. No.	Disparity index	Mean	S.D.	C.V.
1.	Male-female	0.400	0.163	0.4076
2.	Rural (male-female)	0.482	0.211	0.4377
3.	Urban (male-female)	0.222	0.088	0.3946
4.	Rural-urban	0.344	0.161	0.4887
5.	Male (rural-urban)	0.268	0.141	0.5247
6.	Female (rural-urban)	0.519	0.259	0.4997

It may be noted from Table 11.1 that the coefficient of variation for disparity indices involving rural-urban components show high values as compared to those of male-female disparities. With the rural-urban disparities, the highest value of coefficient of variation has been observed for male (rural-urban). It follows from the above that the spatial variations are more marked between different components of rural-urban sector rather than within rural or urban segment.

We now turn to the task of interpreting the underlying patterns that emerge from the factor analysis of the disparity indices using the factor loadings as input variables and factor scores for all 388 districts as the observation set. The examination of the factor loadings for each factor would permit a more succinct description of the territorial patterns in the incidence of intraregional disparities. An examination of the factor scores that represent the behaviour of individual districts with respect to various factors, would bring out the spatial patterns described by more significant factor loadings. The calculation of factor scores weighs the original variables in proportion of the strength of their association with the factor which is measured by their respective loadings. Consequently, the emerging spatial profiles would allow us to bring into sharp focus some of the broad features of the regional structure of educational development in India.

For purposes of the present analysis, factors have been ordered according to their descending contribution to total variance following the method of varimax rotation. The factor structure was limited to two factors which explained more than 91 per cent of the total variance. The communalities vary from 0.77 to 0.99 and are mostly greater than 0.90. The high values of communalities thus clearly show that a large proportion of the variations in each of the six disparity indices is being explained by the two factors. It may also be noted that for both the factors, the factor loadings of the individual variables have the positive sign. This suggests that in the regional space, all the six indicators of disparities covary in the same direction. The detailed rotated factor loadings and communalities corresponding to each variable are given in Table 11.2. The results corresponding to the

TABLE 11.2 : **Rotated factor loadings**

Sl. No.	Disparity index	Factor 1	Factor 2	Communalities
1.	Male-female	0.930	0.240	0.9221
2.	Rural (male-female)	0.884	0.366	0.9154
3.	Urban (male-female)	0.840	0.254	0.7697
4.	Rural-urban	0.309	0.944	0.9863
5.	Male (rural-urban)	0.198	0.960	0.9613
6.	Female (rural-urban)	0.538	0.816	0.9555
	Eigen value	2.78	2.73	
	Cumulative percentage variance explained	46.30	91.80	

analysis of the first two factor loadings are now discussed.

Factor 1 : Disparities in Male-Female Literacy Rates

The first factor on rotation accounted for more than 46 per cent of variance, and the factor loadings for the three variables corresponding to male-female disparities were found to be greater than 0.80. Thus the first factor may be taken to represent the territorial pattern of the disparities between male and female literacy rates.

With a view to examining the pattern of spatial distribution of disparities, the factor scores for all the districts were calculated and classified into six mutually exclusive groups.* The statewise distribution of districts into these categories is presented in Table 11.3. It was also considered desirable at this stage to examine the spatial distribution of these groups. This was done with a view to bringing out the significance of the spatial clusters of districts that have similar patterns of intradistrict disparities, especially when these cut across different state boundaries.

Figure 11.2 presents the spatial distribution of districts based on the factor scores corresponding to the factor representing the disparities in male-female literacy rates. The following points emerge from a close look at the table and the map prepared on the basis of the factor scores corresponding to factor 1 :

(a) It may be observed that even after 30 years of planned development the disparities in male-female literacy in all its facets have continued to be fairly high over a large part of the country so much so that they have significantly high values in nearly half the number of districts in the country.

(b) The statewise distribution of districts in different classes is highly skewed. Normally, the patterns of intradistrict disparities for a large number of districts in a state of a similar nature. For example, in Kerala 10 districts out of a total of 12 have a very low level of disparity which is in contrast to the situation in

* The cut-off points have been determined from the factor scores in the following manner :

	>	
Extremely high		Mean+2 S.D.
Very high	Mean+1 S.D.	Mean+2 S.D.
High	Mean	Mean+1 S.D.
Medium	Mean—1 S.D.	Mean
Low	Mean—2 S.D.	Mean—1 S.D.
Very low	<	Mean—2 S.D.

TABLE 11.3 : Distribution of districts classified by type of male-female disparity in literacy rates—1981*

Sl. No.	States/Union Territory	No. of Districts with disparities						
		Extremely high	Very high	High	Middle	Low	Very Low	Total
1	2	3	4	5	6	7	8	9
1.	Andhra Pradesh		2	14	4	2	1	23
2.	Bihar		7	24				31
3.	Gujarat			2	13	4		19
4.	Haryana			5	7			12
5.	Himachal Pradesh			2	6	4		12
6.	Karnataka			5	9	5		19
7.	Kerala					2	10	12
8.	Madhya Pradesh	1	20	23	1			45
9.	Maharashtra			5	17	3	1	26
10.	Manipur				3	3		6
11.	Meghalaya				1	4		5
12.	Nagaland			1	1	4	1	7
13.	Orissa		3	6	4			13
14.	Punjab				3	9		12
15.	Rajasthan	4	21	1				26
16.	Sikkim			1	3			4
17.	Tamil Nadu				12	2	2	16
18.	Tripura				2	1		3
19.	Uttar Pradesh		15	32	9			56
20.	West Bengal			3	9	3	1	16
21.	Arunachal Pradesh			4	3	2		9
22.	Chandigarh					1		1
23.	Dadra and Nagar Haveli				1			1
24.	Delhi					1		1
25.	Goa, Daman and Diu				2	1		3
26.	Lakshadweep					1		1
27.	Mizoram				1	2		3
28.	Pondicherry				1	1	2	4
29.	Andaman and Nicobar Islands				1	1		2
	Total	5	68	128	113	56	18	388

*Excluding Assam and Jammu & Kashmir.

Rajasthan where 25 out of 26 districts have either a very high or

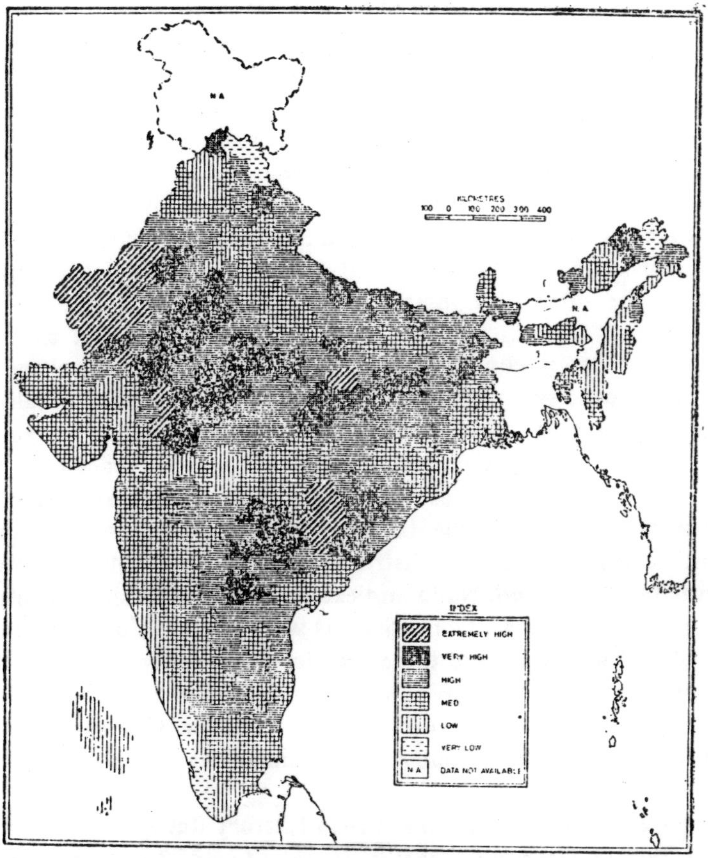

FIG. 11.2 : *India : Disparities in male-female literacy, factor 1*

extremely high order of male-female disparity in literacy rates.

(c) Row-wise examination of Table 11.3 reveals that a large number of districts of the educationally backward states of Andhra Pradesh, Bihar, Madhya Pradesh, Rajasthan and Uttar Pradesh continue to have a high order of male-female disparities. On the other hand, a large number of districts in the states of Punjab, Meghalaya and Nagaland have low levels of disparity.

(d) Columnwise examination of Table 11.3 reveals that out of a total of 73 districts with a very high and extremely high order of disparity, 61 districts lie in the contiguous states of

Madhya Pradesh, Rajasthan and Uttar Pradesh.

(e) An examination of Figure 11.2 reveals that districts corresponding to a particular level of disparity tend to cluster in national space and that these clusters cut across state boundaries. The first cluster of districts corresponding to very high levels of disparities extends from Western Rajasthan to Central Bihar. Contiguous districts from Rajasthan, North Madhya Pradesh, Eastern Uttar Pradesh and South Orissa are included in this cluster. Similarly, the spatial distribution of districts according to the disparities in male-female literacy takes the form of two major clusters that are not contiguous. The first cluster in this category comprises the districts from south-central India and the second corresponds to the districts from central Madhya Pradesh, northern Orissa, northern and southern Bihar and extends all along the Gangetic plains. The cluster corresponding to medium type of disparities forms a U-shaped pattern and extends from the west coast to the districts along the east coast. On the west coast it includes districts from Gujarat, Maharashtra, Karnataka and Tamil Nadu and extends up to Krishna district along the east coast. The districts with low levels of disparities are very few and most of these are along the western coast. The analysis has thus clearly shown that districts with high disparity index do not exist in isolation but take the form of homogeneous clusters that cut across many state boundaries.

Factor 2 : Disparities in Rural-Urban Literacy Rates

The second interaction factor, representing 45.3 per cent of the total variance, represents a pattern involving rural-urban disparities in literacy rates. The significance of rural-urban disparities is indicated by high factor loadings of all the three variables corresponding to rural-urban phenomenon. The explanatory power of the second factor is approximately equal to that of factor 1. The explanation lies in the fact that, in the Indian context, both the factors, that is, male-female and rural-urban disparities, continue to be of a high order and are equally significant.

The factor scores for all the 388 districts were calculated using the factor loadings for the second factor and were classified into six mutually exclusive groups.[1] The spatial distribution of the clusters corresponding of these groups is shown in Figure 11.3

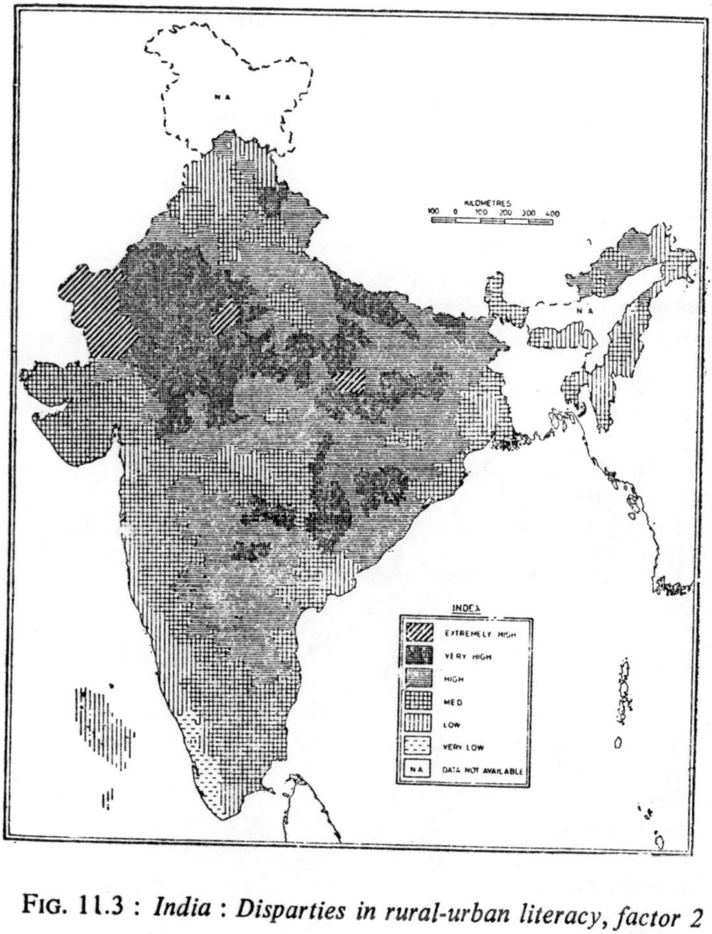

FIG. 11.3 : *India : Disparties in rural-urban literacy, factor 2*

and the statewise distribution of districts is presented in Table 11.4. The following may be deduced from this table :

(a) It may be noted that in the case of rural-urban disparities, while more than 50 per cent of districts have high to extremely high levels of disparity, there are only 22 districts with very low rural-urban disparities. It may be noted that these 22 districts include those districts also that have only either rural or urban population. For these districts the rural-urban disparity would naturally be zero. Such districts hence have been included in the category of districts with a very low disparity index ; the

TABLE 11.4 : **Distribution of districts classified by types of rural-urban disparity in literacy rates—1981***

Sl. No.	State/ Union Territory	No. of Districts with Disparity						Total
		Extremely high	Very high	High	Middle	Low	Very low	
1	2	3	4	5	6	7	8	9
1.	Andhra Pradesh		6	11	5		1	23
2.	Bihar		7	20	4			31
3.	Gujarat			2	12	4	1	19
4.	Haryana			7	5			12
5.	Himachal Pradesh		1	3	3	3	2	12
6.	Karnataka			5	11	3		19
7.	Kerala					3	9	12
8.	Madhya Pradesh	3	18	21	3			45
9.	Maharashtra			4	18	3	1	26
10.	Manipur				1	5		6
11.	Meghalaya		1	1	2	1		5
12.	Nagaland		1	1		3	2	7
13.	Orissa		2	5	4	2		13
14.	Punjab				6	6		12
15.	Rajasthan	6	13	7				26
16.	Sikkim			1	3			4
17.	Tamil Nadu			2	10	3	1	16
18.	Tripura			1	2			3
19.	Uttar Pradesh		7	28	21			56
20.	West Bengal		1	5	8	1	1	16
21.	Arunachal Pradesh		2	3		3	1	9
22.	Chandigarh					1		1
23.	Dadra and Nagar Haveli			1				1
24.	Delhi					1		1
25.	Goa, Daman and Diu				2	1		3
26.	Lakashdweep					1		1
27.	Mizoram				2	1		3
28.	Pondicherry				1	1	2	4
29.	Andaman and Nicobar Islands					1	1	2
	Total	9	59	128	123	47	22	388

* Excluding Assam and Jammu and Kashmir.

number of such districts being 14. In the true sense, only the remaining 8 districts have very low rural-urban disparities.

(b) The broad patterns of rural-uuban disparities are almost the same as those observed for disparities in male-female literacy. In this case it is the rural population that suffers from the inequities in the spread of literacy. A comparison of the maps representing the two components of disparities reveals that there are some districts that have shifted from one category to another because of their being on the margin of cut-off points. The other reason for a slight shift in spatial patterns could be the variations in the composition of the rural-urban population. The analysis clearly reveals that most of the districts of Bihar, Andhra Pradesh, Haryana, Madhya Pradesh, Rajasthan and Uttar Pradesh happen to have a high level of rural-urban disparities.

In view of the above discussion, it would be permissible to conclude that more than 50 per cent of the districts in India still continue to have high level of disparities in the rural-urban as well as male-female literacy rates. This is so in spite of the fact that balanced regional development of the country and equitable sharing the benefits of development among different social groups has been one of the objectives of planning during the last 30 years.

REGIONAL DISPARITIES AND CHARACTERISTICS OF THE ECONOMIC BASE

It is evident from the foregoing analysis that most of the districts in the country are characterized by the existence of a significantly high magnitude of intraregional disparities in literacy rates between different binomial elements therein. It may be recalled from the earlier discussion that the factors which give rise to the disparities do not operate in isolation from each other but are intrinsically intertwined with the historical processes, which have resulted in a distorted socio-economic structure of the national space.

Planning in independent India has been aimed at altering this structure through massive public investment in the vital sectors of the economy. However, even after 30 years of planned development, the inherited regional structure of underdevelopment has continued to persist in many of its essential characteristics. A systematic understanding of these relationships isof crucial significance not only for regional planners of the educational sector alone but of economic development as a whole. This is particularly so if we recognize that education is bidirectionally linked with the development process. We shall make some exploratory investigations into the nature of these relationships in this section. The relationship of disparity indices with the levels of literacy, as well as with economic base indicators, would be examined in some depth. The choice of variables has already been discussed in Section 2. A preliminary exploration into the nature of these relationships has been attempted through the examination of a correlation matrix which is presented in Table 11.5. Some important conclusions that emerge from a study of this table are as follows :

(a) It is obvious that the level of literacy is significantly and negatively associated with all the six types of disparity indices. In view of the negative association, the districts with low levels of literacy tend to have high level of disparities and vice versa. These interrelationships further reveal that in the Indian context, the diffusion of literacy has been accompanied by a corresponding reduction in the magnitude of disparities. It follows that there is no contradiction between equity and growth. In the case of educational development, growth has thus been by and large accompanied by equity.

(b) In view of the rural dominance in the population of the country, rural literacy emerges as the major determinant of the level of overall literacy and is also significantly related to different types of disparity indices. The correlation coefficient between total literacy and rural literacy is 0.843 as compared to 0.552 with urban literacy. It may thus be noted that in the Indian context, the growth of overall literacy has been essentially constrained by the slow progress made in the diffusion of literacy in rural areas.

(c) A close association has also been observed between the indices of literacy and the economic base of a district. An examination of Table 11.5 shows that the correlation coefficient between the literacy rate and the proportion of cultivators is both negative and significantly high (—0.605). Thus, the districts that are mainly dependent on agriculture have low levels of literacy. This evidence clearly shows that literacy among the

TABLE 11.5 : **Intercorrelation table**

	X_1	X_2	X_3	X_4	X_5	X_6	X_7	X_8	X_9	X_{10}	X_{11}	X_{12}	X_{13}
X_1	1.000												
X_2	—0.805	1.000											
X_3	—0.744	0.911	1.000										
X_4	—0.622	0.757	0.728	1.000									
X_5	—0.695	0.519	0.612	0.498	1.000								
X_6	—0.667	0.420	0.492	0.457	0.957	1.000							
X_7	—0.718	0.686	0.816	0.606	0.934	0.856	1.000						
X_8	—0.040	—0.003	—0.062	0.165	0.008	0.037	—0.052	1.000					
X_9	—0.605	0.526	0.439	0.275	0.361	0.324	0.389	—0.484	1.000				
X_{10}	0.713	—0.605	—0.468	—0.461	—0.422	—0.399	—0.416	—0.251	—0.707	1.000			
X_{11}	0.843	—0.674	—0.668	—0.540	—0.664	—0.635	—0.697	0.063	—0.393	0.398	1.000		
X_{12}	0.552	—0.455	—0.392	—0.236	0.044	0.034	—0.064	0.002	—0.290	0.316	0.440	1.000	
X_{13}	0.523	—0.449	—0.262	—0.215	—0.240	—0.244	—0.200	—0.115	—0.622	0.771	0.096	0.204	1.000

X_1=Percentage total literacy
X_2=Disparity index of male-female literacy
X_3=Disparity index of rural (male-female) literacy
X_4=Disparity index of urban (male-female) literacy
X_5=Disparity index of rural-urban literacy
X_6=Disparity index of male (rural-urban) literacy
X_7=Disparity index of female (rural-urban)
X_8=Percentage agricultural labourers
X_9=Percentage cultivators
X_{10}=Percentage other workers
X_{11}=Percentage rural literacy
X_{12}=Percentage urban literacy
X_{13}=Percentage urban population

rural masses is of crucial significance in the process of modernization of the agricultural sector andd iversification of the rural economy.

(d) The correlation coefficient between proportion of workers engaged outside the traditional combination of primary activities and household industries is significantly and positively related to the levels of literacy on the one hand, and negatively associated with the different types of disparities on the other. The inherited economic base of low technology production processes is essentially inadequate and incapable of either generating and supporting processes of social change through increase in literacy or the narrowing of the inequity gaps therein.

(e) Urbanization contributes to the raising of literacy rates and the lowering of the disparities therein. This may be attributed to the fact that there is a stronger tendency amongst the literates to migrate from the rural to the urban areas and to the strong urban bias in the provision of facilities for formal as well as non-formal education. The exact nature of the magnitudes of these relationships, however, needs to be examined in greater detail.

(f) The value of the correlation coefficient between different indices of disparities and rural literacy is higher than the corresponding values for urban literacy. This may be attributed to the fact that the intensity of social inequities particularly those of the caste system and of male dominance is higher and more deeply entrenched in the rural ethos than in the urban.

It follows from the above that there are strong associations between the indicators of literacy : disparities and the economic base characteristics of a district and are intrinsically intertwined. On the basis of these findings it was considered desirable to probe deeper and to examine the causal relationships between the indicators of regional disparities and those of economic base with the help of simple linear regression analysis. The disparity indices and literacy rates have been considered as the dependent, and the indicators of economic base as the independent variable. The causal relationship between literacy rate as the independent, and the disparity indices as the dependent variable has also been examined in a similar manner. The main findings of the analysis are presented here :

Literacy, Disparities and Economic Base Characteristics

The percentage share of workers engaged in activities other than those of the traditional sector of primary production in symbiosis with household industry, and particularly that of the modern manufacturing sector, constitutes the economic base indicator that significantly explains variations in total literacy. It is the strength of the non-primary sector of economic activity that emerges as the major determinant of the literacy rate. The estimated regression equation is as :

$$X_1=19.23+0.544X_{10} \qquad R^2=0.509.$$

The total explanatory power of the equation described above is about 51 per cent and the positive sign of the regression coefficient indicates that the level of literacy and the proportion of workers in non-primary sector of the economy covary in the same direction. It may, however, be noted that the relationship between literacy and the economic base is quite complex. After all, literacy alone may not be an important measure of all the technical skills required in the different sectors of the economy. Nevertheless, the regression results clearly show that there is a high degree of correspondence between the processes of modern organized manufacturing and the level of literacy.

The relationships between the variables relating to the magnitude of the intradistrict disparities and the corresponding levels of literacy in a district have been examined in this section. For the purpose of the present analysis, literacy has been considered as an independent, and the disparity indices as the dependent variable. The analysis resulted in two significant regression equations which bring out the relationship of male-female and rural-urban disparity index with the level of total literacy. The estimated regression equations are as :

$$X_2=0.74997-0.010\,X_1 \qquad R^2=0.648$$

$$X_8=0.64229-0.009\,X_1 \qquad R^2=0.484.$$

It may be noted that the regression coefficient in these equations is negative. This indicates that the character of both male-female and rural-urban disparities are essentially similar in nature. In view of the fact that the literacy rates and the disparity indices always take positive values, the districts with high levels of literacy tend to have low disparities. Stated in other words, the analysis has shown that with the spread of literacy, disparities tend to narrow down. This may be so in view of the fact that the new gains would mainly accrue to the lower strata as the others would have already achieved higher rates of literacy in the earlier phase.

In view of the negative value of the regression coefficient in both the equations, it is possible to find a critical value of overall literacy when the inequities in male-female and urban-rural literacy rates tend to converge to zero. This is evident from the following regression equation that shows a significant fall in male-female disparities as a consequence of a rise in rural and urban literacy.

$$X_2=0.65293-0.009\ X_{11} \qquad R^2=0.454\ ;$$

$$X_2=0.70538-0.006\ X_{12} \qquad R^2=0.207.$$

It is interesting to observe that male-female disparity is reduced to zero level when the rural literacy approaches 72.5 per cent, while it continues to persist even when the urban literacy reaches 100 per cent. This clearly indicates the relative importance of rural literacy in reducing the intraregional disparities between male and females to a zero level. While interpretting these results, it is significant to note that the results are based on the cross-sectional data and has limited use for projection purposes. In spite of its limitations, the equations have broadly captured the trend.

The foregoing results clearly reveal that the existence of a high order of inequities has considerably constrained the diffusion of literacy across different districts of the country. It is in this context that the concept of growth with equity has acquired greater significance in the development strategies in the developing economies. In the Indian context the objectives of growth with equity would essentially mean a more effective strategy oriented towards the education of the females, especially in the rural areas and among those of the deprived sections of society.

THE VICIOUS CIRCLE OF UNDERDEVELOPMENT

With a view to understanding the nature of interrelationships between inequities and economic base indicators as well as between literacy and economic base indicators, use has been made of Stepwise Regression Analysis to analyse the districtwise data. The analysis has shown that the economic base of a region is significantly related to the levels of literacy, which in its own turn is associated with different segments of inequities in literacy rates. Some of the important results are :

$X_1=15.35+0.572\ X_{10}+0.145\ X_8$ $R^2=0.53$

$X_2=0.48973-0.004\ X_{13}$ $R^2=0.20$

$X_2=0.61788—0.006\ X_{10}$ $R^2=0.37$

$X_{13}=49.38—0.460\ X_8—0.542\ X_9+.212\ X_{10}$ $R^2=0.61$

The regression results presented above, give ample and unambiguous evidence to confirm the hypothesis that with social modernization and economic diversification of the economic activities (non-primary activities), the levels of literacy tend to rise and, as a consequence, the inequities at the basal stratum of the multilevel educational pyramid tend to narrow down.

The regression results also reveal that urbanization is the composite effect of the economic base characteristics of the district. The value of R^2 in this case is 0.61. An examination of the regression equation reveals that the processes of urbanization are accompanied with the modernization and the industrialization of the economy.* It is thus evident that the persistence of inter-regional disparities and the attributes of fragmentation of the socio-economic structure are intrinsically intertwined. The above regression results further reveal that urbanization to some extent also helps in reducing male-female disparities.

Following the gross features emerging from the interrelationships between economic base characteristics and educational development indicators, as reflected through inequities and levels of literacy rates, we now proceed a step further in our analysis and propose a model of the vicious circle of under-development, incorporating the basic elements and characteristics of the regional structure of the Indian economy which continue to inhibit the development processes even after more than 30 years of planned development. The vicious circle of under-development is presented in Figure 11.4.

A close examination of the proposed model reveals that in view of the predominantly agricultural base of the Indian

*It may be noted that the three variables relating to the economic base of a district (i.e. X_8, X_9, X_{10},) do not add to give the total workforce. We have not considered the percentage of workers in the household sector as a separate variable.

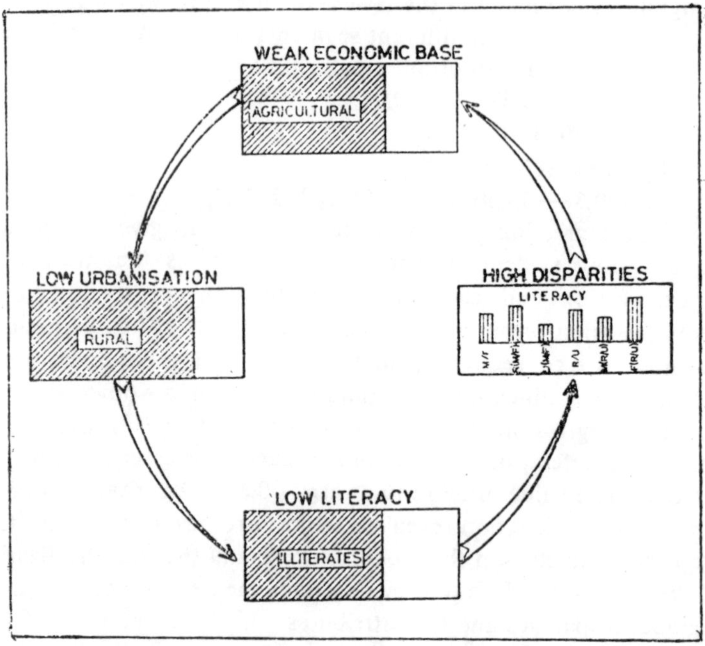

FIG. 11.4 : *Vicious circle of underdevelopment*

economy, a large part of the population lives in rural areas, constituting a hierarchy of the system of settlements which reflects the continuance of some distortions of the inherited fragmentation of the settlement continuum along the rural-urban interface, and which in its turn is associated with the emergence of disparities between rural and urban areas. It is through these infirmities of the spatial organization of underdevelopment that the economic space gets transformed into a highly differentiated field, and as a consequence imbalances and inequities emerge across different regions as well as between different social groups. These mechanisms are crucial, particularly in the developing economies, as the persistence of high inequities and regional imbalances constrain and are themselves constrained by low industrialization and low urbanization leading first, to a weak, inadequate and distorted non-primary sector and second, to a primary sector that is characterized by low productivity and mass poverty. Thus, the reasons for educational backwardness in

India are essentially rooted in the inadequacies and infirmities of its economic base, which in its turn is characterized by the persistence of the distortions embedded in the system during the colonial period, and of the infirmities and limitation of the current development strategies.

CONCLUSIONS

With a view to unravelling the complexities in the process of diffusion of literacy, the foregoing analysis was essentially concerned with the identification, measurement and explanation of inequities in the territorial distribution of literacy. In this endeavour we have proposed a new method for measuring intra-regional inequities between different binomial elements of the regional space. The analysis was based on districtwise data on literacy for the year 1981, and an attempt was made to explain its relationship with disparities in literacy rates and economic base characteristics. From the foregoing discussion, we can conclude the following :

a. The continuing low levels of literacy, particularly in rural areas. and the persistence of high level of inequities partly reflect the failure to achieve the goals of universal primary education and calls for a system of territorial resource distribution of a fundamentally different kind than the present one.

b. In order to pin point the regions that require special remedial measures to raise their levels of literacy, the study has identified the districts with varying degrees of inequities in the distribution of literacy. The analysis has shown that more than half the districts in India still continue to have significantly high level of disparities.

The examination of casual relationships corroborates the finding that the economic base of a region exerts a strong impact on the spread of literacy, and the processes of urbanization and industrialization strongly influence the levels as well as the inequities in the distribution of literates. A vicious circle of underdevelopment incorporating these characteristics has been proposed.

The findings of this study have serious policy implications for the present educational policy, and evidently point to the need for a different strategy for educational development. The analysis has highlighted the fact that universalization of alphabetization is intrinsically linked with the development process as a whole and that a narrow sectoral approach to educational planning would not be able to achieve the national objectives.

REFERENCES

Ahmad, A., Inter-regional 'Inequity in Literacy Levels of Tribals and Caste Segments of Population in India, 1971' (Paper presented at Regional Workshop on *Long-term Educational Planning*, NIEPA, New Delhi, January 1982).

Carron, Gabriel and Chau Ta Ngoc. (ed.), *Regional Disparities in Educational Development : A Controversial Issue* (Paris : International Institute for Educational Planning, UNESCO, 1980a).

Carron, Gabriel and Chau Ta Ngoc (ed.), *Regional Disparities in Educational Development : Diagnosis and Policies for Reduction* (Paris : UNESCO Press, 1980b).

Census of India, *Provisional Population Tables*, Series 1, Paper 1 of 1981.

Census of India, *Provisional Population Totals : Workers and Non-Workers*, Paper 3 of 1981, New Delhi.

D'Souza, S. Victor, "Urbanization and the Rural Urban Differences in Literacy" (Paper presented at the Conference of the Indian Association for Population Studies, New Delhi, December 1982).

G.S. Gosal, 'Literacy in India : An Interpretative Study', *Rural Sociology* 29 (1964).

Kundu, A. and J.M. Rao, 'Inequity in Educational Development—Issues in Measurement, Changing Structure and Its Socio-Economic Correlates with Special Reference to India,' Paper presented at Regional Workshop on *Long-term Educational Planning*, NIEPA, New Delhi, January (1982).

National Council of Educational Research and Training, *Fourth All India Educational Survey*, New Delhi : NCERT (1980).

Raza, 'Moonis, Future of Education : A Vision', Memeographed, National Institute of Educational Planning and Administration, New Delhi, 1983.

Raza, Moonis, and Y.P. Aggarwal, 'Some Aspects of Literacy Among Scheduled Castes and Non-Scheduled Population in India : A District-wise Analysis' (Paper presented at the workshop on *Indicators of Equity in Education*, NEIPA, 24-28 October 1983.

Raza, Moonis, A. Habeed and Y.P. Aggarwal, 'Urbanization in a Fragmented World : Some Aspects of International Inequities,' *Urban India* 3 (1983).

Ryba, R., 'Territorial Patterns of Diversity in Education,' *Comparative Education* 15 (October 1979).

Sen, A.K. *On Economic Inequality*, New Delhi, Oxford Press (1973).

Shah, Maya, *Economic Factors Explaining Variations in Literacy Rates in Rural Areas : A Case Study of Gujarat* (University of Baroda, 1981).

Sopher, D.E., 'A Measure of Disparity', *The Professional Geographer*, 26

(1974).

UNESCO, *Estimates and Projections of Illiteracy* (Paris : Division of Statistics of Education, UNESCO, 1978).

UNESCO, *Comparative Analysis of Male and Female Enrolment and Illiteracy* (Paris : UNESCO 1980).

Usha Rao, N.I., 'Intra-State Disparities in Literacy Rates in Karnataka with Special Reference to Scheduled Castes', in *Regional Inequalities in India*, ed. L.S. Bhatt et al. (New Delhi : Society for Study of Regional Disparities, 1982).

Williamson, W., 'Patterns of Educational inequality in West Germany', *Comparative Education*, 13 (1977) : pp. 29-44.

TWELVE

MEHDI RAZA AND FARKHUNDA BURKE

REGIONAL DIMENSIONS OF INCOME DISPARITIES
A Case Study of Uttar Pradesh, India

INTRODUCTION

IN SPITE of the latest trends in the definition of income, the word somehow reminds (even the layman) of gains measured in terms of money, and of the age-old proverb 'money makes the mare go'. Although income has been defined by academicians in the most sophisticated terms and the Oxford Dictionary defines income as the periodical (usually annual) receipts from one's business, land, work, investments, etc. per capita income may be termed as money income per person or income in the form of finance, or in other words, hard currency—the term in which it is measured. Per capita income is a reflection of the combination of such variables as density of population and economic development of an area measured in terms of national, state or regional income, and it is a known fact that the reflection is rather uneven no matter on how micro a scale it may be observed.

In addition to the marked spatial diversity in the quality of life, the persistent disparities in these spheres is one of the major problems facing the modern world. Although an equal society is unrealistic, yet to achieve general social progress, matters of major societal concern ought to be thoughtfully examined and speedily minimized in order to attain equity.[1] Measurement of quality of life and social well-being, however, is fraught with many conceptual and methodological difficulties as it depends upon a whole range of interrelated variables. Since the quality of life is the reflection of the satisfaction of human needs and wants, income is a very important criterion. It is true that social well-being is not solely a matter of monetary income, but as it provides access to such other critical variables as health, education, housing, nutrition, recreation and status, it might well be regarded as an adequate indicator of societal well-being. Many of the variables mentioned can be adequately measured in terms of monetary income. A number of empirical studies have shown that per capita income indeed correlates positively with some of these variables. A per capita monetary income might, therefore, be an adequate surrogate for the general condition of social well-being.

In developing countries the level of living appears to be closely related to per capita income, as it provides the wherewithal to make purchases ; although different amounts and types of effort and time are required to obtain the same income at different places.

The present study aims at drawing attention to the regional variations in the distribution of per capita income in the state of Uttar Pradesh (U.P.) India, and identifying the factors responsible for the marked spatial disparity therein. Identification of the causative factors may facilitate policy formulations for taking remedial measures.

Spatial Pattern of Income Distribution in Uttar Pradesh

In terms of per capita income, U.P. is one of the poorer states of India. The average per capita income of India in 1970-71 was Rs. 989.* Only six states—Punjab (Rs. 1482), Maharashtra (Rs. 1270), Haryana (Rs. 1217), West Bengal (Rs. 1065), Gujarat (Rs. 1038) and Himachal Pradesh (Rs. 1037) had per capita incomes above the all-India average. Uttar Pradesh with a per capita income of only Rs 336 occupied the 13th rank.

The average per capita income of U.P. for 1970-71 at current prices was Rs. 336. (It increased to Rs. 598 by 1978-79.) But considering only 48 districts for which data are available, the average state per capita income stands at Rs. 345.25. Only 20

* The study relates to the year 1970-71 as data for all the selected variables were available only for this year.

districts have per capita incomes above the state average. The remaining 28 districts, mainly concentrated in eastern U.P. have per capita incomes below the state average. There are, however, pockets of below average districts in the above average income areas. Naini Tal has the highest per capita income (Rs. 857), followed by Meerut (Rs. 474), Pilibhit (Rs. 474) and Muzaffarnagar (Rs. 468). The districts of north-western and western U.P. have comparatively higher incomes compared to the central and eastern parts of the state. indicating that eastern U.P. is the state's major region of economic and social deprivation (Fig 12.1). Ballia district has the lowest per capita income of Rs. 221, followed by Azamgarh (Rs. 223), Jaunpur (Rs. 224). Lucknow (Rs. 227) and Faizabad (Rs 245).

In terms of variations of per capita income, the districts of

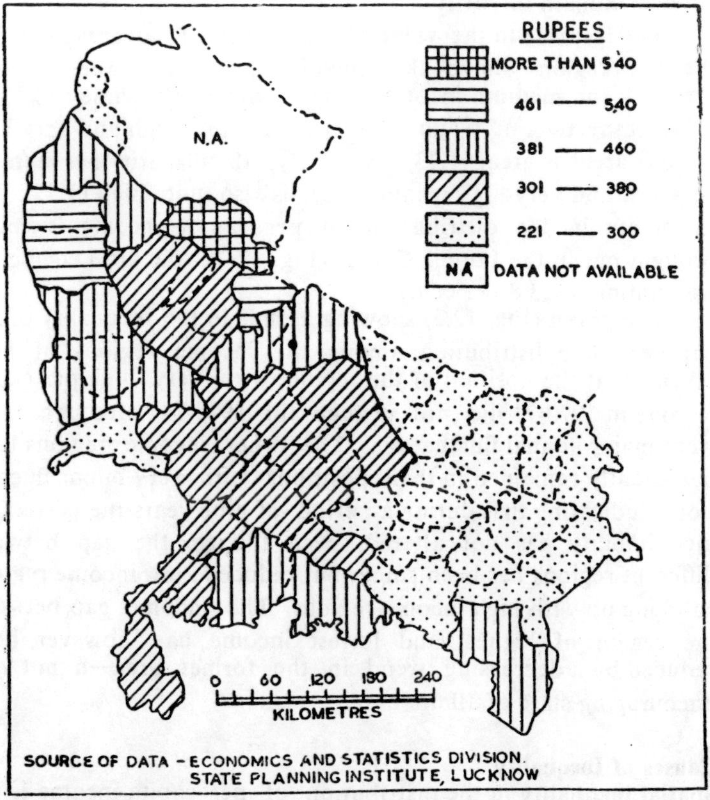

FIG. 12.1: *Major region of economic and social deprivation*

U.P. have been divided into five groups (Table 12.1).

A perusal of the table reveals that population concentration and per capita income are inversely proportional. The districts of eastern U.P. which have the lowest per capita income have the highest population concentration, which is lowest in the area of bighest per capita income (i.e. Naini Tal). The fact that the percentage of population living in the lowest per capita income districts is higher than the income percentage explains the lowest position of these districts in the scale of coefficient of advantages. Similarly, the district with highest per-capita income (Naini Tal) has the highest co-efficient of advantage. The small difference between income and population here shows the equitable nature of distribution of income in contrast to the districts with very low per capita income, where the negative value (−9.74) indicates maximum disparity.

The increase in the value of coefficient of advantage among the five regions shows that income disparity between the very low, low and medium income areas is almost the same (0.23 and 0,24, respectively), while that between the high and very high income areas is great (1.33). Similarly, the disparity between the very low and very high income areas is also quite high (1.77). The inequality in the distribution of per capita income is clearly brought out in the Lorenz Curve (Fig. 12.2), the Gini coefficient amounting to 13.8 per cent.

The graph (Fig. 12.3) showing trends in the inequality of per capita income distribution among the income regions of U.P. reveals that the position of the region with very high per capita income in 1970-71 did not change till 1978-79 although there were major annual fluctuations. The other income regions have persistently remained in their positions with very minor fluctuations, indicating that during a period of nine years the pattern of inequitability has not altered, even though the gap between different regions has been somewhat reduced ; low income regions catching up with high income regions. The income gap between the region of highest and lowest income has, however, been reduced by a decreasing trend in the former area—a not very encouraging state of affairs.

Causes of Inequality

Spatial inequality in the distribution of per capita income is the

TABLE 12.1 : **Showing coefficient of advantage, Uttar Pradesh (1970-71)**

Income/District/Groups	No. of Districts	Range of Per Capita Income (Rs)	Total population	Per capita income (Rs)	Total Income (crores)	Percentage		Cumulative Percentage		Difference	Coefficient of Advantage	Increase in Coefficient of Advantage
						Population	Income	Population	Income			
Very low	18	221—300	37,644,380	259.17	965	43.83	34.09	43.83	34.09	—9.74	0.78	—
Low	15	301—380	2,445,119	339.73	818	28.47	28.89	72.30	62.98	0.42	1.01	0.23
Medium	11	381—460	17,075,035	416.55	704	19.88	24.87	92.18	87.85	4.99	1.25	0.24
High	13	461—540	5,921,356	472.00	277	6.89	9.78	99.7	97.63	2.89	1·42	0.17
Very high	1	More than 540	790,080	857.00	67	0.93	2.39	100.00	100.00	1.44	2.55	1.13

Note : Population of 48 districts of U.P. (1970-71)=85,885,970 and total income in crores (1970-71)=Rs. 2,831.

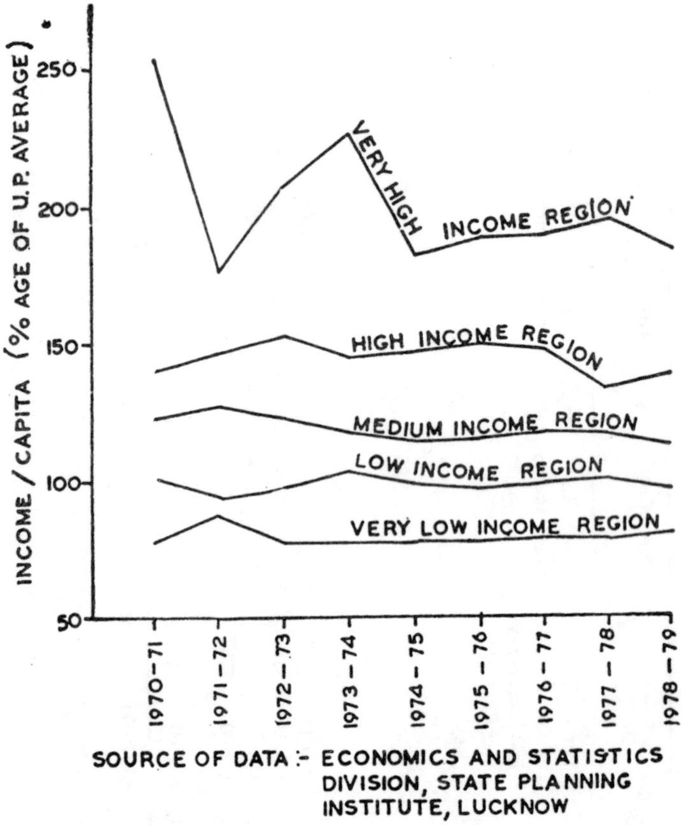

FIG. 12.2 : *Trends in the inequality of per capita income distribution among regions of Uttar Pradesh (1970-71 to 1978-79)*

outcome of a number of contributory factors of varying significance. In the present study an attempt has been made to determine the relative importance of the factors that may be deemed to play an important role in income disparity in the state of U.P.

Methodology

To isolate the dominant factors, the method of Stepwise Multiple Regression Analysis has been used. The per capita income (in Rs)

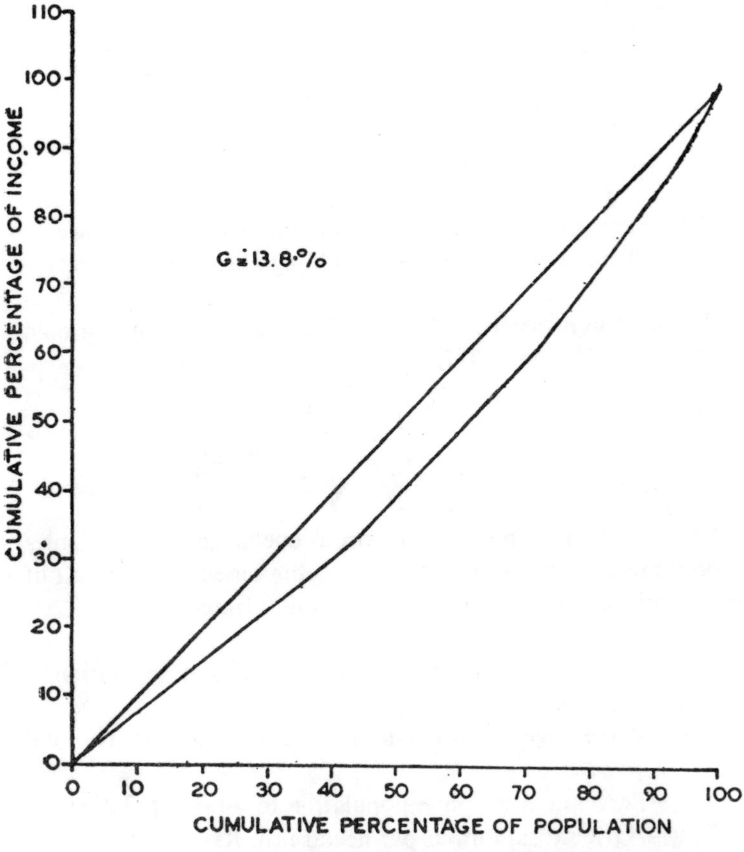

SOURCE OF DATA :- ECONOMICS AND STATISTICS DIVISION
STATE PLANNING INSTITUTE, LUCKNOW

FIG. 12.3 : *Uttar Pradesh Lorenz curve : Showing inequality in the distribution of per capita income 1970-71*

is the dependent variable (y) and 11 arbitrarily selected variables are x_1 to x_{11} :

y = Per capita income in rupees.
x_1 = Percentage of district population to total population.
x_2 = Percentage of workers to total population.
x_3 = Percentage of cultivators to total population.

x_4 = Percentage of workers in tertiary sector to total population.
x_5 = Percentage of industrial workers to total population.
x_6 = Percentage of scheduled caste to total population.
x_7 = Percentage of urban population to total population.
x_8 = Agricultural output per hectare (in Rs).
x_9 = Agricultural output per worker (in Rs).
x_{10} = Percentage share of landholding below one hectare.
x_{11} = Percentage share of landholding above 10 hectares.

Simple correlation matrix (r) (Table 12.2), based on the formula given below (Mahmood 1977) was prepared.

$$r = \frac{\Sigma xy - \frac{\Sigma x \Sigma y}{N}}{\sqrt{\Sigma x^2 - \frac{(\Sigma x)^2}{N}} \sqrt{\Sigma y^2 - \frac{(\Sigma y)^2}{N}}}$$

On the basis of the simple correlation coefficient, the variables showing high multicollinearity were eliminated as a result of which 8 independent variables were finally selected :

x_1 = Percentage of district population to total population.
x_2 = Percentage of cultivators to total workers.
x_3 = Percentage of scheduled caste population to total population.
x_4 = Percentage of urban population to total population.
x_5 = Agricultural output per hectare (in Rs).
x_6 = Agricultural output per worker (in Rs).
x_7 = Percentage share of landholding below one hectare.
x_8 = Percentage share of landholding above 10 hectares.

The Stepwise Multiple Regression approach involves adding one independent variable at a time, and generating a series of intermediate regression equations. The first independent variable considered is the one that has the maximum R^2, that is, the highest simple correlation with the dependent variable. In the first step R^2 is the square of 'r'. In the second step the variable that gives the next highest value of R^2 is added on to the previous regression.

The formula (Johnston 1963) for calculating R^2 from the

TABLE 12.2 : **Correlation matrix**

y	x_1	x_2	x_3	x_4	x_5	x_6	x_7	x_8
1.0	–0.32092	0.02144	–0.31873	–0.00114	0.12880	0.54154	0.61504	0.47426
	1.0	0.28219	0.79309	0.50097	0.23626	–0.00048	0.35153	–0.42263
		1.0	0.18722	–0.02672	0.31994	0.34354	0.09791	–0.10960
			1.0	0.33069	0.00547	–0.13580	0.29811	–0.30111
				1.0	0.09527	0.25311	–0.21127	–0.08559
					1.0	0.52018	0.32269	–0.28907
						1.0	–0.22890	0.08453
							1.0	–0.70864
								1.0

correlation matrix is

$$R^2_{1\cdot 23} : 1 - \frac{R}{R_{11}}$$

After the selection of variables for the stepwise procedure, the Regression Coefficient, S.E., t, $\bar{R}^2$ and F ratio have been obtained by computerising the raw data by the Multiple Regression Programme.

The results (Tables 12.3 and 12.4) show that the share of landholding below one hectare (x_7) explains the maximum variation (i.e. 37%) in the distribution of per capita income; followed by agricultural output per worker (in Rs) (x_6) (i.e. 17%); percentage of urban population to total population (x_4) (i.e. 5%); agricultural output per hectare (in Rs) (i.e, 1%) (x_5); percentage of cultivators to total workers (x_2); percentage of scheduled caste population to total population (x_3); share of landholding above 10 hectares (x_8) and finally percentage of district population to total population (x_1). The contribution of x_3, x_8 and x_1 are, however, quite insignificant in increasing the value of R^2, as is clear from the column showing the increase in R^2. A study of $\bar{R}^2$ also shows that in addition to not having any role in the contribution to R^2, the value of $\bar{R}^2$ gradually decreases after step 5. The value of R^2 decreases as x_7, x_8 and x_1 are added into the model. This shows that their contribution in increasing the value of R^2 is not strong enough to counterbalance the reverse effects on the explanatory model owing to increase in the degrees of freedom (n-k). Hence, the analysis beyond step 5 has not been carried out in later discussions.

The regression coefficients from step 1 to step 3 consistently show significant values for landholding below one hectare (x_7) and agricultural output per worker (in Rs) (x_6) at 1 per cent level of significance. The regression coefficient of percentage of urban population to total population (x_4) is significant at 5 per cent level of significance.

The value of F ratio, however, shows a steady decrease after the second step, indicating that x_7 and x_6 are the most significant variables by all means. All the values of F ratio are however significant at 1 per cent level of significance indicating linear regression.

The average change in per capita income caused by a unit change in percentage of landholdings below one hectare, and in

TABLE 12.3 : Results of stepwise regression analysis

Variables	Regression Coefficient	SE	t	R^2	Increase R^2	$\bar{R}^2$	F
1	2	3	4	5	6	7	8
Step 1							
x_7	4.776	0.903	5.291*	0.378	—	0.365	27.989*
Step 2							
x_7	3.758	0.749	5.020*	0.548	0.170	0.528	35.308*
x_6	10.012	10.002	5.185*				
Step 3							
x_7	4.022	0.739	5.446*	0.600	0.052	0.573	26.285*
x_6	10.012	10.002	5.493*				
x_4	64.414	32.971	1.954**				
Step 4							
x_7	4.265	0.931	4.579*	0.613	0.013	0.577	19.397*
x_6	10.011	10.003	3.579*				
x_4	66.816	33.736	1.981**				
x_5	0.013	0.030	0.434				
Step 5							
x_7	4.014	0.925	4.337*	0.626	0.013	0.582	16.701*
x_6	10.013	10.003	3.989*				
x_4	69.55	33.109	2.101**				
x_5	0 014	0.029	0.467				
x_2	—15.689	9.454	—1.660				

TABLE 12.3 : (*Contd.*)

1	2	3	4	5	6	7	8
Step 6							
x_7	4.061	1.002	4.053*	0.628	0.002	0.574	13.595*
x_6	10.013	10.003	3.942*				
x_4	71.941	38.018	1.892**				
x_5	0.014	0.030	0.477				
x_2	—16.049	9.941	—1.614				
x_3	8.625	64.807	0.133				
Step 7							
x_7	3.734	1.317	2.834*	0.628	0.000	0.563	11.433*
x_6	10.013	10.003	3.915*				
x_4	67.113	40.383	1.662				
x_5	0.014	0.030	0.462				
x_2	—15.884	10.055	—1.579				
x_3	8.135	65.501	0.124				
x_8	4.698	12.087	0.388				
Step 8							
x_7	3.609	1.349	2.676**	0.628	0.000	0.5517	9.867*
x_6	10.013	10.003	3.889*				
x_4	54.474	46.779	1.165				
x_5	0.017	0.031	0.532				
x_2	—14.6210	10.400	—1.406				
x_3	51.520	102.929	0.501				
x_8	4.009	12.257	0.327				
x_1	—16.042	29.181	—0.549				

* Significant at 1 per cent level of significance.
**Significant at 5 per cent level of significance.

TABLE 12.4 : **Analysis of variance for the regression**

	Source of Variation	Degrees of Freedom	Sum of Squares	Mean Squares	F value
	Attributable to Regression (Explained)	1	196,410.516	196,410.512	
Step 1	Deviation from Regression (Residual)	46	322,796.281	7,017.311	
	Total	47	519,206.797		27.989
	Explained	2	317,122.125	158,561.063	
Step 2	Residual	45	202,084.656	4,490.770	
	Total	47	519,206.781		35.308
	Explained	3	333,252.594	111,084.195	
Step 3	Residual	44	185,954.188	4,226.231	
	Total	47	519,206.782		26.284
	Explained	4	334,064.563	83,516.141	
Step 4	Residual	43	185,142.219	4,305.632	
	Total	47	519,206.782		19.397
	Explained	5	345,458.250	69,091.648	
Step 5	Residual	42	173,748.531	4,136.869	
	Total	47	519,206.781		16.701
	Explained	6	345.533.219	57,588.871	
Step 6	Residual	41	173,673.563	4,235.94	
	Total	47	519,206.782		13.595
	Explained	7	346,184.531	49,454.934	
Step 7	Residual	40	173,022.250	4,325.556	
	Total	47	519,206.781		11.433
	Explained	8	347,514.938	43,439.367	
Step 8	Residual	39	171,691.844	4,402.355	
	Total	47	519,206.781		9.867

agricultural output per worker, are Rs 4.02, and Rs 10.01, respectively. Another significant regression coefficient (at 5% level) is that of the urban population, which increases till the 6th step, indicating that urbanization is significant with increasing per capita income.

The most significant factor, therefore, in explaining the spatial disparity in per capita income is the percentage of land-holding below one hectare (x_7), followed by agricultural output per worker (in Rs) (x_6) and percentage of urban population to total population (x_4).

Since agriculture is the principal occupation of U.P. (57.43% of the total workers are engaged in agriculture), the question of landholdings is of utmost significance. Eastern U.P. is characterized by predominantly marginal and submarginal holdings which are being further fragmented under the pressure of a growing population. According to the 1976-77 agricultural census report, the average size of holdings in eastern U.P. was 0.78 hectare, compared to 1.05 hectares for the whole state and 1.27 hectares for western U.P. Such small holdings are economically unviable and lack even the minimum facilities for an average level of cultivation. The highest percentage of landholdings below one hectare is found in the districts of Gorakhpur, Deoria, Faizabad, Azamgarh, Sultanpur, Pratapgarh, Jaunpur and Varanasi. As a result, the output per worker in these districts too is the lowest (between Rs. 600-3,679). Coupled to these factors, eastern U.P. has the lowest percentage of urban population (0.1%). This is another significant factor responsible for putting the region in the lowest per capita income category (Rs. 221-300 range).

Naini Tal is one of the districts with the smallest percentage of landholdings below one hectare (36.34%), therefore, the output per worker is comparatively high (Rs. 13,271.9). The level of urbanization also is relatively high (22.13%). All these factors put together have resulted in the highest per capita income of Rs. 857 recorded there.

Although it is a known fact that urbanization is positively correlated with economic development, yet Dehradun district in spite of having 47.08 per cent urban population, has a low per capita income (Rs. 253). The benefits of high level of urbanization are undone because the percentage of landholding less than one hectare is quite high (65.60%), as a result of which output per worker is quite low (Rs. 4271.7). The same is the case with the district of Agra. This brings into focus the fact that a balanced urban-rural development is essential for economic progress in a country such as India.

The present trend of thinking on the lines of a welfare approach towards which a study of disparities and inequities leads us, would prove futile if the malaise of society were not removed. Although we know that it is always easier to raise problems than to solve them, in our present capacity we can only make a few suggestions, which may not necessarily be both rational as well as feasible.

In terms of our empirical data, small size of landholdings is the most important single factor responsible for low per capita income in U.P. Logically, as a necessary first step this situation must be rectified. Some form of collectivization seems to be the answer in the long run. What seems most feasible as a short term measure, is the modification of Indian inheritance laws, so that further fragmentation of land subsequent to consolidation may be prevented. This measure coupled with a proper implementation of ceiling and tanancy laws may, to a certain extent, ameliorate the dismal situation.

Minimizing of the gap between rural and urban areas and reconstructing rural society to ensure self-sufficiency, for exmiple, by providing employment during lean periods or by proper implementation of the food-for-work programme and thus lessening somewhat an explosive migration to metropolitan areas, are essential prerequisites for a balanced development of the economy and equitable distribution of the means of livelihood.

However, these are just stop-gap measures. What is needed is to attack the fundamental cause of inequality—inequity. This can be achieved by combining growth with proper redistribution, which can be brought about through suitable taxation and by the provision of social services, promoting the extension of education and ensuring employment.

REFERENCES

Johnston, J., *Econometric Models*, McGraw Hill Book Co., New York, 1963, pp. 159-68.

King, L.J., *Statistical Analysis in Geography*, Prentice Hall, Inc. Englewood Cliffs, New Jersey, 1969, pp. 143-48.

Mahmood, A., *Statistical Methods in Geographical Studies*, Rajesh Publications, New Delhi, 1977, pp. 145-53.

Smith, D.M., *Where the Grass is Greener : Geographical Perspectives on Inequality*, Croom Helm, London, 1979, pp. 364-68.

Census of India, Uttar Pradesh, Population Tables, Series 21, Part I, IA (i), II A, II C (i) and IX, 1971.

THIRTEEN

AIJAZUDDIN AHMAD

SOCIO-CULTURAL CHANGE IN A DEVELOPING SOCIETY
The Case of India

In a society, change comes in the natural course of social evolution. It may be rapid if the social group has generated an internal dynamism of its own which encompasses all spheres of material culture and affects all segments of the population. Such an endogenetic change may be lasting as it gets reinforced by exogenous contact and a continual renewal. On the other hand, it may be slow, even imperceptible, if it gets stagnated by historically determined or socially caused inertia which thwarts internal dynamism and arrests progress. In any case, change has to be seen in a time-space frame. Denuded of its time context and space milieu, it would be as incongruous as it would be unintelligible.

It is necessary to emphasize that conceptually, change is not unidirectional ; it may be progressive or regressive. The concern for direction of change, however, raises the inevitable question of the value-system that will provide the necessary logic for discriminating between positive and negative dimensions of change. The former, as the cumulative human experience shows, can be attained through institutionally directed intelligent interventions in the situation. Then it is called 'development' as it brings about a qualitatively new and irreversible sequence of changes eventually leading to total transformation. In the ultimate analysis what is desirable is not social change but social transformation.[1]

Contemporary Indian society presents a case of utter

confusion. To an observer of the Indian scene the society appears to be a strange mix of the old and the new, the conservative and the modern, the stagnant and the dynamic, the virile and the anaemic. An elitist section of society has undergone at least a century of modern education and has remained exposed to rational thinking, secular beliefs and a scientific world-view. Its exposure to media has contributed towards its consciousness of human values that transcends the limits imposed by the prejudices of caste, creed, language, religion or ideology. The democratic experience spread over three-and a-half decades has partly weakened the dogma of social hierarchy and inequality bred by centuries of casteist philosophies and brahmanical doctrine. The performance in the field of science and technology is baffling. India is a developing country that is known to have produced its own nuclear reactor. It has set up prestigious institutions that are engaged in research on Antarctica, space science, medicine, agriculture and engineering. India's cities try to project a facade of what they consider modernity is their night clubs, discotheques, high-speed cars, colour television sets, five-star luxury hotels, sports stadia, revolving restaurants, race courses, health clubs and the like. And yet each of these cities is a blend of the old and the new, the outmoded and the modern. New Delhi retains within its confines dozens of mediaeval villages where hardly anything has changed. Cow-dung cakes dry on the street pavements and the television set is decorated by an image of lord Ganesha. The layer of modernity is superficial and only skin deep.

The national scene is characterized by contradictions of a very high order. India has achieved self-sufficiency in foodgrain production and yet some 300 million people live below the poverty line. The nation has developed an impressive infrastructure for primary school education. However, millions of parents in rural as well as urban areas prevent their children from attending schools as they are needed at home as 'earning hands'. India jealously guards its hard-earned political independence and sovereign status and is the chief spokesman of the freedom of the South, although it is crucially dependent on the North for development aid, technology, armaments and expertise. Millions of Indians cannot either read or write in any of the regional languages, yet imported video-tape recorded films in English dealing with pop music, Crazy games, Western humour, detective stories and science fiction take up a substantial proportion of the national television time. India is a strange country.

PARAMETERS OF SOCIO-CULTURAL CHANGE

It is difficult to conceptualize the parameters of socio-cultural change in a society as complex as that of India.

What is social change ? Is it independent of technological and economic change ? Why is diffusion of technology so much constrained by the institutional framework ? What processes—social, cultural, politico-institutional—retard the uniform spread of economic development ? These and similar other questions can be answered only if one acknowledge the logical relationship between change and social structure. In trying to answer these questions one is inevitably drawn into the basic issue raised so admirably by the Indian sociologist, Desai (1969) :

> . . . increasing spread of the modern means of communication, the introduction of the British system of administration and laws, and the growth of modern, capitalist, competitive economy, which shattered the subsistance economy of . . . the village, undermined more and more the functional basis of caste. However, the transformation of self-contained rigid castes into modern mobile classes has taken place in a peculiar manner. Certain castes have been monopolising the position of the privileged upper classes of modern society . . . . This development has created a peculiar social structure in modern India, with the result that within the existing Indian society, class struggles have been often assuming the form of caste struggles.

What role has the 'doctrine of immutable casteism' played in the continuation of a stagnant rural social order in a society that essentially remained unaltered owing to unchanging technology and the consequent rigid and institutionalized division of labour (Desai 1969).

As a normative process of change one may assume that man's interaction with nature and his cognition of the laws of nature lead to the emergence of new technology which enhances further his level of interaction and expands his area of freedom. All technological innovations eventually promote man's capacity to attain greater skills, and re-define the framework of production relations leading to an economic change. The mechanism of the distribution of surplus thus generated determines the pace and direction of social change.

Rural society in India has been changing ever since its emergence. Changes in technology have brought about changes in economy, which in their turn have produced changes in social institutions, ideology, art, religion and social outlook. However, owing to the constraints imposed by the social structure, all these processes have neither been uniform in social coverage nor in spatial extent—the latter being determined by the former. As soon as an innovation, a technique, a social amenity reaches the doorstep of the rural countryside, the paramount question that emerges is which castes and which social strata will gain initial advantages and will eventually monopolize the gains and thus alienate the disadvantaged sections. Each innovation, therefore, leaves the disadvantaged even more disadvantaged and the deprived more impoverished.

SOCIO-CULTURAL CHANGE AND RURAL SOCIETY

There are intrinsic links between socio-cultural change and the rural space. However, there continue to be serious gaps in geographical research on spatial structuring and functional linkages of the rural settlements in India. It is felt that no rational policy on problems of rural development and social change can be evolved or satisfactorily implemented in the absence of such crucially important knowledge. In the final analysis, all decision-making in such matters as the location of basic social facilities such as a primary school, a primary health centre and a seed depot, hinges on an intimate knowledge of the settlement structure, its spatial linkages and its capacity to function as a chain in the framework of social circulation.

It is important to keep this point in mind : that the rural society in India is far from homogeneous. In fact the social formations are so rigidly defined that they determine everything else including the pace and direction of economic change. The relationship between the disparate social groups has been determined in the course of long-drawn historical processes, and the value-system that links them together in a hierarchical social order is universally respected. These caste relationships are accepted as given and are believed to be lying beyond the purview of change. The preponderance of a social system that engenders inequality acquires nuances of meaning in an essentially agrarian setting such as rural India. It is this social inequality which forms the basis of agrarian organization, besides providing a framework for an analysis of the causes of rural poverty. As Beteille (1966) has put it :

> . . . India represents more sharply than any other country the extremes of privilege and deprivation which are so characteristic of the agrarian societies of Asia. The Hindu caste system which formed the basis of social divisions in traditional India may be viewed as the prototype of a hierarchical system. Caste has been particularly important in the social system centring around agricultural production, especially the control and use of land.

Whatever may have been the origins of the caste system, it became extremely hierarchical and rigid, encouraging high castes to exploit the masses of toiling peasantry, artisans and the like. Very often high castes could live as parasites on the entire community which was directly responsible for production and rendering of needed social services. Since 1947 the parasitical character of the high castes has been slightly modified in the wake of land reforms, but its effects have not entirely been wiped out. An alarming feature of the social situation is the continuing coercion of the bottom stratum in the caste hierarchy comprised mostly by the harijans, or the scheduled castes, and other disadvantaged groups. The spatial concentration of the scheduled castes and their economic correlates merit special mention because they promote an understanding of one of the important facets of the regional dimension of rural poverty, which is a major constraint on rural transformation.

The harijan work-force is by and large concentrated in the following occupational categories, (i) landless agricultural labourers, (ii) cultivators with small landholdings, (iii) small commodity producers or artisans, and (iv) services, particularly 'polluting' or 'dirty' services. An earlier study revealed a *prima faci* correspondance between the concentration of scheduled castes and agricultural underdevelopment. The fact that these castes are associated with agriculture mainly as landless labourers explains their main concentration in the northern and the coastal plains of India. In as many as 70 districts, lying mainly in Uttar Pradesh and Tamil Nadu, these castes account for about one-fifth to one-half of the rural population. Although at the national level about one-third of the agricultural work-force of the scheduled castes is composed of agricultural labourers, the phenomenon is highly variant in space. For example, the share of agricultural labourers is as high as 96 per cent in Palaghat and 80 per cent in another five districts of Kerala. A scatter diagram showing distribution of scheduled castes on the one hand and agricultural labourers, on the other, indicates high positive correlation. Thus the very nature of social and economic deprivation in rural India suggests their interpenetrative character and operates as a major constraint on the homogenous distribution of developmental gains in the rural areas (Raza and Ahmad 1977).

ROLE OF EDUCATION IN SOCIAL TRANSFORMATION

Perhaps education plays a crucial role in the process of social transformation as it paves the way for the attainment of the objectives of social development.

Assuming this as a premise and further assuming that agricultural development in a predominantly agrarian society is expected to create favourable conditions for the spread of literacy (and education), an attempt was made to explore the pattern of statistical correlation between agricultural productivity[2] and rural male literacy.[3] The exercise did not prove helpful as is evident from Table 13.1. See also Fig. 13.1.

A study of Table 13.1 leads to the following broad generalizations :

1. In the traditional green revolution areas of Punjab, Haryana and Upper Doab there is no significant relationship between agricultural productivity and spread of literacy. Indeed, in the case of Punjab the correlation is negative although insignificant (Table 13.2).

TABLE 13.1 : **Coefficients of correlation between rural male literacy and agricultural productivity**

	Region	Correlation Coefficients
1.	Western Himalaya	—
2.	Punjab-Haryana upper Doab	0.134
3.	Middle Ganga plain-east	0.066
4.	Middle Ganga plain-Bihar plain	—
5.	Bengal-south	—
6.	Sikkim and north Bengal	—
7.	North-east	—
8.	North-eastern plateau and Orissa coast	−0.12
9.	Northern plateau	0.169
10.	North-western plateau	0.502+
11.	Rajasthan Gujarat dry region	0.59+
12.	Maharashtra	0.338*
13.	Peninsula-interior	0.604+
14.	Peninsula-littoral	0.331*
15.	Islands	—

+ Significant at 50% level.
* Significant at 10% level.

2. Literacy is significantly correlated with productivity in the regions within the peninsula which, a few exceptions apart, are generally marked for their low productivity.

3. There is a negative correlation, although insignificant, between literacy and productivity in the north-eastern plateau region and the Orissa coast.

The patterns of this relationship as displayed by data for the selected states union are given in Table 13.2.

These generalizations do not take us very far. However, it is evident that neither the levels of literacy nor of productivity display any homogeneous spatial pattern. Both the phenomena are fragmented in space, although for different reasons.

It may be inferred that higher income generated from the farming sector does not necessarily release the work-force to avail of the educational facilities or of the skills that can be acquired only through formal education. The child population remains bogged down with ancillary work at home and on the farms and fails to return to school. The skewed distribution of

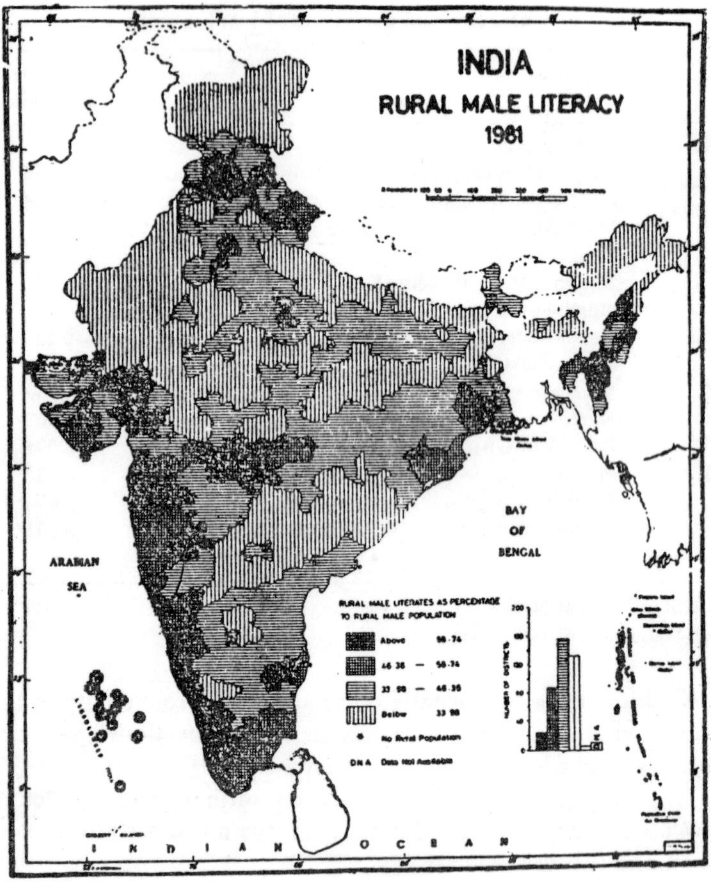

FIG. 13.1 : *India* : *Rural male literacy*

agricultural land and of assets leaves a vast section of the rural population unaffected by the winds of change.

Likewise, the problem of the spread of literacy (and education) should be seen in its proper historical perspective. It is generally agreed that there were historically determined constraints on the universal spread of literacy and the imparting of education in Indian society before independence. The initially stunted and highly fragmented growth of literacy stemmed from the primary socio-economic inequalities and the stratified

TABLE 13.2 : **Coefficients of correlation between agricultural productivity and male literacy for selected states**

States	Correlation Coefficients
Karnataka	0.51*
Maharashtra	0.34*
Madhya Pradesh	0.45*
Rajasthan	0.312
Punjab	−0.025
Uttar Pradesh	0.147

*Significant at 10% level.

structure of society. The institutionalized framework of social inequality coupled with variations in income distribution engendered vast disparities not only in access to the institutions of formal learning but also in the levels of educational attainments of the different segments of the Indian people.] As noted by Cippola (1969), while the art of 'writing spread around the world, across geographical and cultural borders, and adopted itself to a variety of languages, it hardly filtered down through social stratification' (Cippola 1969). For ages, education in India was regarded as a preserve of the uppermost stratum in the traditionally caste-ridden society. Those who had access to education eploited it as a means of making the individual accept and conform to the hierarchic structure of society and completely subordinate his individuality to it (Desai 1966). The social system thus operated as a constraint on the universal spread of education and curtailed the capability of education to make an impact on the masses who remained by and large outside the framework of institutionalized learning.

India at independence carried on her shoulders this heritage of inequities in all facets of social development. The pattern of inequities was most striking in the field of education.

One of the major challenges facing India today is the need to impart education to the teeming millions of the deprived communities and to bring them to that level in terms of attainment of knowledge and skills at which they can assume their responsibilities as useful citizens of an independent state.

A good deal of progress has been made in this field since

independence. Special efforts have been made to reduce inequality in all facets of social development. The lagging sections of the population have received the attention of governmental and quasi-governmental agencies and have been introduced to formal learning on a special footing. However, the problem of educability of the scheduled castes and tribes is much more complex and represents essentially an extension of the social inequality. It can be resolved only by persistent and committed efforts.

Second, the educability of the tribal communities is not analogous to that of the general population or even to that of the caste segment. By and large, the spread of education among the tribal communities is contingent on their stage of social evolution, ecological setting, mode of economy, their exposure to a language other than their own dialect, the pattern of economic interaction with the non-tribal groups in their neighbourhood, the role of tribal elites and of the exogenous forces that operated in the tribal regions before and after independence.

The Indian experience in the pre-independence era shows that the spread of literacy in tribal societies has been constrained by a complex set of factors operating in close conjunction with each other.

These observations leave little scope for optimism. Part of the inequity in the development of education flows from social inequality. Partly, the explanation lies in the physical accessibility to schools of different levels for different age-groups of school-going children.]

Accessibility to Schools

Accessibility has been one of the major constraints in achieving the objective of universalization of elementary education. The location of schools in rural areas—in the early phase of planned development—was such that they were beyond the manageable reach of children. It was in this context that the Kothari Commission on Education proposed that schooling should be provided to each and every child within a 'walkable' distance. The recommendation was considered as one of the significant stepping stones and attempts were made to remove this constraint. Efforts made in this direction are evident from Table 13.3, which depicts progress during the period 1957—78.

TABLE 13.3 : **Population served by schools**

(*as percentage*)

Type of Schools	First Survey 31.3.1957	Second Survey 31.12.1965	Third Survey 31.12.1973	Fourth Survey 30.9.1978
Primary schools	74.38 (0.5 miles)	86.33 (0.5 miles)	81.86 (0.5 km)	85.13 (0.5 km)
Middle schools	50.34 (3.0 miles)	72.58 (3.0 miles)	57.26 (3.0 km)	78.83 (3.0 km)
Secondary schools	36.44 (5.0 miles)	61.20 (5.0 miles)	54.65 (5.0 km)	57.01 (5.0 km)
Higher secondary schools				19.64 (4.0 km)

The concept of walkable distance varies from state to state and from region to region depending upon the terrain conditions. For the purposes of this study the following distances are considered as walkable for children in the different age-groups and may be considered as norms.

Primary schools	0.5 km
Middle schools	2.0 km
Secondary and Higher Secondary schools	4.0 km

Primary School

According to the Fourth All India Educational Survey 78.53 per cent of rural population was able to avail of the primary schooling facility within the area of habitation. For another 6 60 per cent of population primary schools were situated within a distance of 0.5 km. Altogether, the proportion of the population with access to primary schools amounts to 85.13 per cent. However, there are variations from district to district in regard to the population covered by primary schools as is evident from Table 13.4.

The geographic pattern is such that the better served areas are situated in Punjab, Haryana, Gujarat, Maharashtra (excluding the Konkan coast) Karnataka (excluding the southern and coastal districts), Andhra Pradesh, Nagaland and Manipur. Areas in the moderate category include parts of the mid-Indian tribal belt,

TABLE 13.4 : **Population served by primary schools**

Percentage Population Served within 0.5 km.	Number of Districts
Above 91.13	150
76.21—91.13	146
Below 76.21	99

Bihar, Assam, Tamil Nadu and Kerala. Districts with the lowest share of population served by primary schools within the specified distance are concentrated in the northern region of the country. The majority of the districts of Jammu and Kashmir, Himachal Pradesh and Uttar Pradesh fall in this category. This pattern suggests that apparently the accessibility to primary schools is not linked with any developmental indicator as such.

Middle School

Speaking in physical terms, education becomes more inaccessible as one moves from the level of primary schools to middle schools. It is noted that only 46.57 per cent of the population is covered by middle schools within a range of 2 km. This suggests that at the national level a major effort has to be made to induce the process of change at the middle stage of school education. However, there are areas in which the population covered is quite significant as is evident from Table 13.5.

TABLE 13.5 : **Population served by middle schools**

Percentage Population Served within 2.0 km	Number of Districts
Above 88.23	28
70.73—88.23	102
52.51—70.73	128
Below 52 51	137

Regions such as Haryana, Punjab, Bihar plain, deltaic West Bengal, the coastal districts of Orissa and Andhra Pradesh, Gujarat, Maharashtra, Karnataka and Kerala are fairly well served. On the contrary, accessibility to middle schools is very poor in Rajasthan. Madhya Pradesh, North Bengal and the

North-east, excluding Assam and Mizoram. Districts of western Uttar Pradesh, east Rajasthan, Telangana, Tamil Nadu, Assam Valley, Tripura and Mizoram fall in the middle category.

Secondary School

Schooling at the secondary level in rural areas according to the Fourth All India Educational Survey is available only for 58.12 per cent of population within a distance of 4 km. However, the situation varies from district to district as is evident from Table 13.6.

TABLE 13.6 : **Population served by secondary schools**

Percentage Population Served within 4.0 km	Number of Districts
Above 86.61	41
65.29—86.61	90
43.97—65.29	113
Below 43.97	151

Areas ideally served are in Punjab, Haryana, north Bihar, deltaic Bengal, coastal Andhra Pradesh and Kerala. On the other hand, the population in the hilly districts of Uttar Pradesh and Himachal Pradesh, central Uttar Pradesh, Rajasthan, Madhya Pradesh, Bihar plateau, Arunachal Pradesh, Nagaland, Mizoram and Meghalaya suffers from poor accessibility to secondary schools.

Higher Secondary School

Accessibility to higher secondary schools in the rural areas is very poor. It is evident from the fact that only 19.64 per cent of the population is served by higher secondary schools within a distance of 4 km (Table 13.7).

Uttar Pradesh and deltaic Bengal are better served by higher secondary schools. It is significant to note that the situation in Uttar Pradesh in regard to lower stages of schooling is generally bad. Areas of poor accessibility are comprised by western Rajasthan, interior Andhra Pradesh, Orissa, Bihar, Arunachal Pradesh, Manipur, Nagaland and Meghalaya.

The social transformation of the rural communities in India

TABLE 13.7 : **Population served by higher secondary schools**

Percentage Population Served within 4.0 km	Number of Districts
Above 39.54	35
24.90—39.54	65
10.26—24.90	151
Below 10.26	144

is not being thwarted by the inconsistencies in the location of the schooling facility alone. The rural school is yet to emerge as the harbinger of social change and as a centre of diffusion of the qualitatively new ideas among the rural masses—ideas that may inspire hope and not dispair, ideas that are not incongruous and do not lead from one darkness into another.

APPENDIX—A

REGIONAL DIVISIONS

Region	Constituent NSS Regions
1. Western Himalaya	Himachal Pradesh, Jammu and Kashmir (mountainous, outer hills Jhelum valley).
2. Punjab-Haryana upper Doab	Eastern Haryana, western Haryana (excluding Mahendragarh district), northern Punjab, southern Punjab, west Uttar Pradesh, Delhi, Chandigarh.
3. Middle Ganga plain east	Central Uttar Pradesh, eastern Uttar Pradesh (excluding Mirzapur district).
4. Middle Ganga plain—Bihar plain	Northern Bihar, central Bihar.
5. Bengal South	West Bengal (eastern plains, western plains, central plains).
6. Sikkim and north Bengal	Himalayan West Bengal, Sikkim.
7. North-east	Assam (eastern plains, western plains, hills), Manipur, Nagaland, Meghalaya, Tripura, Mizoram, Arunachal Pradesh.
8. North-eastern plateau and Orissa coast	Sidhi and Shahdol districts of Vindhya, Chhatisgarh, south central Madhya Pradesh, Mirzapur district of U.P., southern Bihar, Orissa (coastal, southern, northern), Srikakulam and Vishakhapatnam, districts of coastal Andhra Pradesh.
9. Northern plain	Southern Uttar Pradesh, Vindhya (except Sidhi and Shahdol), northern Madhya Pradesh, central Madhya Pradesh.
10. North-western plateau	Malwa plateau, south-western Madhya Pradesh, North-eastern Rajasthan, southern Rajasthan, South-eastern Rajasthan, Mahendragarh district of Haryana, eastern Gujarat, Gujarat plains-northern Gujarat plains-southern Saurashtra, Dadra and Nagar Haveli, Daman and Diu.

11. Rajasthan Gujarat dry region	Western Rajasthan and Gujarat dry.
12. Maharashtra	Maharashtra (coastal, inland western, inland northern, inland southern, inland eastern, eastern).
13. Peninsular-interior	Karnataka (coastal, ghats, inland eastern, inland southern, inland northern), Andhra Pradesh (inland northern, south western, inland southern).
14. Peninsular-littoral	Coastal Andhra Pradesh, coastal northern Tamil Nadu, coastal Tamil Nadu, northern Kerala, southern Kerala, Pondicherry.
15. Islands	Andaman and Nicobar Islands, Lakshadweep.

APPENDIX—B

GENERAL LITERACY, 1981*

	State/Union Territory	Number of districts in each category					
		Above 53.72	40.74—53.72	27.76—40.74	Below 27.76	N.A.	Total
1	2	3	4	5	6	7	8
1.	Andhra Pradesh	1	1	8	13	—	23
2.	Assam	—	—	—	—	10	10
3.	Bihar	—	—	8	23	—	31
4.	Gujarat	2	9	7	1	—	19
5.	Haryana	—	2	9	1	—	12
6.	Himachal Pradesh	—	6	5	1	—	12
7.	Jammu and Kashmir	—	1	2	11	—	14
8.	Karnataka	—	7	9	3	—	19
9.	Kerala	12	—	—	—	—	12
10.	Madhya Pradesh	—	3	16	26	—	45
11.	Maharashtra	3	14	9	—	—	26
12.	Manipur	—	3	3	—	—	6
13.	Meghalaya	—	2	2	1	—	5
14.	Nagaland	1	3	2	1	—	7
15.	Orissa	—	3	5	5	—	13
16.	Punjab	—	6	6	—	—	12
17.	Rajasthan	—	—	4	22	—	26
18.	Sikkim	—	1	2	1	—	4
19.	Tamil Nadu	3	7	6	—	—	16
20.	Tripura	—	2	1	—	—	3
21.	Uttar Pradesh	—	3	23	30	—	56
22.	West Bengal	1	6	6	3	—	16
23.	Andaman and Nicobar Island	—	1	1	—	—	2
24.	Arunachal Pradesh	—	—	—	9	—	9
25.	Chandigarh	1	—	—	—	—	1
26.	Dadra and Nagar Haveli	—	—	—	1	—	1
27.	Delhi	1	—	—	—	—	1
28.	Goa, Daman and Diu	1	2	—	—	—	3

1	2	3	4	5	6	7	8
29.	Lakshadweep	1	—	—	—	—	1
30.	Mizoram	2	—	1	—	—	3
31.	Pondicherry	2	2	—	—	—	4
	All India	31	84	135	152	10	412

* National Average=36.12 ; Mean=34.25 ; Standard Deviation=12.98.

Note : Data have been categorized on the basis of mean and standard deviation.

APPENDIX—C

RURAL MALE LITERACY, 1981*

	State/Union Territory	Number of districts in each category: Above 58.74	46.36—58.74	33.98—46.36	Below 33.98	N.A.	Exclusively Urban Dist.	Total
1	2	3	4	5	6	7	8	9
1.	Andhra Pradesh	—	—	10	12	—	1	23
2.	Assam	—	—	—	—	10	—	10
3.	Bihar	—	—	16	15	—	—	31
4.	Gujarat	2	9	7	1	—	—	19
5.	Haryana	—	3	9	—	—	—	12
6.	Himachal Pradesh	1	7	4	—	—	—	12
7.	Jammu and Kashmir	—	—	2	12	—	—	14
8.	Karnataka	—	6	9	4	—	—	19
9.	Kerala	12	—	—	—	—	—	12
10.	Madhya Pradesh	—	—	22	23	—	—	45
11.	Maharashtra	—	18	7	—	—	1	26
12.	Manipur	—	3	3	—	—	—	6
13	Meghalaya	—	—	1	4	—	—	5
14.	Nagaland	1	4	—	2	—	—	7
15.	Orissa	—	4	7	2	—	—	13
16.	Punjab	—	4	5	3	—	—	12
17.	Rajasthan	—	—	6	20	—	—	26
18.	Sikkim	—	—	3	1	—	—	4
19.	Tamil Nadu	2	9	4	—	—	1	16
20.	Tripura	—	2	1	—	—	—	3
21.	Uttar Pradesh	—	7	29	20	—	—	56
22.	West Bengal	—	4	8	3	—	1	16
23.	Andaman and Nicobar Islands	—	2	—	—	—	—	2
24.	Arunachal Pradesh	—	—	—	9	—	—	9
25.	Chandigarh	—	1	—	—	—	—	1
26.	Dadra and Nagar Haveli	—	—	1	—	—	—	1
27.	Delhi	—	1	—	—	—	—	1

1	2	3	4	5	6	7	8	9
28.	Goa, Daman and Diu	1	2	—	—	—	—	3
29.	Lakshadweep	1	—	—	—	—	—	1
30.	Mizoram	2	—	1	—	—	—	3
31.	Pondicherry	2	1	—	—	—	1	4
	All India	24	87	155	131	10	5	412

• Standard Deviation=12.38, Mean=40.17, National Average=40.62.

Note : Data have been categorized on the basis of mean and standard deviation.

APPENDIX—D

URBAN LITERACY : TOTAL (1981)*

	State/Union Territory	Number of districts in each category						
		Above 68.48	58.96—68.48	49.44—58.96	Below 49.44	N.A.	No. Urban Population	Total
1	2	3	4	5	6	7	8	9
1.	Andhra Pradesh	—	—	11	12	—	—	23
2.	Assam	—	—	—	—	10	—	10
3.	Bihar	—	2	14	15	—	—	31
4.	Gujarat	—	10	7	23	—	1	41
5.	Haryana	—	4	8	—	—	—	12
6.	Himachal Pradesh	4	6	—	—	—	2	12
7.	Jammu and Kashmir	—	2	4	8	—	—	14
8.	Karnataka	—	8	5	6	—	—	19
9.	Kerala	10	1	—	—	—	1	12
10.	Madhya Pradesh	—	5	23	17	—	—	45
11.	Maharashtra	1	16	7	2	—	—	26
12.	Manipur	—	1	2	3	—	—	6
13.	Meghalaya	—	2	2	1	—	—	5
14.	Nagaland	—	4	2	—	—	1	7
15.	Orissa	—	2	6	5	—	—	13
16.	Punjab	—	3	6	3	—	—	12
17.	Rajasthan	—	1	10	15	—	—	26
18.	Sikkim	—	—	4	—	—	—	4
19.	Tamil Nadu	1	11	4		—	—	16
20.	Tripura	3	—	—	—	—	—	3
21.	Uttar Pradesh	2	5	10	39	—	—	56
22.	West Bangal	1	10	4	1	—	—	16
23.	Andaman and Nicobar Islands	—	1	—	—	—	1	2
24.	Arunachal Pradesh	—	—	4	1	—	4	9
25.	Chandigarh	—	1	—	—	—	—	1
26.	Dadra and Nagar Haveli	—	—	1	—	—	—	1
27.	Delhi	—	1	—	—	—	—	1

1	2	3	4	5	6	7	8	9
28.	Goa, Daman and Diu	—	3	—	—	—	—	3
29.	Lakshadweep	—	—	1	—	—	—	1
30.	Mizoram	2	1	—	—	—	—	3
31.	Pondicherry	1	2	—	1	—	—	4
	All India	25	102	135	130	10	10	412

* Standard Deviation=9.52, Mean=54.20, National Average=57.19.

Note : Data have been categorized on the basis of mean and standard deviation.

FOOTNOTES

1. 'Social change' in India was systematically analysed by the Indian Sociologist M.N. Srinivas. He developed the two concepts of 'Sanskritization' and 'Westernization', to explain some features of religious, cultural and social change. There is no evidence to show that Srinivas considered either technology or economy as crucial phenomena related to any such change or explaining it in any way. However, both the concepts were inadequate measures of social and cultural change that overtook Indian society in different historical situations—the first smacked of high-caste chauvanism and the second of loyalty to western imperialism.

   It can, therefore, be concluded that in the writings of Srinivas the word social change has been used in a narrow, parochial sense, and is essentially devoid of any rational content of meaning.

   It is interesting to note that if Sanskritization is the word for organized change that is a concomitant of social evolution in India, then the social and cultural change that has overtaken Europe during the last two and a half millennia should be described as Greekization, Romanization and Arabicization.
2. Agricultural productivity has been measured as the produce from the farming sector expressed in monetary value. Interdistrict variations in agricultural productivity were measured by considering only 19 crops for the trienneum 1962-65 and 1970-73 (courtesy G.S. Bhalla 1979).
3. Data for literacy pertains to the census year 1981.

REFERENCES

Ahmad, A., 'Integrated Rural Development in India : Social Implications of Planning since Independence', *Proceedings of the Second International Symposium on Asian Studies* (Hong Kong : Asian Research Service 1980), pp. 905-922.

Beteille, André, *Caste Class and Power* (Berkley : California University Press, 1966), p. 120.

Beteille, André, 'The Social Framework of Agriculture', in *Regional Development Experiences and Prospects in South and Southeast Asia*, Ed. Louis Lefeber and Mrinal Dutta Chaudhuri (Paris, Mouton 1971).

Bhalla, G.S. and Alagh, Y.K. : *Performance of Indian Agriculture : A Districtwise Study* (New Delhi : Sterling Publishers Pvt. Ltd., 1979).

Cippola, Carlo M., *Literacy and Development in the West*, London, 1969, pp. 7-8.

Desai, A.R., *Social Background of Indian Nationalism*, Bombay, 1966, p. 138.

Desai, A.R., *Rural Sociology in India*, Bombay : Popular Prakashan, 1969, pp. 38-44.

Dubey, S.C., 'Social Structure and Change in Indian Peasant Communities', in *Rural Sociology in India*, Ed. A.R. Desai (Bombay : Popular Prakashan, 1979), pp. 201-205.

Goody, Jack, *Literacy in Traditional Societies* (Cambridge : Cambridge University Press, 1975).

Haldipur, R.N., 'Sociology of Community Development and Panchayati Raj', in *A Survey of Research in Sociology and Social Anthropology*, Vol. 2 (Bombay : Popular Prakashan, 1974).

Raza, Moonis and Ahmad, A., *India—A General Geography* (New Delhi : National Council of Educational Research and Training, 1977), p. 99.

Singh, Yogendra, *Modernization of Indian Tradition : A Systematic Study of Social Change* (Delhi, Thomson Press (India) Ltd., 1973).

Srinivas, M.N., *Social Change in Modern India* (Berkeley : University of California Press, 1973)

Wertheim, W.F., 'Betting on the Strong', in *Rural Sociology in India*, (Bombay : Popular Prakashan, 1969).

FOURTEEN

MOHAMMAD SHAFI

ECONOMIC DEVELOPMENT AND SOCIAL TENSIONS WITH SPECIAL REFERENCE TO INDIA

ECONOMIC development, in the course of history, has very often led to the conflict of cultures. One of the earliest examples comes from the Industrial Revolution which took place in England by 1760. Prior to this England basked in the summer of village industry and pleasant overseas trade. The society was by and large rural. The commodities that were consumed were local and marked their imprint on the culture of the region. By 1760 the summer faded and trade grew more comparative. The Industrial Revolution, brought a cardinal change in the economic development of England by shifting industry from the home to the factory. Before 1760 it was customary in the villages to work at home, but after 1760 it was more usual for everyone to go to work in a factory.

The great drive in the Industrial Rovolution was for a new organization of production : the factory system. This was the result of a number of technical inventions. The interplay of methods and machines of economic and techniques was constant and mature.

The Industrial Revolution was followed by the Agricultural Revolution when new methods of farming were introduced. Rotation of root crops with grass and grain crops began to be practised. There was a new enclosure movement to grow food on the common land and about six million acres of common lands were enclosed. This made possible the growing of large, scientifically cultivated crops. The whole economic climate of Britain had changed.

This economic development produced social tensions and conflict of cultures. An elite class arose in the rural areas with large holdings of land, better housing facilities, and plenty to eat and to spare. The goods of village industries were no longer in demand as the elite class could afford better quality goods produced in the cities, or even imported ones. The enclosure movement helped to destroy village industries. The home worker was earlier able to eke out his living by grazing his few animals on the common land and getting his fuel from there ; but the enclosure of the commons drove him from the village into the industrial towns where he provided cheap labour. Oliver Goldsmith sums up the situation in his Deserted Village. Thus, 'where wealth accumulates and men decay'.

Economic development was the result of scientific development, and scientific progress was instrumental in bringing down the death rate, with the result that after 1760 the English population rose by leaps and bounds—the Demographic Revolution had set in. The migration of people from the rural to the urban areas led to their concentration in the latter. The crowding of the people in the towns led to the development of slums resulting in two distinct cultures : the culture of the elite and that of the common man. In the rural areas the culture of the elite was the culture of the feudal lords, while in the towns it was the culture of the factory owner or the capitalist.

The economic development that followed the Industrial Revolution brought about a change in the system of production. The main industrial growth was in wood and cotton textiles, and in the complex linkages of iron, coal and steam. The factory culture that emerged was distinct from the rural culture : men spent their working hours in the factory and their leisure hours in the crowded streets.

The twentieth century is the age of the technology. Technology has made amazing progress in the fields of agriculture, industry and means of transportation and communication. In the developing countries the application of new agricultural technology has doubled and trebled the yield per hectare. The new technology is based on the investment of heavy doses of inputs such as chemical fertilizers, adequate irrigation facilities, high yielding varieties of seeds and application of insecticides and pesticides. In India, by the application of new technology the production of wheat rose from 12 million tons in 1966-67 to 20 million tons in 1971, and in 1979-80 it stood at 36 million tons. The peasant who could afford the modern inputs became rich overnight, while those who could not afford them found themselves at a disadvantage. Some of those whose farm acreage was small, sold their farm to the rich farmers and became wage-labourers, while others continued with their poor existence. The uneven application of technology resulted in two distinct rural social groups : one class of people with well-built masonary houses, adequate food, enough clothing and with facilities of modern education, and the other class—the rural poor—ill-nourished, ill-clad, and illiterate. This state of affairs has generated social tensions and cultural conflicts.

If technology is not applied in the right perspective, social tensions and cultural conflicts are bound to arise. The introduction of technology without adequate understanding of its impact on employment and other social consequences leads to distress. People are displaced from traditional occupations. Millions of persons earn their living by selling fruits and vegetables, processing farm produce, and drying and storing grains. They also sometimes look after the areas of energy management based on the collection and utilization of agricultural and animal wastes. Energy intensive and labour displacing techniques are inimical to their economic interest.

Economic development with the attendant cultural conflict is indeed a complex matter. On the one hand, in view of the rising population (it is estimated that the population of the world by the turn of the century may be 8-9 billion, and half of the population of the world to-day is believed to be undernourished), economic development is a necessity. Economic development cannot take place without the application of technologies, and those who possess the means of applying technologies will accumulate capital while those who are small and marginal farmers and petty artisans will remain poor and will progressively become poorer. The basic question is how, along with economic development, social tensions and cultural conflicts can be avoided.

THE INDIAN EXPERIENCE

Realizing the generation of social tensions along with economic development, India has embarked upon a scheme of Integrated Rural Development (IRD). The focus of the programme is on target groups consisting of small and marginal farmers, landless labourers, share-croppers and rural artisans.

The primary objective of the IRD programme is to generate additional employment and raise the income level of the identified target groups. In order to achieve this objective, the identified beneficiaries are provided subsidies to undertake various activities in the primary, secondary and tertiary sectors. While the intensification of development effort in agriculture and other allied, activities will create additional employment opportunities, the secondary and tertiary sectors will also receive the subsidies and thus, generate further sources of income for the rural poor.

One of the most important features of the programme is to adopt the family as a unit rather than the identified beneficiary as an individual. The subsidy given is in proportion to the size of the family and the family is supposed to undertake productive investment on various schemes, the returns of which are large enough to lift it above the poverty line. The IRD programme is linked with such special programmes as the 'Small Farmers Development Agency', the 'Drought Prone Area Programme', and the 'Command Area Development Programme'.

The country has been divided into 50,004 blocks for rural development. Each block consists of a group of villages. In 1978-79, 2,000 out of 3,000 blocks covered by one or other of the three special programmes were selected for taking up the IRD programme. In addition, 300 blocks were taken up from those areas where no special programme was in operation. Starting from 1978-79, 300 new development blocks were expected to be covered each year from areas not earlier selected for any special programme.

Of the 2,000 blocks earmarked from the special programme areas, 1,000 blocks were allocated Rs. 500,000 per block, and the other 1,000 blocks, Rs. 1,000.000 per block. In addition, a provision of Rs. 60,000 per block was made for conducting the baseline survey of the households in the blocks selected from the non-special programme areas.

Under the integrated rural development programme, an attempt has been made to integrate the 'area' and the 'beneficiary' development approaches. Ten per cent of the outlay provided under the programme is devoted to the development of the infrastructure.

It is estimated that the number of families that can be raised above the poverty line during one year will be about 300 in blocks for which this outlay is Rs. 500,000 and 600 for which the outlay is Rs. 1,000,000. The programme is closely linked with the availability of bank credit.

Similarly, the planning strategy for the rural industries aims at providing a reasonable income to the rural artisans and a sustained increase in employment opportunities.

Again, the National Scheme of Training of Rural Youth for Self-Employment has also been initiated. The main thrust of the scheme is on equipping rural youth with necessary skills and technology to enable them to seek self-employment. It has been proposed to train about 200,000 rural youth every year in various skills. The schemes envisage training through local servicing and industrial units, master craftsmen, artisans and skilled workers. The scheme provides for suitable stipends for training expenses.

Economic development accompanied by social tensions is bound to erode the structure of parliamentary democracy. Social tension and cultural conflicts can lead to upheavals, but if measures are adopted which promote economic development, a balanced healthy and durable economic development will take place and strengthen the foundations of parliamentary democracy.

FOOTNOTES

1. Micheles, Dean Vera, *West and Non-West* (New York : 1963).
2. Swain, Joseph Ward, *History of Civilization*, Vol. 2 (New York : 1958).
3. Government of India, *India* (New Delhi : Ministry of Information and Broadcasting, 1980).

SECTION FOUR

PLANNING AND DEVELOPMENT

FIFTEEN

SATYESH C. CHAKRABORTY

ON DECENTRALIZED REGIONAL PLANNING IN INDIA

INTRODUCTION

IN ORDER to conduct the affairs of a society through planning there must exist an assurance, explicit or implicit, of public commitment for it. Without people's participation, the objective of planning, be it economic, administrative or social, cannot be fulfilled. Such a commitment for participation is necessary to create an organization for and to sustain the process entailed in planning. It is not enough to claim that such a commitment exists. Continual verification of this commitment is a necessary condition in order to sustain the process of planning. For this reason, when doubts get expressed about the utility of a given planning practice that may be currently in use. it becomes necessary to settle the issue through public debate or a similar socially accepted mode.

Doubts about any given practice may emerge from widely different circumstances. Failure to achieve the declared targets of growth may upset some. However, doubts emerging thereof are relatively easy to alleviate by upgrading organizational efficiency. The more common sources of doubts emerge from the anxiety regarding the very objective of planning that a given political authority may have decided to pursue. The reason is basic, because the object of planning is to invoke interventions upon a given course of social development. These interventions when exercised cause upsets or inconvenience to some segments of society which had otherwise expected to profit. Unless the socially held belief-system that imparts rationality to the invoked interventions is clear and appears just, public protests tend to reduce the validity of the object of planning and also of the *modus operandi* used to pursue the desired process of change. Public debates help to clarify the belief-systems of a given society.

In recent years questions have been raised on the usefulness of continuing with the current planning system in India. This has created a need for public debate. In fact a number of debates have already been organized by the different political parties that hold political power in the states. A number of social scientists have also participated in these political debates, and there is reason to believe that many more of such debates will follow, as they are expressions of the doubts held at some level of Indian society.

In these debates a consensus seems to be emerging in favour of branding the current practice of planning in India as an instance of centralization. The idea of centralism is disliked by the participants as they claim to have experienced it as arbitrary and hence oppressive. According to them the regional (state) priorities on social and economic change are hardly accommodated within the nationally adopted policies. This has been held as a factor contributing to the disparity in the achieved levels of economic development between the regions. Therefore, the consensus calls for greater autonomy at the state level than what has been provided for within the present framework of the Constitution of India. Running concurrently is another consensus for decentralized planning. Since the objective of decentralized planning is to accommodate the varying regional priorities, these debates may be considered as implicitly suggesting the need for an institution for regional planning in India.

Needless to mention, these feelings have produced counter-views in other quarters. Some consider it unjust to describe the current planning practice as an instance of unmitigated centralism. They argue that India had invoked the concept of multilevel planning while initiating the Fifth Five-Year Plan and that as a result we may now have planning boards for all the states or district planning committees for all the districts of India if we want to. The original idea was to create an institutional mechanism to identify the planning priorities for the districts, the states and the nation-state in sequence or through iteration. The Planning Commission of India, as an apex organization, exercises its authority by preparing the final plan as comprehensively as possible in response to the so identified needs at sub-national levels. In other words, according to this point of view our present planning practice is just one variety of decentralized planning with some order of centralism within it.

The existence of divergent views about the same reality is not something unusual. This is a common feature of any society, as all its constituent social groups may perceive the same reality in many different ways. The differences in their perceptions merely suggest that some among them are unhappy about the reality they face. They are obviously worried about the way the current process of planning is being regulated in India. Such anxieties are being expressed in almost all the states. It is the territorial spread of this feeling that makes it an important issue. Obviously, the cognitive distance between the contending perceptions requires careful handling.

Those who are looking for an alternative approach are not against planning *per se*. They have not at any stage questioned the need for planning for the various regions. Given these factors the substantive issue is to examine the proper meaning of centralism within an acceptable mode of decentralized regional planning for India. The problem here is to define a structure for decentralized regional planning that could resuscitate public commitment for guided social change in India.

CONTENT AND CONTEXT OF PROTEST

While it is important to be aware of protests, it is injudicious to support them before knowing what the protest leaders want to achieve. Such caution is necessary when the protest about a current institution is unmatched by any alternative suggestion. There are many who are unhappy about the current system of planning, but it is difficult to believe that all of them are unhappy about the system for the same reason and in the same way. Therefore, there may be general agreement on the need to join in a protest against a current situation, while disagreeing on what to substitute in its place.

The fact that the proponents of decentralized planning constitute a mixed group is significant. There are professional academics as well as professional politicians among the protest leaders. Until it is known that these two groups of professionals are unhappy about the current system of planning for the same set of reasons and are agreed on the need to evolve a common format of decentralized planning, it is well to be wary of opportunistic compromises that emanate from a possible combination of unexamined motives. It is, therefore, necessary to examine the content as well as the context of the protests in some detail.

Taking a look at the reality that confronts professional politicians, it may be recalled that planning in India was instituted at a time when the political powers at both the levels of the union and the state governments were held by a single political party. Through its own disciplines and procedures this political party maintained an image of a monolithic polity by proxy. However, the situation has changed a great deal since then. Different political parties are now holding political power at different levels of government and their placements in this hierarchy vary from region to region. The professional politicians now acting as protest leaders represent the parties that have come to acquire power in some of the states in recent years, and which together constitute the opposition group at the level of the union government. They are asking for greater autonomy for the states than what has been provided for in the Constitution of India.

Autonomy refers to the right of self-determination. Since the demand is not for absolute autonomy, the question arises as to how the absolute autonomy now enjoyed by the nation state is going to be shared in future between the states and the union government. The answer should not only specify where the residual power will be in the relationship between a state and the union government, but should also indicate the rules that will govern the relationship between the states themselves. Needless to say such a classification has not yet been forthcoming from the professional politicians. What is apparent from their expressed dissatisfaction with the present situation is that they would prefer a better form of federalism that what now obtains in India.

Federalism, as a political philosophy, is needed to provide standards of norms for managing interrelationship among diverse centres of political authority. This requires the contenders jointly to participate in an exercise to formulate the rules of business that will govern their future role. This joint exercise of course needs the involvement of the professional politicians. While it is true that such explorations are now being conducted only by the political parties holding power in the states, it is hoped eventually perhaps they will involve the party in power at the centre, that is, the Union Government. Perhaps also a time will come when the results of such explorations will eventually get incorporated within the framework of the constitution of India.

It is not clear what kind of autonomy the states would like to have. Neither is there any indication of how the change is to be brought about. Yet the desire for changing the current practices of planning is being seen as a corollary to the demand for autonomy of the states. It has already been mentioned that the expressed desire for decentralized planning has been mainly to enable the states to put appropriate emphasis on regional needs. It is this objective of the professional politicians that seems to have attracted a large number of professional academicians to their cause.

It is not suggested here that the wish to include regional needs within a planning process is mistaken. On the contrary, for a multiracial, multiethnic and a multireligious polity as we have in India, a format of decentralized planning is a necessity. Also, co-operation between professional politicians and academicians is desirable, for when a polity faces the need to redesign the formal rules of interaction between its hierarchically structured components such a coalition is indeed essential. However, so far none of the parties has cared to spell out why and how decentralization of planning can be used as a means of restructuring the form or process of the government.

The demand for decentralized planning can come from different quarters to fulfil different objectives held therein. For example, one such objective perhaps held by the political parties in charge of the government in the states, may be to have greater access to tax funds now accruing to the central government. As long as the national commitment for investment in the public sector remains unquestioned, any mechanism of decentralized planning will direct the flow of funds towards the state governments. Some social scientists on the other hand might have quite a different goal : that of social involvement in the planning process. They may look upon decentralization of planning as a mechanism to uphold the felt needs of the different regions. As long as funds become available for investment in such areas, they are not likely to worry much about the source from which these are made available. Co-operation between the proponents of these two objectives is possible only when the procedures for admitting the felt-needs of the regions and for guaranteeing the flow of funds for them are spelt out.

There exist just too many unexamined beliefs which if left so would prevent us from arriving at any consensus in the near future and may eventually lead to confusion. For example, there is a widely held belief that interregional disparity at the level of economic development has been increasing over the past planning decades. Although a large array of data have been marshalled to prove this point, hardly any attempt has been made to examine whether the emergence of this disparity is a natural consequence of the planning process itself. Nevertheless, this belief is obliquely used to prove the incompetence of the Planning Commission of India and, therefore, of the current planning procedures. It is also used as evidence of neglect of the so afflicted regions. There hardly exists any comprehensive analysis of the decision-making process to show where it went wrong and how such mistakes came about. Instead, this belief is used to arouse strong regional feelings, which is easily done in a multiracial, multilingual and multireligious society such as India's. Unless the experiences of our polity are fully examined, it will be difficult to disentangle the procedural issues of planning for proper evaluation.

Before examining the need to change the procedures of planning, the following questions will have to be answered : First, in what way does the current planning process produce disparity between regions? Second, how will decentralized planning reduce the chance of disparity emerging between regions? Finally, can decentralized planning be carried out without a centralized control mechanism? This last issue is related to the question of autonomy of the states—a base upon which a future federalism can be built in India.

REGIONAL DISPARITY AND CENTRALIZED PLANNING

It is useful to remember that India had inherited a very high order of interregional disparity at the level of economic development from the past colonial period. There were a handful of industrial towns in India and most of them were located in and around the port cities. The general level of agricultural productivity was dismally low. The mining economy was desperately dependent on overseas demand; metallurgy was poorly developed. Set against this situation, the Planning Commission was given the task of achieving a fast rate of economic growth. It was also given the task of reducing the levels of disparity between regions and between social groups within any given region. This indeed was a formidable task to be confronted with.

The planners, however, did not get a blank canvas to work upon. There were diverse social processes for them to deal with. For example, there were the forces of communalism and racism; again, the staff available for administering social change had their orientation in the imperial order. Among all the social forces, the power of the landed gentry, embedded as it were in the ascriptive authority by tradition, and that of the industrial monopoly houses sustained as it were by the rules of truncated capitalism, were something to reckon with. These two power groups appeared as disjunct sets in so far as unfolding the productive potentials of the society was concerned. To overcome these disjunctions was the specific task for the planners.

The problems confronting the nation were far from simple. The Planning Commission, prescribed the norms of public sector investments and designed the policies by reserving certain sub-sectors of the economy for this purpose. The private-sector was granted the freedom to invest in the remaining sub-sectors of the economy guided as it were by the norms set by the nation state. Thus the country was committed to a policy of mixed economy.

The task of managing the process of development has not been at all an easy one. For example, the proposals for public-sector investment do not originate from within a single institution. These emerge from decisions taken by the different departments of the government and there are three tiers of government in India, namely, the national, the provincial and the local.

For the management of a set of resources or sub-sectors of the economy, the Constitution of India confers upon the national and the state levels of the government certain exclusive and separate rights. There are also some other resources or sub-sectors of the economy, whose management responsibility is concurrently shared by these two levels of government. The relationship between the local and the state governments is also designed in a somewhat similar fashion. Such a partitioning as also sharing of responsibility has produced numerous problems of interface management. The growth of the sub-sectors is dependent upon the growth of subsidiary and ancillary sectors, provided commensurate investments are made in all the inter-related areas. However, such investments cannot comprehensively be estimated by any individual department that is engaged in the work of developing the sub-sector that it is specifically charged with. In the absence of an institution to tackle the problem at interfaces between the sub-sectors of the economy, suboptimal resolutions threaten the implementors of approved projects. As a result, anxiety mounts at the lower tiers of the government charged with the execution of the projects through the successive years.

There is yet another institutional process through which tensions built up between the different tiers of the government, resulting in differences of opinion on plan priorities. This process is generated by the relative paucity of investible funds available at the lower tiers of the government. At the lower tiers, projects get designed on the assumption that the bulk of the required funds will flow from the higher tiers of the government. Considering that the total resource available with the nation as a whole has always been less than what is required for investment, the prerogative of restraining the ambitions of the project designers has always been seen as that of the upper tiers of the government. The lower tier institutions, as far as possible, gain access to resources without any strings attached. The higher institutions on the other hand, urgent on allocating grants tied only to specific purposes which they feel they alone have the right to define. Undoubtedly, such tensions are inherent within the institutionalized processes of planning in India.

The various conflicts between the different sub-sectors of the economy have an inhabiting effect on the productive potentials of the country's economy. To remove these disjunctions is the specific task of planning. As the specificities of these disjunctions vary between regions, it is necessary to design specific strategies for their removal, which, for obvious reasons, are bound to vary from region to region. The chances of discovering the correct strategy are higher where the observant planners are located closest to the domain of action and have the necessary authority to do so. However, within the current structure of institutionalized planning this authority has been retained by the upper tier organizations. Inevitably, the prescribed strategies are more likely to reflect the generalities of the situation obtaining across the country and to appear somewhat unrelated to the regionally varying situations. Since all general strategies of planning affect only the macro-sphere of the society, their effectiveness at the local level remains doubtful. The local power groups retain considerable powers of manipulation as the specific sources of their politico-economic authority remain largely untouched. Such a problem confronts all forms of centralized planning institutions.

It would be incorrect to claim that the lower tier institutions do not enjoy any freedom to design strategic interventions. They do, but they get constrained in other ways. Any intervention upon a given disjunction, in order to be effective, requires the support of a package of sub-programmes related to the ancillary and subsidiary sub-sectors of the economy. For example, to encourage the cultivation of high-yielding variety of cereals, there is need, simultaneously, to mobilize irrigation facilities, distribution mechanisms for fertilizer and pesticide, and infrastructural outlays on marketing. It is rarely possible to obtain such a co-ordinated package of investments, as the jurisdictions over these supporting activities are held by different departments at any tier of the government. As a result, the substantive issue of managing the spatial conjunction of activities tends to fall between the departmental stools at any given level of governmental machinery.

A similar problem exists between the various levels of government. Since the upper tier institutions enjoy a right todeal with the sub-sector of the economy in their charge, the responsibility of providing the supportive investments at any given location rests with the lower tier institutions. For example, when the union government decides to locate, say, a petro-chemical industry in a given state, the responsibility of providing the related infrastructure is assigned to that state government. The demand is pre-emptive, for there may be some other state in the line which is also angling for such a bait. It is true that the multiplier effects of the investment on a petro-chemical industry would largely be available to the state that gets it. It is also true that the differences in the gestation periods of alternative investment schedules can appear irksome. In the context of higher orders of scarcity of investible funds faced by the lower tier institutions, such pre-emptive demands may also condition suboptimal returns from investments within a descrete plan period.

The above description of the institutionalized processes of planning in India may explain not only how resentment gets generated at the lower tier organizations, but also why it becomes difficult for them to amend the consequences of centralized planning decisions. It is proposed to touch upon only a few facets of the problem that arises from decisions on location of investments taken by any centralized planning machinery.

Decisions on location are taken with two important assumptions in mind. In the first instance it is assumed that at a chosen location of investment, men and materials will flow. Second, it is assumed that for every input there exists a discrete cost surface across the country as there exists a cost surface for the disposal of final outputs. By following the second assumption, the procedure for location decision is designed. Normally it is usual to compute the aggregate costs for mobilizing inputs and for dispcsing of outputs for a number of initial locations. Among the initially selected sites, that which offers maximum return on investment is then chosen.

It is not suggested that the Planning Commissson in India in every instance actually follows such a procedure meticulously. Quite probably diverse non-economic considerations influence the decision-making process thus producing suboptimal results. It is now necessary to examine the impact of the procedure mentioned even when correctly followed upon the Indian social reality. This examination can be gainfully conducted in the context of the first assumption that the economic optimization models are based on location decisions. It is assumed that the model would help to select a location for the creation of certain types of jobs and that the job seekers would flow towards such a site to secure benefit of employment from the investment.

It may be noted that all investments do not create the same types of jobs. The complexities of technology in use decide the structure of the organization of production and thereby lay down specifications for the jobs created. The more specialized the job is, the more difficult it is for the people coming from the backward regions to qualify for it. Although a job seeker can be given the requisite training before absorption, this practice is hardly ever followed by Indian industrial establishments in real life. As a result, investments for high order technology are advantageous to people belonging to regions with a highly developed social base of technology. This will perhaps explain why the communities living in and around the newly developed mines do not easily get absorbed in the industries that utilise the products of such mines. For this reason, such industries are either located in places where a skilled labour force is readily available or they tend to recruit such labour from technologically advanced regions and in the process exclude the backward regions from the advantages of economic development.

The significance of the effect of the level of technology in determining labour migration should not be underestimated. Where the level of technology is low, primary economic activities such as food-gathering, fishing, animal husbandry, agriculture and mining are pursued through labour-intensive technology ; here the component of capital in the form of tools and machinery is low. When an industry that requires a great many tools and machinery for its functioning gets located in such a region and recruits labour locally, the impact upon non-industrial activities is far from simple. It is difficult even for a willing labourer now absorbed in an industrial establishment completely to detach himself from the tiers of the economy that he has come from. This is manifest in many of the older industrial centres of India that experience seasonal homebound migrations of their industrial workers. The workers return to their villages because they are needed to help in the agricultural production activities that are still pursued by their family members. Until their social base of technology is improved to the level where the workers can be permanently released from such labour, the seasonal homebound migration is likely to continue. The lower the level of social base of technology of the region, the greater is the strength of such ties. Until the social base of technology is improved to a level capable of permanently releasing such labour, the seasonal homebound migration will continue. This is indeed irksome for the industries. As a result the industrialists gradually turn to such areas to recruit labour, where the social base of technology is high enough to enable the immigrant to stay permanently in the industry. The falling rate of participation of the tribal communities in urban activities is an example of this state of affairs.

There is another social force to reckon with in regard to the flow of unskilled labour to the places where investments are made. This type of labour is abundantly available in our country in almost all regions. However, the spatial variation of the socio-cultural base in our country appears as a constraining factor in obtaining free mobility of labour migration. Ours is a multiethnic, multireligious and multilinguistic society, where tolerance for aliens has hardly been cultivated. For this reason, any sizeable transfer of population between regions often produces social tension within the labour absorbing areas.

These two processes operate simultaneously in India across all regions and thereby disprove the assumption that labour flows freely where jobs get located. Nevertheless, the other assumption that materials flow freely to places where investment are made remains unimpared. Since such flows of materials are designed to obtain higher value-addition through fabrication, a net transfer of value takes place from the raw material exporting areas to the finished goods producing centres. A commensurate flow of labour from the former to the latter regions does not necessarily occur. As a result the country gets reclassified into overprivileged and underprivileged segments. This is basically the process that sustains interregional disparity in levels of economic development. Thus it can be seen that a purely centralized mechanism of planning in India, can produce disparity in the levels of development between regions owing to the peculiarities of the socio-technical base of regional economies.

REGIONAL DISPARITY AND DECENTRALIZED PLANNING

It has already been mentioned that the demand for decentralized planning in India has come as a reaction to centralised planning. As a result many are prone to see the merit of decentralized planning just because the other mode appears inappropriate. It is essential, however, to appreciate the fact that the philosophy of decentralized planning is quite different from that of a centralized one. Unless this is done, the modes of decentralized planning cannot be gainfully applied in India. It is necessary to find out the positive strengths of this method of planning in the given context of India before accepting it as desirable.

Decentralized planning is a technique that recognizes and admits the need for decentralized decision-making centres for allocating investments. This requires the creation of multiple organizations of planning, each having a distinct territorial jurisdiction or domain of action. How the boundaries of the territories in charge of the different planning organisations are to be delineated need not concern us here. Instead, what should concern us is the fact that the need for delineating definite territorial domains of action arises from the very reality facing the Indian polity.

On the one hand, obvious differences exist in the social base of technology, say between the cities and the villages, and between industry and agriculture. On the other hand, there exists a complex system of interdependence between the metropolitan cities in India, based as it is upon an equally complex system of interdependence between diverse types of industries. Large segments of society are engaged in pre-industrial forms of production in the primary sector of the economy, pursuing their livelihood in a relatively independent state. This has produced inequalities between industry and agriculture as also between cities and villages.

The city-based industries are mainly producing a large array of goods that are meaningful primarily to the urban economy in India. These industries hardly produce any meaningful goods for the rural economy to improve its productivity status. Nevertheless, the country's rural economy has been made to meet the demands of the metropolitan economy by supplying the nutritional requirements of the city dwellers and by producing cash crops for fabrication in city-based industries. As far as meeting their own demands is concerned, rural society has been left to fend for itself. Such disparities make the Indian polity decidedly vulnerable.

The objective of planning is to take appropriate measures to overcome these inequalities. In order to do this, there is need of a planning machinery that is adequately decentralized, because the nature of the problem varies from region to region. In some areas the farmers may be willing to take up the cultivation of cash crops, but are restrained owing to the non-availability of a marketing infrastructure. In some other areas the factor restraining increased productivity may be the non-availability of industrial inputs. In yet other areas an initial investment may be needed for a warehouse, or the functions of an existing warehouse may be required to be further strengthened by introducing an industry to process the available primary goods. The contents of developmental work would thus vary according to the patterns of livelihood pursued by the different communities. It is almost impossible for a single centralized planning machinery to assess the situation in each and every region and take appropriate action. This task can better be handled if a number of planning organizations are given independence of action to conduct their affairs.

It may then be said that the philosoply of decentralized planning is to bring jobs to where the people are. This is indeed a commendable ideal, for it would permit the people of all regions to derive advantages from planned investments. It would at least minimize the social costs of large-scale labour migration. However, a planner while pursuing this philosophy has to take care of certain other objectives if he wishes to alter the given level of disparity in economic development between regions. He may not opt for mere increase in the number of such jobs that are already there in the region. His basic endeavour should be to obtain a new production-function by altering the composition of capital in favour of fixed assets such as tools and machinery. The reason is simple. His task is to increase the productivity of labour, which he can do by acquiring a greater number of tools. By opting for such a production function, he will be able to elevate the social base of technology in his domain of action. His competence in this regard will be shown by his ability to identify and to absorb such tools and machinery that will make the new production function economically viable.

It is often argued that with the use of more tools and machinery, reduction of labour in the production process emerges as a distinct possibility. The planner has to guard against such a possibility by redirecting more and more of the redundant labour to newer resources available within the region. He may do this by extracting such resources locally, by fabricating the so extracted materials locally and even by consuming the thus fabricated material locally. In this way, he may initially minimize the redundance of labour.

For all these reasons, the planner has to give serious thought to the orientation of the natural resource endowment of the region ; but this should not lead him to believe that it is his task to make that region economically autonomous. To give an example.

In order to improve productivity in agriculture, the use of diverse intermediate goods such as fertilizer, pesticide, irrigation-implements, etc. is required. An industrially backward region has to obtain these goods from an industrially advanced one. Similarly, to obtain an economic outlet for all kinds of surplus from one region, markets have to be sought in other regions. This is a simple example, demonstrating the need for developing regional interdependence. We may note that no society will continue to produce more goods than it needs unless willing consumers are found elsewhere to absorb the surplus. Without consumption, the object of production can never fructify and the anticipated value addition can never be realized. Trade facilitates the process by institutionalizing negotiation between consumers and producers.

In developing interdependence between regions, the planner may not entirely depend upon the market forces. Especially in countries such as India, where market forces are distorted by monopsonistic and monopolistic powers, free-trade exists merely as a dream. If demands are placed in markets without preparing the producers of the relevant goods in advance, then the resultant short supply of materials leads to the strengthening of speculative trade. This means that the planners have to take pre-emptive actions by working out as carefully as possible the nature of interdependence that they wish to obtain between regions. If this is not done, then unregulated demands would strengthen only the economically powerful groups and allow the economically advanced regions to retain their grip over the rest of the society and regions. The current pattern of interregional disparity in the levels of economic development will then persist.

OBJECT OF INTERREGIONAL NEGOTIATIONS

Interregional negotiations between the respective planning organizations need not be seen as a matter of pure political convenience. It is an economic necessity governed by the compulsion of pricing of goods. Unless the meaning of pricing of goods is properly understood in the context of the need of the region for which planning is intended, even the choice of technology may become irrational. To demonstrate the issue, we may look into the case of apatite deposits of the district of Purulia in West Bengal.

In the drought-prone district of Purulia, which is also one of the poorest regions of India, there exists a fairly good deposit of apatite or calcium phosphate. To use this natural resource gainfully, a planner in India has a choice of three technologies. First, the rock can be pulverized and used directly in the field with green manure as a low grade fertilizer. Second, Nitrogen may be bound with phosphate to produce ammonium superphosphate, which is a high grade fertilizer. Third, Phosphoric acid may be extracted from apatite and used as a base for producing diverse types of pesticides. Now the problem before the planner is which of these three technologies to choose.

Measured in terms of attainable value-addition as determined by the prices now current in India for the three final products mentioned, pesticide production will appear as the most profitable and pulverized apatite as the least profitable. Amonium superphosphate will find a place somewhere in between. If the planner is asked to respond only to market forces, he will naturally select the technology of pesticide production. It would be interesting to examine the consequences of such a decision against the philosophy of decentralized planning that has earlier been discussed.

Pesticides are expensive. Therefore, only those crops succeptible to attacks by pests and incurring substantial losses thereby will justify the use of pesticides. High value crops such as tea, coffee, tobacco, etc. will deserve the use of pesticides. In contrast, low value crops such as cereals and grains, may not accommodate the cost of pesticides. The district of Purulia does not produce any high value crop. This means that the users of pesticides will be found outside the district, where the value addition that is derivable from the use of pesticides will take place. Therefore, the selection of this technology will be seen as another instance of transfer of potential value from a poor region to a richer one. This will upset the philosophy of decentralized planning. Given the regional orientation of the market, the district of Purulia may not qualify even as a good location for the pesticide factory.

The second alternative, that of conversion to ammonium superphosphate, which we know is a valuable fertilizer for the Indian agricultural economy, may now be considered. Compared to pesticides, ammonium superphosphate appears more meaningful to a far larger segment of the Indian peasant society. However, its application demands a guaranteed supply of water to the farms. Now water is precisely the resource that the drought-prone district of Purulia does not have in abundance. Naturally, the decision to opt for this technology would mean a commitment to permit value addition to take place in tracts that are well-endowed with water and are located in other than the region where apatite is found to exist. This will also impair the philosophy of decentralized planning.

The technology of pulverized apatite will thus appear in concordance with the philosophy of decentralized regional planning. Since the product is a low-grade fertilizer, it cannot travel over long distances in search of markets. It has to be used within the region and in places not far from the mine-head. Its utility within the region will be seen through the ability of calcium in apatite to reduce the level of acidity in soils. The district of Purulia has, on the whole, acid soils. The problem of locally procuring the green vegetative materials to assist the soil microbes in the conversion of inorganic phosphate into plant food can be readily solved. The labour force and machines required to pulverize apatite are also easy to procure locally. On all these considerations, the planner for this region will naturally find this technology as most appropriate to adopt. But the question is, can this planner opt for this technology in response to the retionality of market forces ?

The above example demonstrates that a regional planner practising his art within the framework of decentralized planning, may have to opt for suboptimal resolutions in many instances. The technology of pulverized apatite as fertilizer for direct use in the adjacent areas is decidedly a suboptimal choice when seen against the other available opportunities to maximize value addition for the given concern from the given resource. Yet it is the most relevant action for the district where the basic resource is available. It is relevant because the output can be consumed locally to increase the level of productivity of the agricultural farms. It is relevant also because it would widen the range of jobs in the locality and because the types of jobs it can offer can be taken up by the local people. The plant-machinery is also locally available. When all these factors taken into consideration it will be found that such a choice would lead to an enhancement of the social base of technology in the region.

However, all these gains will be obtained by the society at large and not by the organization that opts for the technology of pulverized apatite. For this reason, the planner would find it extremely difficult to mobilize the necessary investment to set up this concern within the ambit of pure market forces. To get the funds, he would need policy support from the nation as a whole. The nation can give this support through a national-level institution that is duly authorized to adjudicate upon such proposals. The national-level institution, as its primary task, shall have to accept the philosophy of decentralized planning, to compute the net social gains from each type of proposed investments, and to allocate priorities between such investments. The institution can address itself to this exercise by facilitating negotiations between the different units of the decentralized planning machinery.

ROLE OF CENTRALIZATION

The role of a central institution is not to defend sub-optimal investments in regions. This institution is needed for the purpose of discovering the alternative means of reconciling diverse and conflicting economic interests. Such conflicts can be resolved through the choice of alternative technologies. The conflicting economic interests may also be resolved in terms of alternative designs of intraterritorial flow of factor-endowments and products. The primary task of this national-level institution of planning is to examine the implications of each proposal and through negotiations to arrive at a common resolution of the conflcts. It would be interesting to examine how the case under reference, that is, the apatite deposits of Purulia, can also admit of a common resolution between the three contending claimants.

It may be recalled that there are three contending claimants of the apatite deposits, each one of whom has a specific final product in mind. In order to reconcile their contentions, the first task would be to assist them to recognize that for each final product there exists a need for a specific grade of apatite as raw material. For example, apatite with the highest concentration of phosphate is needed for pesticide production. The superphosphate industry can make do with intermediate grade apatite. The lowest grade is suitable for the production of pulverized apatite.

Now, it is known that in the humid tropics all apatite deposits are subject to leaching. This means that the surface deposits have proper phosphate content than the deeper layers. If this situation is found to prevail also with the Purulia deposits, then all the three objectives can be met by organizing mining operation in a special way. That is, to open up the mine in sections and to excavate the vertical length of each section at any given time. This will allow for the materials to be graded and allocated in accordance with the varying needs of the users. The overall advantage of this method of mining is that there will be much less waste, thus upholding the value of conservation of resources.

Such an exercise on resource allocation for multiple-objective projects can surely be worked out by any competent planner. But that is not the most important issue under consideration in the context of decentralized planning. The basic concern here is to institutionalize the process of allowing the contending regions to fully review the reality across the table so as to arrive at an appropriate consensus, thus upholding the value of negotiation. If such a value is endorsed then a great deal of discontentment can be avoided. For example, if in the given instance it is found that the size of available reserve of a certain grade of apatite is inadequate to support a given industry, then there will be little room left for feeling ill-used.

Incidentally it may be mentioned that neither the Planning Commission of India nor the National Development Council offer such opportunities for holding negotiations between the contending regions regarding the options exercised for the allocation of investments. The participants are not given an accurate picture of the probable impact on any one kind of investment designed for one region upon the other regions. As a result, the participants in the meetings of the National Development Council, after approving the plan, carry back a number of unresolved feelings upon which rides the arrogance of regionalism.

The point to consider, therefore, is not whether to take a stand exclusively for decentralized regional planning against centralized national planning. There is need to strive for smaller areas of focus within the national planning frame in such a manner as to preserve the strength of the identified objective and overall welfare of the nation-state, with the advantage of dealing positively with micro-level features and resource attributes.

The case for decentralized planning should not be seen from a narrow point of view—as observed by its protagonists in India today ; it has to be seen from an overall perspective. Only then can the failures, largely administrative but not excluding technical, be highlighted in such a way as to allow for the development of the alternatives concurrent with central plan formulations. This clarity seems to be missing from the statements of many of the advocates of decentralized planning, so that one does not know what happens to the idea of centrality of regional planning within a region.

SIXTEEN

J. T. COPPOCK

PLANNING FOR COUNTRYSIDE RECREATION IN GREAT BRITAIN

THE purpose of this chapter is to examine how planning for countryside recreation has evolved in Great Britain, to outline its characteristics and to evaluate its effectiveness in the light of current trends. It will focus primarily on day recreation from home, though it is virtually impossible, within the confines of such a small country, to isolate the recreational requirements of tourists from those of residents (except in so far as the former require accommodation, whether in-built facilities or in-camp grounds). It will also be confined mainly to public recreation, excluding that enjoyed by landholders and their friends or provided by them privately on a commercial basis. Such private recreational use of land takes place over a very large proportion of rural land, usually as a subordinate activity to farming or forestry ; but much use of this kind is outside the purview of either statutory planning or public policy and is not generally considered in recreational planning (Coppock and Duffield 1975).

PERSPECTIVES

It is essential for a proper perspective on recreational planning in Great Britain to appreciate the context in which outdoor recreation occurs. First, land is at a premium, with approximately one acre of land of all kinds per head of population to serve all purposes. Second, virtually all the land is used for some purpose, almost 80 per cent of it for farming and a further 9 per cent for forestry, and very little of the countryside is not man-made or man-modified. Third, only in north-west Scotland there are large areas that are relatively remote (i.e. more than 3 hours drive) from large centres of population. Lastly, only about a tenth of the countryside is publicly owned and little of this is used primarily for public recreation, much of it having been acquired since the Second World War for specific purposes such as afforestation. It is true that a considerable part of this public land is used incidentally for recreation, but the distinctive feature of outdoor recreation in Great Britain is that much of the privately owned land also serves a public recreational purpose, whether by virtue of its legal status or with the agreement of the owner, or by default, or because such use cannot easily be prevented. Thus, although the extent of (mainly public) land used primarily for recreation has been increasing, the greater part of recreational activity in the countryside takes place on land used primarily for other purposes, most of it in private ownership.

It is also important to see planning for recreation in the context of planning generally. Effective statutory planning was introduced in 1947 by the Town and Country Planning Acts ; but despite the epithet 'country', the emphasis in such planning has been primarily urban and on control over built facilities. Countryside planning has thus tended to be urban planning writ small and the statutory planning process has been ill-suited for dealing with the resource-based planning which is characteristic of much rural activity. Changes in the machinery of local government and in planning now provide a better basis for taking account of resource-based activities, though the urban bias of planning remains. Recreational planning in the countryside has focused mainly on public provision and there has been some tendency, in the absence of effective rural planning, for recreational planning to be conceived in isolation from other aspects of rural life (Palmer and Bradley 1971). Indeed, because of the increasing attention given to recreation, it has been suggested that there is a danger of rural planning becoming synonymous with recreational planning (Davidson and Wibberley 1977).

Despite the small size of the country, it is also necessary to consider recreational planning at three different levels—nationa l, regional and local. In a British context, planning is primarily a local activity. Yet the framework for recreational activities is often a regional one, and both local plans and regional strategies should be consistent with national policies and strategies. In practice, national policies, in so far as these exist, have tended to be permissive, local plans restrictive and partial, and regional strategies, with few exceptions, limited and unenforceable.

COUNTRYSIDE RECREATION BEFORE 1949

Planning for recreation has evolved in piecemeal fashion, beginning with a growing appreciation of the rural recreational needs of an increasingly urban population in the middle of the nineteenth century, when legislation was passed to prevent the disappearance of the remaining common land around London (Blackie, Coppock and Duffield 1979). Public access for recreation was also granted by subsequent legislation to common land in England and Wales that lay wholly or partly in an urban district, and to common land for which a landowner had made a deed of access, although most of this land continued to be used primarily for grazing. In total, such land represents only a small part of the total area of rough land and its distribution was to a large extent fortuitous (Burton and Wibberley 1965). Another legacy of the medieval economy, a network of over 160,000 km of footpaths and other rights of way interwoven throughout the agricultural land, also survived as a potential recreational resource, though it was probably not of major importance until the following century (Countryside Review Committee, 1977). In a few localities rural land was acquired for recreational purposes, well-known examples being Epping Forest and Burnham Beeches ; but the principal interest of the local authorities lay in creating urban parks and other recreational facilities in the towns.

Nonetheless, there was also an increasing interest in countryside recreation and from the end of the nineteenth century several abortive attempts were made to secure public access as of right to open land, generally, both in Scotland and in England and Wales. Frictions were developing with landowners, notably in the Pennines, and eventually the Access to Mountains Act of 1939 did grant permissive powers in England

and Wales, though it was never implemented, being overtaken by subsequent legislation (Cherry 1975).

Until 1947 there was nothing that could properly be called planning for rural recreation and the role of the central government was either permissive or restrictive, providing local authorities with powers to make provision if they wished, and preventing a further loss of recreational resources. There was, however, one national agency, the Forestry Commission, which sought to play a more active role in respect of the resources it controlled. The commission had been formed in 1919 both to encourage private forestry and to create new forests and from early on it permitted access on foot to its forests where this was consistent with its other objectives and obligations. More important, it broke new ground in 1935 by creating the first of a number of forest parks where outdoor recreation was facilitated, though it had no statutory obligation to do so (Ryle 1969).

THE NATIONAL PARKS COMMISSION AND THE NATIONAL PARKS AND ACCESS TO THE COUNTRYSIDE ACT 1949

Before the Second World War, pressure had been growing to create national parks, articulated after 1936 by a Standing Committee on National Parks representing many of the voluntary organizations. This movement had both conservation and recreational objectives, and these were given legislative form in 1949 by the National Parks and Access to the Countryside Act, which established a National Parks Commission with the dual and (as it was subsequently to appear) sometimes conflicting aims of encouraging both the conservation of the landscape and the promotion of public enjoyment of the countryside. The National Parks which the commission was to designate (but whose running was to be primarily a local matter) were quite different from those that had been created in North America. They were predominantly areas of privately-owned land, chiefly in agricultural use (much as rough grazing for hardy sheep and cattle) and contained large populations of permanent residents numbering over 40,000 in the Lake District National Park (LDNP). They were described by John Dower (1945), whose seminal report helped to shape thinking on national parks, as extensive areas of relatively wild country, though the Act defined them only by reference to their natural beauty and the opportunities they afforded for open-air recreation.

With one exception (the Pembroke Coast NP) they were predominantly upland areas, containing extensive tracts of moorland, open land under semi-natural communities of grasses, sedges and low shrubs. Again with one exception (the Peak District NP) the parks were relatively remote from the main centres of population and, in the minds of those who shaped the legislation, they were to be used primarily by walkers and cyclists—'a people's charter for the open air, for the hikers and the ramblers, for everyone who likes to get out into the open air and enjoy the countryside,' as the Act was described by the responsible minister.

The Act also gave the commission powers to create long-distance footpaths (trails) and required local authorities in England and Wales to prepare definitive maps of public rights of way in their areas. In addition, it also required them to make surveys of access to open country, defined as 'mountain, moor, heath, down, cliff or foreshore' and to consider whether any action was required to secure public access, whether by concluding access agreements or making access orders where agreement was not possible.

Between 1951 and 1957 ten parks were designated, covering nine per cent of England and Wales, but access agreements were largely confined to the Peak District NP which had been the scene of conflicts between recreationists and landowners in the 1930s. Most authorities found that de facto access satisfied public needs in their areas and that no action was necessary. The creation of long-distance footpaths, like the surveys of footpaths, was a slow process and it was not until 1965 that the first long-distance footpath, the Pennine Way, was actually opened, though it had been approved in 1951 (Patmore 1970). No subsequent parks were designated, nor was there any significant increase in the area under access agreements. However, by 1982 13 long-distance footpaths had been opened and a further 33 named walks identified (Patmore 1983). The Act thus provided the first components of a national plan for public recreation in the countryside, based primarily on blueprints provided by the Dower (1945) and Hobhouse Reports (National Parks Committee, 1947), though there was no explicit plan as such.

The first government committee to investigate the possibility of establishing national parks, the Addison Committee (National Parks Committee, 1931), considered Great Britain as a whole, and similar investigatory committees to those which preceded the 1949 Act were established in Scotland in the 1940s (Cherry 1975). It was expected that a parallel Act for Scotland would follow but no such legislation was introduced, probably because public pressure on the countryside was much less in Scotland (with 10 per cent of the population but 70 per cent of the rough land) and because both landowners and local authorities were opposed to legislation. It was not until 1967 that an Act relating to recreational use of the countryside in Scotland was passed. As a result, 'Scotland is the only civilised country without National Parks' (Select Committee on Scottish Affairs, 1973, A1820). There was little that could properly be called recreational planning in the countryside in Scotland before 1967, although a circular from the Scottish Development Department in 1962 did request local authorities to appraise their scenic resources and to make plans for tourism.

The National Parks Commission, which was established in 1949, was an advisory body, understaffed and underfinanced. It was resented by civil servants and by local councillors who were opposed to the creation of ad hoc bodies in fields they regarded as properly their own (though paradoxically they had for the most part little enthusiasm for countryside recreation). Even in the commission's work the main emphasis was on preservation of landscape rather than on provision for leisure, and its focus was the national parks rather than the countryside at large. In the remaining countryside, only golf courses and public open space were shown on the development plans prepared under the 1947 Town and Country Planning Acts, that contained few (if any) proposals for new facilities. For the most part countryside was shown on these plans as 'white land' where no statutory change of use was envisaged and for which no policies were, therefore, necessary.

What had not been appreciated in the 1940s was the great increase in leisure time and in personal mobility that was to come in the postwar period and the effects that these would have on the countryside. The rapid growth in car ownership (now eight times higher than in 1939) and the widespread adoption of the five-day week and of progressively longer holidays with pay have provided both the means and the opportunity for growing numbers of the predominantly urban population to participate in countryside recreation. In the 1960s and early 1970s, participation in such recreation was believed to be growing at 10-15 per cent compound per annum (Select Committee on Sport and Leisure, 1973, Q1571), and by 1977 it was estimated that over 10 million people were visiting the countryside on a summer Sunday (Fitton 1979). Increasing pressure on farmland and woodland both around towns and in the popular holiday areas (notably the National Parks) was believed to have led to widespread damage to soil and vegetation and sometimes property, and to conflicts between recreationists and farmers, foresters and other primary land users and between different kinds of recreationists. It became increasingly clear, notably in the influential *Countryside in 1970* conferences, that additional powers and facilities would be required and that a more comprehensive view of countryside recreation was necessary.

THE COUNTRYSIDE ACTS 1967 AND 1968

Such additional powers were provided by two pieces of legislation, the Countryside (Scotland) Act 1967 and the Countryside Act 1968. The former established a Countryside Commission for Scotland, with similar duties and powers to that created in 1968 in England and Wales, except for those relating to national parks which (at least on the English model) were seen as inappropriate to the Scottish situation (Countryside Commission for Scotland, 1974). The latter Act replaced the National Parks Commission by a stronger Countryside Commission, with a wider remit covering the countryside as a whole and with powers to grant-aid whatever it considered conducive to the purpose of the Act. Both Acts contained powers for greater public provision, particularly through the creation of picnic sites, car parks and country parks, and widened the concept of 'open land' to include woodland. Country parks had been proposed in a White Paper on *Leisure in the Countryside* (Cmnd. 2928, 1966) to facilitate public use of rural areas for leisure purposes, ease pressure on more remote places and reduce the risk of damage to the countryside. They were defined in the Act as 'a park or pleasure ground for the purpose of providing or improving opportunities for the enjoyment of the countryside by the public.' Such parks could be owned and run by local authorities, by other public bodies and by private individuals or organizations, though they would require the approval of the appropriate countryside commission, which could give grants of up to 75 per cent of the costs of creating them. Though the Countryside Commission has indicated priorities in England and Wales, the initiative for creating them lies largely at a local level and authorities have varied greatly in their enthusiasm for doing so. The pattern of provision is thus uneven. A similar situation obtains in respect of picnic sites. So far 167 parks have been created in England and Wales (Countryside Commission, 1982) and 10 in Scotland (Countryside Commission for Scotland, 1979), most of them within motoring distance of large towns, although relatively few lie in the urban fringe. They vary considerably in both size and character, from under 25 ha to over 500 ha, though most give opportunities for informal recreation in the open air in semi-natural environments in which recreation is the primary purpose. Although some parks were already used for recreation, increasing numbers are new creations ; but no information is available to show whether they do in fact act as honeypots or intervening opportunities, diverting pressures from elsewhere or intercepting those who might otherwise travel further, or have created a new supply-led demand.

The wide remit of the Countryside Commission and the new powers available to local authorities have certainly improved provision for recreation in the countryside, but the location of new facilities often seems to have been determined intuitively and to reflect local enthusiasms or the availability of land already held by a local authority. Indeed, in some areas existing deficiencies have been aggravated by new provision (White and Dunn 1975). Nor does adequate attention seem to have been paid to the location of the large number of private parks open to the public (often associated with historic houses), even though many of these appear to fulfil a similar function and some have even become country parks.

OTHER NEW AGENCIES

The countryside commissions have played the leading roles in stimulating the provision of opportunities for informal recreation in the countryside, but they are not the only agencies concerned with recreational planning. In the late 1960s and early 1970s statutory sports councils and tourist boards were formed for each of the constituent countries (England, Scotland, Wales), the former to promote participation in sport and physical recreation and the latter to promote holiday making, both by residents and by visitors. The sports councils have concentrated on active physical recreation and their remits cover both town and country, where they have, for example, provided grant aid for facilities for skiling. The tourist boards have primarily been concerned with the provision of accommodation, but they also have an interest in the provision of recreational facilities as tourist attractions. Indeed, some facilities will be viable only if the needs of both residents and visitors are taken into account while others, provided primarily for visitors, will also benefit local people. The encouragement of farm-and of activity-based holidays are other examples of the initiatives of tourist boards that have implications for recreational planning in the countryside.

As well as these promotional and advisory agencies there has also been an increase in recreational provision in the countryside by executive agencies. The emerging role of the Forestry Commission in this respect has already been noted, and in 1967 the commission adopted provision for public recreation as one of its objectives, though a subordinate one to that of timber production. Such provision is primarily for informal passive recreation and for those camping and caravanning (though the commission has retained its prohibition, with few exceptions, on the use of motor cars in its forests) and each of the commission's conservancies (regions) has been required to produce a recreational plan. Since 1973 the commission has also required those private owners seeking grant-aid to consult the appropriate local authorities about the possibility of access agreements to such land. Large stretches of the inland waterways, which are the responsibility of the British Waterways Board, are retained primarily for recreational purposes, but perhaps more important for recreation generally, the 1973 Water Act, which created nine regional water authorities in England and also a Welsh Water Authority, required these to make the best recreational use of the land and water they control. Plans to meet these requirements are in various stages of preparation (see, Welsh Water Authority, 1980), with guidance from the Water Space Amenity Commission which was also established under that Act as a source of recreational expertise and advice (though the Commission was in fact abolished in 1983).

Provision and planning for recreation at a national level has thus become increasingly complex as the number of bodies has multiplied and as some have prepared national and others regional strategies (Gilg 1978). Attempts have been made to achieve better co-ordination between them at both a research and a policy level, the former through the Countryside Recreation Research Advisory Group and the latter through the Chairmen's Policy Group, at which the chairmen and chief officers of the main agencies meet to exchange ideas. No attempt has been made to produce an analogue of the United States National Recreation Plan, although it has been argued that the Countryside Commission should have the responsibility of producing such a strategy (Davidson and Wibberley 1977).

THE NEW PLANNING AND THE NEW AUTHORITIES

At a local level the development of recreational planning has been affected by changes in statutory land-use planning and in the structure and organization of local government. Although recreation planning has been and remains but a small part of statutory planning, the latter provides the frame within which recreational plans are conceived and, at least to some extent, implemented Statutory planning also provides the mechanism whereby permission is given for countryside recreation that requires the erection of structures, whether for public or private use.

By the mid-1960s experience with comprehensive land-use planning since 1947 had aroused considerable criticism, in particular that it was slow to adjust to changing conditions, too concerned with detail and too urban-oriented. Substantial modifications were accordingly made in a series of Town and Country Planning Acts between 1968 and 1972 notably, incorporation of a distinction between structure plans, in which broad strategies not confined to land use would be adumbrated, and local plans, which would be much more specific showing how such policies would be effected on the ground. The latter would also include subject plans on a single topic such as tourism. It is still too early to judge the effectiveness of these changes (which coincided with a major reorganization of local government and a period of severe restraint on public expenditure), though there is a widespread impression that the planning process is still too detailed and too urban. It is certainly true, however, that recreation and leisure, which are not regarded as key issues, have only a small place in such plans.

It was also recognized that the structure of the basic units of local government, which provide the spatial framework for development (and recreational) plans, no longer corresponded with the social geography of the country and that larger units were necessary for the more efficient provision of some services. The reports of two royal commissions, the Redcliffe-Maud Commission for England (Cmnd. 4040, 1969) and the Wheatley Commission for Scotland (Cmnd. 4150, 1969), provided a framework for decision by the central government, which paradoxically accepted more radical proposals for Scotland and rejected the less radical recommendations for England in favour of more limited changes. As a result two rather different systems were created in the two countries and these have had some significance for recreational planning.

In England, counties and districts replaced the various categories of urban and rural areas and there were some changes through amalgamation and reorganization of counties. The role of the conurbations other than Greater London (which had been the subject earlier of separate legislation) was also recognized by their designation as metropolitan counties. Similar arrangements were made in Wales, except that there was a much greater reduction in the number of counties and no metropolitan counties were created. In Scotland, counties and counties of cities were abolished, to be replaced by a system of large regions (for the most part city regions), subdivided into districts, but with special provision for the three island groups (Orkney, Shetland and the Western Isles), which were made all-purpose island authorities. The chief consequence of these changes for planning was that it became a concurrent function between the upper-tier authorities (counties/regions) and the lower-tier authorities (districts), the former being responsible for structure plans and the latter for local plans.

Associated with these changes in area and status were some changes in organization and responsibilities for recreation and leisure in local government, which also became concurrent functions between counties/regions and districts (Veal and Travis 1979). These responsibilities had grown piecemeal since the nineteenth century and committees considering the administration and management of the new local authorities recommended that separate departments of recreation and leisure should be established. While this advice has often been followed (especially in the upper-tier authorities), a great variety of structure has emerged, with the planning function often remaining with the planning department and tourism being separated from recreation. Reorganization might have been more effective if it had not been accompanied by a ban on recruitment, which led to many key positions being filled by senior staff without professional training, and by financial stringency, which has limited initiatives. Nevertheless, for the first time there existed a structure at local level in which leisure (including that in the countryside) could be considered as a whole.

Inevitably, the existence of concurrent functions at different levels for both recreation and physical planning has led to disagreement, especially where there were differences in the political composition of councils. A further complication was that, whereas regional councils in Scotland had a duty to make provision for recreation and leisure, districts in Scotland and local authorities in England and Wales had only permissive powers. In theory, the county/region is concerned with recreational strategy and with other than local provision, but this distinction is not always seen as clear-cut. Such duplication has led the government to transfer the statutory duty in Scotland from the regions to the districts, with the former retaining only permissive powers (Local Government and Planning (Soctland) Act, 1982).

It is not implied that there has been no planning for recreation under the new planning. The report of survey has often included sections on tourism and countryside recreation and considerable efforts has been devoted to developing methodologies for assessing recreational demand at the local scale (Countryside Commission, 1973b). Two studies by the Tourism and Recreation Research Unit (1970, 1971) have been described as models of what should be attempted by local authorities in Scotland (and Select Committee on Scottish Affairs, 1972). The unit (1977) has also produced a type study of local planning for recreation and tourism. The Lothian Region Department of Leisure Services had, before its abolition in 1983, produced a draft Leisure Strategy for Countryside Recreation (1981). The main weakness at the structure plan level is that recreation has not been seen as a key issue. It is also possible that the difficult financial climate since 1973, during which increased spending on recreation was specifically singled out for adverse comment by the central government, has contributed to a neglect of recreational issues.

REGIONAL/NATIONAL CO-ORDINATION

Outdoor recreation is no respecter of local government boundaries, and increasing dependence on the private car has given recreationists much greater freedom in where they go, not least on the 'trip out', which is the most popular form of countryside recreation. There is an increasing recognition that such recreation must be planned on a regional scale, though there are few examples, such as the Lee Valley Regional Park, of facilities conceived on a regional basis and little evidence of joint provision by adjacent authorities. Nevertheless, there is abundant evidence from numerous surveys of the regional character of recreational trips and several regional surveys have been undertaken as inputs to planning, for example Survey of Informal Recreation in South East England (SIRSEE) (Countryside Commission, 1973a), (though it is not known how far their findings have subsequently affected plans at a local level). The Countryside Commission (1974b) has also published a study of planning for recreation at the regional scale. National agencies with responsibilities for encouraging participation in various forms of recreation or with policies for providing recreational opportunities/facilities also require a national strategy (or at least compatible national strategies), although unlike the new departments of recreation and leisure at local level, there is no single focus for outdoor recreation at national level, responsibility being diffused among the various agencies and departments. Various approaches have been adopted to facilitate better co-ordination of recreational policy, provision and planning at both national and regional levels.

In Scotland, the large size of most of the regions has meant that the top-tier authorities can themselves play a regional role, though co-operation with the district councils could undoubtedly greatly be improved. The removal of responsibility for recreation from the regions in 1982 will undoubtedly create problems for effective co-ordination. In Wales, with its smaller population and area, co-ordination is often possible at a national (i.e. Welsh) level, though parts of Wales are within the recreational catchments of conurbations that lie wholly or partly outside the country. In England, the much higher density of population and the generally smaller area occupied by the top-tier authorities make some intermediate scale between the national and local desirable. In some areas, for example, south east England, a degree of regional co-ordination has been achieved by the establishment of standing conferences of planning authorities (though inter-county co-operation in planning recreation seems to be the exception). In 1966-67 regional sports councils were created to co-ordinate provision for sport and these have since been transformed into regional councils for sport and recreation, as indicated in the 1975 White Paper on Sport and Recreation (Cmnd. 6200). They include representatives not only of the various local authorities, but also of the national and regional agencies (though not, surprisingly, the tourist boards). Unfortunately, the regional strategies being prepared by these councils, which will have to rely on their constituent authorities to implement their proposals, are emerging at a time when most councils have already prepared their structure plans (Department of the Environment, 1977) and stringent economies are being imposed. Furthermore, these regions often do not correspond with those used by other agencies preparing regional strategies such as the Forestry Commission, the tourist boards and the regional water authorities, so that integrating plans is complex.

It is increasingly recognized that there are also issues of national importance on which those preparing local and regional plans need national guidance, which is also required from each agency in respect of its own policies and how it sees these being applied locally and regionally. Several agencies have begun to provide such advice and in Scotland there has been some attempt to co-ordinate advice at a national level.

In a sense, the Hobhouse Report (National Parks Committee, 1947) was the first such strategy providing a framework for the National Parks Commission when it undertook its first and principal task of designating national parks. The national tourist agencies have also produced advisory strategies, whether at national or regional level, and the Countryside Commission for Scotland has published a discussion document on a park system for Scotland (1974), which outlines a hierarchical (though not spatially specific) strategy which might be followed when the statutory powers necessary to create the upper levels of the hierarchy have been provided by new legislation. More commonly, national agencies have issued policy statements (e.g. Countryside Commission, 1982b) or adopted schemes of grant-aid which are reactive, in that they depend on individuals, voluntary clubs, local authorities and the like applying for such help. Very general indications have been provided by the central government in the two white papers on Leisure in the Countryside (1966) and Sport and Recreation (1975) and more detailed guidance in circulars to local authorities. The most comprehensive guidance has been that given by the Scottish Development Department (1975) in the advice notes issued to Scottish regional authorities to assist them in preparing regional reports and in the National Planning Guidelines (1977) issued in connection with the preparation of statutory plans, though it must be confessed that the guidance provided in respect of recreation is very limited.

Undoubtedly the most ambitious attempt to achieve both national co-ordination and that between national and regional plans/policies has been provided by the Scottish Tourism and Recreation Planning Studies (Countryside Commission et al. 1978), which were designed to assist the evolution of outline regional strategies for sport, outdoor recreation and tourism, coordinated within a broad national framework. The first major task was to identify the policies of the national agencies and their implications for the regions, and this culminated in the publication of a Strategic Issues Report (Countryside Commission for Scotland et al. 1976a). A subsequent Guide to the Preparation of Initial Regional Strategies then provided a consistent methodology for the assessment of demand and supply and the evaluation of impact (Countryside Commission for Scotland et al. 1976b).

Progress in achieving these strategies has been much slower than had been hoped, largely because of the unfavourable economic climate since 1973. As a result, their appearance (like that of the regional strategies in England) has largely been preempted by the appearance of structure and local plans. Nevertheless, the approach adopted has strengthened co-operation between the national agencies and encouraged them to make any adjustments required to make collective sense of their respective approaches. The local authorities have also been encouraged to think comprehensively about the recreational needs of their electorates and to collect data required for recreational planning in a consistent way that will serve both their own needs and those of the national agencies (Coppock 1979). Unfortunately, the momentum created by this initiative is in danger of being lost by the abolition of regional departments of leisure as a result of the Local Government and Planning (Scotland) Act 1982, though it is hoped that regional planning departments will fulfil some of their planning functions.

COUNTRYSIDE MANAGEMENT

Planning for recreation, as with planning generally, has been primarily concerned with the construction of built facilities or with the provision of public resources exclusively or primarily for recreation. Yet a substantial proportion of outdoor recreation takes place on private land (and must continue to do so whatever the foreseeable scale of public provision), and thereby runs the risk both of environmental damage and of adverse effects about other land users (though Fitton [1979] thinks this is exaggerated). The countryside commissions have statutory duties to promote the conservation of the natural beauty of the countryside, particularly in areas of high amenity, and all public bodies have an obligation, under the Countryside Acts, to have regard to the desirability of safeguarding the natural beauty of the countryside (a phrase that is interpreted in the Act as including not only scenic beauty but also the natural flora and fauna). Largely with the aim of reconciling recreation, amenity and other uses in the countryside which lie outside the formal planning process, there has emerged an approach to countryside (and within this recreational) planning to which the term 'countryside management' has been widely applied (Hookway 1977).

Countryside management is not planning in the accepted sense, but rather involves the adoption of a range of policies that secure such objectives as the conservation of landscape and the provision of facilities for the optimum number of people and cover a wide range of activities. It approaches recreational planning most nearly in the National Park Plans that have been prepared for each of the ten National Parks under the authority of the Local Government Act 1972 (Dennier 1978). These plans include many topics that are not subject to formal planning control and are to be seen within the context of the structure plan(s) covering the area. Essentially they provide a framework for action and guidance, not only for the local authorities but also for private owners and managers of land. The Countryside Commission, which has been particularly active in promoting the concept of countryside management, has advocated that similar plans be prepared for areas of outstanding natural beauty (Countryside Commission, 1980d), and a few experimental plans have already appeared, for example, for the Wye Valley (Gwent County Council, 1980). If this proposal is accepted, such plans (i.e. for national parks and areas of outstanding natural beauty) would ultimately cover a quarter of England and Wales.

Similar approaches have been adopted on an experimental basis in Scotland, for example, in the Torridon area, and in three areas identified as Heritage Coasts in England and Wales. In the latter, as in the areas of outstanding natural beauty (AONB), an essential feature has been the appointment of project officers who can act as the vital link between the local authorities, government agencies, voluntary bodies and owners and occupiers of land. Not surprisingly, this kind of approach has been most effectively applied in an area which is largely publicly owned and which is managed by a public agency, the Forestry Commission, namely, the New Forest, where the commission plays a key role, in consultation with other agencies and with local interest groups (New Forest Joint Steering Committee, 1980). Not only has it the power to make plans, but it can also play a major role in implementing them within the constraints imposed by its own resources and by government financial policies. As a result, the number of access points for vehicles has been greatly reduced, provision made for car parking and camping and advance warning provided at the peak so that would-be visitors can be diverted elsewhere when the Forest is 'full'. Indeed, in many ways the plan for the New Forest is like a national park plan with teeth. More recently, local authorities in Scotland have prepared a local subject plan for the Loch Lomond area, which is of national importance and lies in two districts and two regions (Loch Lomond Planning Group, 1982).

In a more strictly managerial sense the Countryside Commission has sponsored a number of experiments in visitor management in the countryside designed to minimize conflict and maximize recreational opportunities in areas of stress, notably the Upland Management Experiments (Countryside Commission, 1976a), and the parallel experiments in the urban fringes of Manchester (1976b) and London (1981). Such an approach has usually involved the appointment of a project officer with a small budget to facilitate minor improvements and repairs. Also at a local level, the commission has been exploring the concept of management agreements under which individual owners who agreed to manage their land in ways that achieved conservation or recreation objectives could receive public compensation for any resulting loss (Countryside Review Committee, 1976). Although such approaches can hardly be called recreational planning, they certainly contribute to the effective implementation of planning. In an attempt to bridge the gap between individual sites for recreation and broad strategies, Ferguson (1979) has proposed a method of linking the provision of informal recreation opportunities, using the 'strategic choice approach', which is specifically designed for use by the kind of liaison groups of bodies responsible for management of specific resources (as envisaged for AONBs) : he believes that the method might be suitable for the development of recreation opportunities as identified by the regional recreation strategies being prepared by regional councils for sport and recreation.

CONCLUSION

Recreational planning in the countryside in Great Britain is both complex and still evolving and is characterized by much experimentation in finding the best way of satisfying the various interests. Attention was originally concentrated on the location of public facilities, with little regard to their implications for the surrounding countryside. Recreational planning then formed a very minor component of the statutory planning process adopted in 1947, but the explosive growth in countryside recreation led to a variety of procedures, some part of, and others outside, the formal planning process, which has itself undergone major changes. The new planning and new structure of local government provided a formal framework for recreational plans and the Scottish tourism and recreation planning studies have shown how such plans can be related to a wider canvas. Equally important is the essentially exhortatory and consensus planning illustrated by the national park plans, though there is as yet little firm evidence to show how effective such approaches have been.

During the postwar period, and particularly during the past decade, there has also been an increasing awareness that outdoor recreation in the countryside is socially selective. Evidence from the National Countryside Recreation Survey suggests that many of those who do not now make recreational visits to the countryside would like to do so (Fitton 1979), and there is growing concern at both national and local levels to improve opportunities for all sections of the community to enjoy countryside recreation if they wish. There is also an increasing tendency to avoid categorizing recreation as rural or urban and recognition of the need to see a continuum of provision, most explicitly formulated in the Countryside Commission for Scotland's proposal for a park system for Scotland. The Countryside Commission already gives priority in grant-aid to projects in the green Belts and it can be expected that there will be an increasing emphasis on facilities on the urban fringe that are readily accessible by public transport, a tendency that may well be reinforced by the increasing cost of motoring. There is also some evidence of a trend that aims to wean visitors from their cars, exemplified by the Goyt Valley scheme (Countryside Commission, 1972) and by the publication of numerous circular walks intended for the motorist.

Attention needs to be paid to ways of improving recreational opportunities by encouraging multiple use of land, and enlightened local authorities are exploring the right mix between public provision of recreational sites and better use of the remaining countryside, the former through wise capital investment or grant aid, the latter through an appropriate combination of consultation, persuasion and assistance, though there is a long way to go before, as Davidson and Wibberley (1977) envisage, there is such a range of planned and organized choice that people channel themselves, dispersing naturally and unobtrusively. As long ago as 1970 it was asserted that 'a great deal of the machinery already exists ; the task is to make it work effectively' (Royal Society of Arts, 1970, p. 13.23). Perhaps an even greater need is the political will to see countryside recreation less in terms of problems and more in terms of opportunities that can enhance the quality of life of the whole community.

REFERENCES

Blackie, J.A. ; Coppock, J.T. and Duffield, B.S., *The Leisure Planning Process*, State of the Art Review, Sports Council/Social Science Research Council, 1979.

Burton, T.L. and Wibberley, G.P., *Outdoor Recreation in the British Countryside*, Wye College, 1965.

Cherry, G.E., *National Parks and Recreation in the Countryside*, Environmental Planning 1939—1969, Vol. II, HMSO, 1975.

Coppock, J.T., 'Review of Scottish Tourism and Planning Studies', *Recreation News* 5, 1979.

Coppock, J.T. and Duffield, B.S., *Outdoor Recreation : A Spatial Analysis*, MacMillan, 1975.

Countryside Commission, *The Govt. Valley Traffic Experiment*, CCP 55 : 1972.

Countryside Commission, *Study of Informal Recreation in South-East England*, 1973a.

Countryside Commission, *Planning for Informal Recreation at the Local Scale*, CCP 69, 1973b.

Countryside Commission, *Planning for Informal Recreation at the Sub-Regional Scale*, 1974.

Countryside Commission, *The Lake District Upland Management Experiment*, CCP 93, 1976a.

Countryside Commission, *The Bollin Valley : A Study of Land Management on the Urban Fringe*, CCP 97, 1976b.

Countryside Commission, *Areas of Outstanding Natural Beauty : a Discussion Paper*, CCP 116, 1978.

Countryside Commission, *Areas of Outstanding Natural Beauty : a Policy Statement*, 1980.

Countryside Commission, *Countryside Management on the Urban Fringe*, CCP

136, 1981.

Countryside Commission, *Fourteenth Report of the Countryside Commission for the Year Ended 30 September 1981*, 1982a.

Countryside Commission, *Countryside Issues and Action : A Prospectus of the Countryside Commission 1982*, CCP 151, 1982b.

Countryside Commission for Scotland, *A Park System for Scotland*, 1974.

Countryside Commission for Scotland, *Eleventh Annual Report*, 1979.

Countryside Commission for Scotland *et al.*, Planning for Sport, Outdoor Recreation and Tourism, *Strategic Issues*, Vol. 1, 1976a.

Countryside Commission for Scotland et al., Planning for Sport, Outdoor Recreation and Tourism, *A Guide to the Preparation of Initial Regional Strategies*, Vol. 2, 1976b.

Countryside Commission for Scotland et al., *Scottish Tourism and Recreation Planning Studies*, 1978.

Countryside Review Committee, *The Countryside—Problems and Policies*, HMSO, 1976.

Countryside Review Committee, *Leisure and the Countryside*, HMSO, 1977.

Davidson, J. and Wibberley, G.P., *Planning and the Rural Environment*, Pergamon, 1977.

Dennier, A., 'National Park Plans', *Town Planning Review* (49), 1978. pp. 173-183.

Department of the Environment, *Memorandum on Structure and Local Plans*, Circular 55/77, HMSO, 1977.

Dower, J., *National Parks in England and Wales*, Cmnd. 6378, HMSO, 1945.

Ferguson, M.J., 'A strategic choice approach to recreational site resource allocation', *Town Planning Review*, 50, 1979, 325-45.

Fitton, M., 'Countryside Recreation—the problems of opportunity', *Local Government Studies* (5), 1979, 57-90.

Forestry Commission, Evidence to Select Committee on Sport and Leisure (q.v.) 1973.

Gilg, A.W., *Countryside Planning : the First Three Decades 1945—76*, David and Charles, 1978.

Gwent County Council, *Towards a Management Plan for the Wye Valley AONB*, 1980.

Hookway, R., 'Countryside Management, the development of techniques', Proceedings of the Town and County Planning School, Royal Town Planning Institute, 1977.

Leisure in the Countryside : England and Wales, Cmnd. 2928, HMSO, 1966.

Loch Lomond Planning Group, Loch Lomond. A Plan for Tourism, Recreation and Conservation, Dumbarton District Council et al., 1982.

Lothian Regional Council, *Leisure Strategy* (2), *Countryside Recreation*, Consultative Draft, 1981.

National Parks Committee, *Report* (Addison Report), Cmnd. 3851, HMSO, 1931.

National Parks Committee, *Report* (Hobhouse Report), Cmnd. 6628, HMSO,

1947.

New Forest Joint Steering Committee, Conservation of the New Forest, Hampshire County Council, 1972.

Palmer, J. and Bradley, J. 'Planning for outdoor recreation', Ch. 13, in Lavery, P. (ed.) *Recreational Geography*, David and Charles, 1971.

Patmore, J.A:, *Land and Leisure*, David and Charles, 1970.

Patmore, J.A., *Recreation and Resources*, Blackwells, 1983.

Royal Commission on Local Government in England, *Report* (Redcliffe-Maud Report), Cmnd. 4040, HMSO, 1969.

Royal Commission on Local Government in Scotland, Report (Wheatley Report), Cmnd. 4150, HMSO, 1969.

Royal Society of Arts, The Countryside in 1970, *Proceedings of the Third Conference*, 1970.

Ryle, G., *Forestry Service*, David and Charles, 1969.

Scottish Development Department, *Planning for Sport, Outdoor Recreation and Tourism*, Planning Advice Note No. 5, 1975.

Scottish Development Department, *Recreation and Landscape Conservation*, 1977.

Select Committee on Scottish Affairs, *Report on Land Resource Use in Scotland*, HMSO, 1972.

Select Committee on Sport and Leisure, *2nd Report*, House of Lords, HMSO, 1973.

Sport and Recreation, Cmnd. 6200, HMSO, 1975.

Tourism and Recreation Research Unit, *Leisure + Countryside = : A Geographical Appraisal of Countryside Recreation in Lanarkshire*, Research Report No. 1, 1970.

Tourism and Recreation Research Unit, *Leisure + Countryside = : A Geographical Appraisal of Countryside Recreation in the Edinburgh Area*, Research Report No. 2, 1971.

Tourism and Recreation Research Unit, *Tourism and Recreation in the Chichester Area, A Basis for Planning*, Research Report No. 35, 1977.

Veal, A.J. and Travis, A.S., Local authority leisure services—the state of play, *Local Government Studies* (5), 1979, 5-16.

Welsh Water Authority, *Strategic Plan for Water Space Recreation and Amenity*, 1980.

White, J. and Dunn, M.C., Countryside Recreation Planning : Problems and Projects in the West Midlands, Occ. Paper 33, Centre for Urban and Regional Studies, University of Birmingham, 1975.

SEVENTEEN

AKIN L. MABOGUNJE

REFLECTIONS ON LOCAL AND REGIONAL DEVELOPMENT IN THIRD WORLD COUNTRIES
The Special Case of Africa

THE LATE Professor Mohammad Anas will perhaps be best remembered as representing a passing scientific tradition that sees the world from a holistic perspective and considers the obligation of the scholar as one of intellectually trying to come to grips with any section of it in as comprehensive a manner as possible. It is thus not without significance that Professor Anas in his own discipline of geography covered such diverse fields as political, cultural, urban and regional geography, not to mention his interest in the philosophy and methodology of the subject. The problems of social and economic development in many third world countries are thus unlikely to have escaped his attention. Indeed, no geographer worth his salt can, in the second half of the twentieth century, ignore in his research activities and studies the tantalizing paradox of the growing immiseration of a significant proportion of the population of many third world countries with each periodic and heroic, if often misdirected, planned development effort of their governments.

Although in his life time Professor Anas had little direct contact with Africa for reasons largely of material constraints, there can be little doubt that he would have found the current problems of development in that part of the world most challenging and absorbing if he had had the opportunity. It is thus in the spirit of a well-deserved tribute to the memory of this notable geographer that, as President of the International Geographical Union, I feel that reflecting on the experience and main issues in the local and regional development efforts in Africa most appropriate in a *festschrift* in his honour.

The period since the 1960s when the United Nations declared the need for galvanising global society into making determined developmental effort on the basis of decade-by-decade strategy, has been marked in Africa by complex and somewhat dispiriting results. As the Executive Secretary of the Economic Commission for Africa noted :

> Although during the past two decades there has been some structural changes in the African economies, the fact is that today the African economy is still basically underdeveloped : low income per head, a very high proportion of the population in agriculture, low levels of productivity, a circumscribed and fractured industrial base, a high dependence on a vulnerably narrow spectrum of primary export commodities, a transport network geared largely to the export sector, a sharp bifurcation between the traditional and modern sectors, a high degree of illiteracy, low levels of life expectancy, and the predominance of expatriate business enterprise in banking, commerce, finance, industry and management.[1]

The picture, at least on the surface, appears to be one of very little growth or even of retrogression. Yet the fact belies the appearance since it reveals considerable activities on the part of governments and the people to alter the situation and to bring about significant changes especially with regard to the spatial configuration of the development process. It is particularly in this latter respect that the role of local and regional development assumes real importance in Africa. In different countries of the continent, valuable experiences have been gained with regard to the problems and prospects of incubating the development process at these levels and integrating or reconciling them to overall developmental goals and directions. In this paper an attempt has been made first, to review the progress made in local and regional development over the last two decades ; second to examine the main strategies and policies directed at achieving this level of progress ; third, to identify the various problems encountered : and finally, to attempt some discussion of the prospects for local and regional development in the future.

PROGRESS OF LOCAL AND REGIONAL DEVELOPMENT

The idea of local development for most African countries is as old as that of local government or local administration. Local government in most countries of Africa can be said to pre-date the colonial era although the colonial administration was later to try to formalize it and leave on it imprints that are still markedly visible even today. In the British colonial territories the idea was to establish local government based on the concept of community. A community comprises people who are linked together by strong historical ties. This may be as a result of accepting a myth of common origins or of conquest or of long political association. In line with their accepted strategy of administering dependent populations, the British colonial government tended to rule such communities indirectly through their own native administrations. At independence, many of these became local government although the tendency had been for them to be broken down into smaller groupings for this purpose.

In the French colonial areas, no such extension of traditional practices was encouraged. Beyond the level of the village, territorial administration was based on a set of artificially created subdivisions, canton and circles presided over and directed by a member of the French colonial administration. Although in the major urban areas, such territorial units emerged to become full communes with an elected local government, even at the end of the colonial period this system operated in very few urban centres.

At any rate, whether in the British or the French-speaking areas of Africa, it was generally agreed that such units of administration should have a number of developmental functions more relevant to immediate grass-roots needs. The broad criteria determining which functions were so categorized were three : first, whether the function is one of a cluster which needed close local-level integration and co-ordination between themselves, for example, roads, transport, housing and sewerage, or health and personal welfare ; second, whether the function requires considerable adaptation to suit local circumstances and culture, for example, parking and transport services, marketing, reading rooms, community centres or homes for the aged and crippled ; and third, whether a major local effort is needed to start off the function, for example, rural water supply points constructed largely by voluntary labour.

As long as most of these activities were directly the responsibility of staff of the central government or decisions about them could be approved or vetoed by them, much local developmental activities were recorded. However, as elected local government emerged before and particularly after the end of colonial rule, charges of corruption, mismanagement, lack of trained manpower became a useful excuse for undermining the development role of local authorities.

This situation became part of a general tendency at administrative centralization as the bureaucracy in most African countries grew rapidly in the years soon after independence. This growth in bureaucratic omnipotence became more pronounced as country after country in Africa adopted a neo-Keynesian planning model that concentrated in the hands of the central bureaucracy powers not only to oversee the performance of the total economy of the country but also to determine how resources would be mobilized and allocated. In the circumstance, the areas of initiative allowed for local government or local development agencies became increasingly diminished.

Two factors helped to worsen the situation. The first was the adoption in the 1960s in many of these countries of a strategy of development based on the concept of industrialization through import substitution. This meant the acquisition of numerous capital-intensive industries for local production of formerly imported manufactured consumer goods. For countries whose underdevelopment is partly a function of capital shortage, this strategy compelled the administration to evolve a system of central concentration of available national revenue resources. A process of abstracting away to the central authorities, many of the major revenue sources of local authorities began and has had the effect today of leaving vast areas of the country in a precarious state in which they hardly have enough resources to establish or maintain modern infrastructure or social service facilities.

On the other hand, the industrial capacity on which a large part of national resources was being expended was characterized by a strong tendency to concentration in one or a few urban centres usually port and/or capital cities. The result was the phenomenon of the rise of primate cities in many of these countries. Their vast attraction for migrants from all over the countries was only the outward visible sign of the resource-concentration process described above.

Initially this development was conceptualized and described in such benign terms as, the need for industrialization to serve as the "engine" of growth for the development process. This had the effect of obscuring for some time the full negative implication for the rural areas in general and agriculture in particular of this process of starving vital areas of the national economy of funds and raising false expectation of a possible later beneficial "spread effect". The growing appreciation that there were no basis for anticipating an automatic realization of these expectations is, in a sense, a major progress made in the understanding of the development process in most of Africa over the last two decades.

The position was best reflected at the Conference of Ministers of the Economic Commission for Africa which met in Tunis in February 1971. At that conference the ministers noted the failure of development effort in the preceding decade and commented especially on three aspects of it.[2] First, they noted the inadequate integration of the national economies in many countries, especially its physical, organizational, economic and socio-psychological aspects. Second, they observed the limited relevance of the development of science and technology in the developed world to the problems facing African countries, in particular to problems bearing upon the transformation of their socio-economic structures. Third, they recognized that some of the models adopted for formulating and implementing economic development plans in many African countries were either over-elaborate or otherwise unsuitable.

It was in these circumstances that increasing attention began to be turned to local and regional development as an alternative strategy of planning and development. The expectation was that this strategy would enable these countries to :

a. effectively marshall their national and external development resources ;
b. mobilize all sectors of their population for participation in activities that should lead to the integration of the traditional sector, at present the less productive sector,

with the modern dynamic sector ; and

c. promote structural changes to reduce almost exclusive dependence on external factors for the process of transformation and development.

MAJOR STRATEGIES OF LOCAL AND REGIONAL DEVELOPMENT

After a decade of these declarations it is worthwhile examining the main strategies attempted for achieving the new objectives implicit in those recommendations. At the outset it is important to stress that these strategies vary considerably from one African country to another and in many countries more than one strategy was used at any given time. But by and large it is possible to indicate four broad strategies each of which had both a local and a regional dimension.

The first of the strategies simply involves regional disaggregation of national plans. In theory this strategy to be effective requires more detailed knowledge of local conditions especially of the geographic, economic and social structure and, therefore, depends on close involvement of local administrators and planners in the formulation and the implementation process. In practice, however, such individuals tend to have rather limited capacity to influence the selection and design of projects that ultimately have to be implemented and administered by them. Furthermore, the need to ensure that regional or local plans fit the state or national planning framework so as to ensure consistency of overall resource utilization and observe social welfare priorities, often means the vitiation of much local or regional initiatives in the development process.

Regional disaggregation of national plans could be said to have achieved some success only in those countries with a federal constitution where individual states of the federation are treated as regions. The best example of this is Nigeria where the creation first of three, then four, then twelve and finally nineteen states made it possible for resources to be more broadly distributed across the national space. The position has, however, not seriously affected development at the local level, since each state has tended to operate as a microcosm of the nation. The result has been the continued agitation for the creation of more and more states as a means of ensuring broader access to allocatable national reasources.

In other countries in such as Uganda before the Amin regime, regionalization was less formal or constitutional, but involved dividing the country for planning purposes into four regions, each of which contained four to seven districts which in turn were divided into a number of counties. For each of these units, criteria were established for assessing current levels of development and determining priorities. This allowed projects and programmes to be formulated on a regional basis and to be integrated into the Planning, Programming and Budgeting System (PPBS) operated by the Ministry of Finance and monitored by an Inter-Ministerial Planning Committee.[3]

A second strategy of regional and local development adopted during the last two decades emphasizes physical resources development, particularly land and water, on a regional basis. In Zambia for instance, this took the form of a geographic-economic approach to investment programming from the village level through the ward and district levels to the provincial and regional levels.[4] This enabled natural resource endowment of each region of the country to be a major input into consideration of development potential and possibilities. In other countries such as in Egypt and to a lesser extent the Sudan, such resource development approach concentrated on land settlement schemes as a means of achieving a wide range of objectives, including a more balanced distribution of regional population and employment opportunities, evening out inequalities in regional per capita incomes, and improving social and economic infrastructures in different parts of the country.

But by far the more striking form of this strategy has been of the integrated river basin development type. In a number of African countries, the initiative for this approach has been the need to augment the installed capacity for electricity generation through damming large rivers and creating huge man-made lakes with great potentialities for multiple uses. The regions around these lakes notably the Aswan (Egypt), Kariba (Zimbabwe), Volta (Ghana) and Kainji (Nigeria) were seen as presenting developmental opportunities for irrigation farming, large-scale inland water fishing, tourism and improved navigation. Although this type of strategy was always treated as part of the economic benefits that tilt the scale in favour of the huge investment on hydro-electricity, it is seldom that the resultant regional development rises to the level claimed for it in the proposal document. Part of the reason for this is the very poor data base on which the decision to embark on this type of regional development investment is made. Nonetheless, a few countries notably Nigeria, has gone ahead to elaborate on the strategy of integrated river basin development. The country has been divided into 11 major river basins, each under a River Basin Authority which has the responsibility of planning and implementing a phased programme for the over-all development of the basin. Many of these authorities are generously provided with funds which they are being encouraged at present to devote to the expansion of land cultivated to cereals, notably rice and maize, to meet the country's rapidly growing demand for these food crops.

The concept of "the growth pole" provided the theoretical underpining for the third main strategy of regional development attempted in many African countries during the 1970s. As a strategy it essentially encouraged governments to locate industries in a few centres other than the capital-cum-port cities that had attracted most industries up till then. It also coincided with effort to spread infrastructural development to other parts of the country. Furthermore, it provided a basis for looking at conditions in urban centres where generally were concentrated the more vociferous minority of the national population. In all of this, therefore, the growth pole strategy had definite political advantages in providing a plausible intellectual rationale for what political leaders wanted to do or in any case found more convenient to do.

Within a relatively short time, however, it became clear that some of the basic conditions necessary for this strategy to exert beneficial effects on regional development did not exist in many of the countries. In the first place, the consumer goods industries which dominated the portfolio of investments had very limited capacity to stimulate regional economies. In the second place, the traditional and untransformed characteristics of the regional economies made it difficult for them even to derive full benefits from the increasing urbanization arising from the growth in industrial capacity. Third, the institutional, legislative and pricing protection necessary to enable the so-called "growth-pole" or "growth centres" to be installed in each country turned out to be inimical to the continued development of regional economies especially the rural sector.

For many African countries, therefore, it was more the "backwash" rather than the "spread" effects of the growth centre that was felt. Particularly for the rural areas, the pursuit of this strategy increased wage and opportunity differentials between urban and rural areas, discouraged investment in agricultural production, depressed social and economic conditions in rural areas, and led to massive migration out of them to the cities. Progressively, the strategy created a situation in which most of these countries were no longer able to feed themselves and were starting to use their hard-earned foreign reserves to import vitally needed food.

In the meantime a few countries, notably Tanzania and Guinea, were attempting a different strategy based on the structural transformation of their economy. In Tanzania, for instance, emphasis was to seek self-reliant development through restructuring production organization especially in the rural areas. Socialist ideology provided the conceptual framework for the transformation of productive resources and people were expected to be persuaded about the superiority of this strategy of development through concrete examples of Ujamaa villages. Unfortunately, progress on a voluntary basis proved very slow and not totally satisfactory. By 1973 it was therefore decided to expedite action on the matter by phasing the process. The first phase known as villagization, was expected to see the physical nucleation of isolated farmsteads, hamlets, farmlands and fields into standard size villages which would be more appropriate for the second phase of institutional development, during which effort would be concentrated on building various co-operative social structures within which the activities of the population could be guided and directed.

The first phase was virtually completed within the short space of three years. Although many villagers were quite co-operative, a good number particularly in areas of relatively prosperous agricultural production were most recalcitrant. Varying degrees of coercion had to be employed to achieve compliance among this group. The whole effort, however, coincided with a period of global economic crisis and recession and the effects on the restructuring process had been far from favourable. This has encouraged rather glib criticisms of the process especially on the part of groups committed to showing that socialist ideology as a framework for organizing regional or national development in developing countries is impractical.

Clearly, the last two decades in Africa have been noteworthy in regional development terms not so much in concrete achievements of economic growth but rather in the manner in which irrespective of the strategies of development adopted, the results have been virtually the same. Indeed, 42 out of the 53 countries in the Africa region, achieved a rate of growth of per capita GNP (real) between 1970 and 1978 of less than 3 per cent per annum with more than half of them registering negative growth rate.[5] The eleven countries that achieved higher rates of growth comprised seven small countries and islands (total population 10 million), the North African countries of Egypt, Tunisia and Morocco and the rather special case of Nigeria with vast petroleum resources. In this circumstance, it is pertinent to enquire into the nature of the problems encountered in the various countries as a prelude to discussing the prospects for local and regional development in the future.

PROBLEMS OF IMPLEMENTING LOCAL AND REGIONAL DEVELOPMENT

It is possible to identify four major problems that have militated against greater achievement in local and regional development in many African countries in the last two decades. These are : the global economic situation, data inadequacy, manpower shortage and bureaucratic centralization.

The global economic situation was particularly significant in the last decade. Indeed, the 1970s had opened with a greater global consciousness of international interdependence. Initially this had taken the form of persuading leaders in both developed and developing countries of the need to treat global environment and natural resources with greater circumspection. The Stockholm Conference on the Human Environment in 1972 was the high-water point of this effort but no one would assert that its achievements were at best more than socially conditioning. It therefore, took the Arab-Israeli War of 1973 and the decision of Arab countries to use their petroleum resources as an instrument of War for the implications of international interdependence to be more widely recognized. The rising prices of petroleum and petroleum products hit not only the Western industrialized countries but virtually all developing countries other than perhaps the oil producers themselves. But even the latter felt the inflationary pressure of mounting bills for manufactured imports from the industrialized countries. The result was to distort the pattern of sectoral resource allocations in most African countries and to push into a position of secondary significance for their governments the question of regional inequalities. National problems of reduced foreign exchange earnings, of declining rural productivity, of serious food shortages and massive emigration from rural areas into urban centres—all came to be more pressing and to be accorded national instead of regional attention.

A second reason militating against active implementation of local and regional development is the inadequacy of data presented in sufficiently disaggregated form to portray the urgency of the problems. This situation was more serious in some countries than in others. In the eastern and northern African countries, the data situation was fairly good compared to those in the western and central African regions. In some countries, the National Office of Statistics is just not organized in a manner to provide on a continuous basis either cross-sectional or time-series data for units of administration smaller than the nation. To do this often involves mounting national surveys. The frequency of such surveys tends to be irregular nor are their observational units always the same. In some other countries, fear of the political implications of portraying differential developmental performance at the local and regional levels has meant an indifference to publishing such figures even when they exist. In still others, the decline in the efficiency of governmental activities generally implies in itself a loss of timeliness in presenting data and similar reports concerning the social and economic situation in different parts of the country. In all of this, the result has been a certain loss of visibility for local and regional problems and of the need for political pressure to deal with them.

The shortage of the requisite manpower can also be shown to have been a factor. Although in most African countries, the last two decades had witnessed a remarkable increase in the output of educational institutions at all three levels, the economic situation has tended to discourage many of these seeking or securing employment in the public services. In some African countries such as Nigeria and the Ivory Coast, the greater and better attraction of private sector employment is of considerable significance. In others, political instability and mismanagement had postured senior and experienced civil servants as critics of the regime and resulted in many of them being removed from office or emigrating from the country. This type of "administrative haemorrage" has resulted in both a qualitative and quant.tative loss of experienced personnel and has further weakened the over-all effectiveness of government. In such circumstances not only would the questions of how to deal with and resolve problems of regional inequalities escape attention but the overall national economic situation is likely to be so mismanaged as to further compound the regional problems. The result, as is so obvious in many African countries today, would be a pervading immobility of action and a debilitating poverty most evident in rural areas but also in many cities, especially those away from the capitals.

But perhaps the most serious danger from an impoverished and immature civil service is the loss of confidence which is often compensated for by an elaborate tendency to concentrate and centralize all powers in the bureaucracy. The less the government can actually do on its own to get things moving at local and regional levels, the more it tries to concentrate powers in itself to do everything. This increasing bureaucratization of decision-making in a country takes different forms. In some countries, it is the administrative response to the indicative political preference either for a socialist form of government of the Tanzanian type or a mixed economy as in the Ivory Coast or Nigeria in which the government attempts to take control of the "commanding heights" of the economy.

In the Tanzanian case, for instance, the role of bureaucracy had come in for serious censure in many of the recent assessments of why governmental effort at regional and rural development had been so severely frustrated. Mapolu and Phillipson, for instance, noted that bureaucrats, because of educational background, experience, and expectations were ill-equipped to deal with the complexities of rural development. They had little or no training in agricultural development, and to vindicate their role they devoted more and more effort in areas where they could compete with the peasants whom they considered stubborn and resistant. Consequently, there was a "... growing commandist trend in the bureaucratic implementation of the *ujamaa* policy."[6]

Even where effort had been made at so-called administrative decentralization as a means of involving the population in the decision-making process, it is often found that it is the civil service that is decentralized and not the process of decision making. The structure of the service and its system of rewards and up-ward mobility runs counter to the type of decentralization which would concede considerable initiatives to local and regional levels of government. Indeed, the over-riding impression that one gets in country after country in Africa is of a bureaucracy that has difficulty in distinguishing between the tasks of co-ordination and execution. Finding that the latter gives a more visible impression of activity, even if not of effective action, the civil service in many African countries have extended government interest in various local and regional matters into an umbrella for wide-ranging bureaucratic interference.

Where such interference has been resisted through political means, it has been easy to starve local and regional initiatives of any real meaning through scanty or indifferent resource allocation. Sometimes, this is coupled with denying local or regional bodies of the powers to raise their own revenue even if they could get the local population to respond effectively. The sceptre of the corruption of local politicians and officials is often raised to dampen any enthusiasm in this direction.

The significance of the bureaucratic factor in the gradual weakening of the push for local and regional development in many African countries must also be seen against the succession of *coup d'etat* and of military regimes that became prevalent during the last two decades. Generally lacking popular support especially among the political elites, these regimes came increasingly to rely heavily on a captive bureaucracy for advice and guidance as to how to run the country. Not unexpectedly, these advice led to an increasing concentration and centralization of power and a growing neglect of the needs and expectations of the masses of the population in their different localities and regions.

CONCLUSION : PROSPECTS FOR LOCAL AND REGIONAL DEVELOPMENT

The above sweeping survey of experiences with local and regional development in many African countries may paint a very negative picture of little having been achieved. It is true, for instance, that hardly any of the strategies that have been adopted has resulted in any substantial growth in the local, regional or national economies. It is also true that social and regional disparities in income have become pronounced in many countries. Furthermore, it has become increasingly patent that the backwash effects of the various regional policies and strategies have rolled on themselves and crashed with reverberating resonance on rural life and agricultural production in many of these countries.

Yet for all that, some positive achievements may be recognized especially in terms of the lessons which these experiences are teaching. Perhaps the most important of these lessons is the fact that local and regional development is nothing if it is not about power sharing in a country. Talks about 'disaggregation of' national plans, of growth centre development or of decentralization are meaningless unless they are backed by actual devolution of decision-making powers to the people actually involved in this process. Anything else simply plays into the hands of the bureaucracy to reinforce its stifling hold on the overall development of a country.

Power sharing, however, does not imply total autonomy. It involves legally defining the limits of regional and local initiatives within the larger collectivity of the nation. It also is not confined to sharing of national resources. At its best, it should enable both local and regional leaders to indicate how they intend to raise revenue and increase or enhance assets and capacities for higher output and productivity. Such a pre-condition for success implies a willingness on the part of national political leadership for a plurality of approaches to local and regional development. Moreover, since the political leadership must still depend on using bureaucrats to achieve results, it requires that a new set of output-oriented criteria for determining career prospects in the civil service must be established.

Another lesson from the failures of the past is the recognition that if local and regional development is to succeed, it must entail a high degree of popular mobilization. For this reason, serious attention needs to be put on the territorial framework for achieving such mobilization. The issue here relates to the optimal size of territory for achieving and sustaining local participation. Given the fact of the poverty of many African countries and the certainty that much asset and capital formation especially in rural areas will depend on voluntary labour inputs and other locally generated resources, this factor is going to bulk large in future concern about local and regional development in many of these countries.

It must of course be understood that the division of a country into districts and of the districts into smaller territorial units of organization is not all that is hereby advocated. Relevant and necessary institutions would have to be fostered and developed. These are needed to provide the basis for participatory decision making, for implementing economic projects and for undertaking social and political actions. Such institutions will be the crucial factor in planning future local and regional development, especially their co-ordination over the total national space.

The acceptance of both power sharing and territorial reorganization as crucial elements in the future progress of local and regional development will be possible only in countries where the leaders exhibit a high degree of political will in wanting to bring lasting changes to the economic circumstances of the majority of their population. Such political will is not very evident in many African countries at the moment. But the progressive denudation of capital resources formerly wasted on prestigious and diversionary investments and the increasing difficulty of securing such resources through loans or aids from external sources are bound to enhance the prospects of an alternative style of development based on the mobilization of the population for self-reliant development. The exercise of political will in this direction will then recognize the crucial importance of seeing and approaching local and regional development as a learning process with emphasis on patiently guiding and helping local and regional populations to help themselves. Centrally allocated resources in these circumstances will consequently be seen not as the be-all and end-all of development but only as an enabling gesture to motivate and actively stimulate local participatory reaction.

Such a new perception of local and regional development as a learning process has implications for the training and education of regional planners. It should help to underscore the fact that whilst past strategies and techniques are not necessarily irrelevant, their viability in achieving their objectives will come only if the planner also appreciates how to incubate these ideas in the minds and consciousness of local and regional leaders and popla-tion and not simply in those of central government officials. This requires an emphasis in development strategy on the role and importance of collective action and of the political dynamics needed to nurture and sustain it. For in many African countries, local and regional development will succeed only when adequate attention is paid to its rural component. And for this to happen, it will be necessary increasingly to appreciate the aptness of President Nyerere's observation that :

> Governments by themselves cannot achieve rural development. They can only facilitate and make it possible. They can only organise, help and guide ; they cannot do. For rural development is people's development of themselves, their lives and their environment. The people cannot do it if they have no power.[7]

For African countries, the 1980s are going to be of critical significance for local and regional development. Given the experiences of the past and the changed international economic situation, it is becoming increasingly clear that there are limited options open to these countries and that the degree of freedom of their choice is considerably circumscribed. Indeed, for some of these countries the choice is already between accepting as a permanent fact of life immiserizing poverty, lightened occasionally by foreign aid and technical assistance, or making a supreme effort to mobilize their population to undertake their own development. It is the hope that many of these countries will soon come to appreciate the need to redefine real development not simply in terms of increases in gross domestic product but more importantly in the slow, steady but necessary changes in the conceptual images that the majority of their citizens have about their role and contribution to the progress of their local district in particular and the whole country in general.

FOOTNOTES

1. Statement by Professor Adebayo Adedeji, Executive Secretary of the Economic Commission for Africa on 'Perspectives of Development and Economic Growth in Africa upto the Year 2000', at the OAU/ECA Colloquium, Monrovia, Liberia, 12-16 February, 1979.
2. See United Nations Economic Commission for Africa, *Africa's Strategy for Development in the 1970s* (E/CN. 14/RES/218/x), 1971, p. 3.
3. R.O. Lukwiya, "Regional Planning and Budgetary Priorities in Uganda" in *Regional Planning and National Development in Tropica Africa*, ed. A.L. Mabogunje and A. Faniran (Ibadan ; 1977), pp. 51-59.
4. L.S. Chivuno, "Regional Economic Planning in Zambia", pp. 39-59.
5. World Bank, *1980 World Bank Atlas* (Washington D.C., 1981), p. 12.
6. H. Mapolu and G. Phillipson, "Agricultural Co-operation and the Development of the Productive Forces : some Lessons from Tanzania", *Africa Development* 1, 1976 : 46.
7. Julius Nyerere, "On Rural Development", *Preceedings of the World Conference on Agrarian Reform and Rural Development* (Rome, 1979), p, 1360.

EIGHTEEN

SUNIL KUMAR MUNSI

ENCLAVES AND GROWTH POLES
Some Observations on the Indian Situation

I. ENCLAVES

MOST scholars today agree that the spatial organization of the Indian economy during the period of colonial rule was marked by a kind of enclavization within the broad framework of macro-economic regions that had emerged around the port-based metropolises of the country.

Going by dictionary definition an enclave is a piece of foreign territory surrounded by territories of another state. In pre-independence India all centres of major British investment were like enclaves of foreign territory without normal linkages and relations with surrounding regions. These were the so-called centres or areas of 'growth' and 'prosperity' where economic activities were controlled and conducted solely or almost entirely for foreign interests.

The tea gardens of Bengal, Assam and the Nilgiris, the mining areas of Bihar, Orissa and Bengal could be taken as typical examples of such enclave development with varying degrees of significance. Irrigation was another area of major British investment. The irrigated areas of Punjab, Sind and Madras bore characteristics akin to enclavization.

Plantations

Tropical plantations the world over have their origin in the colonial period. The term 'plantation' was initially used to designate a plot of land set with plants. During the period of British colonization of the West Indies, North America or Asia, it came to denote a group of settlers or even the political unit constituted by such a group. When in 1768 the Office of the Secretary of State for the Colonies was created by the British government, it inevitably came to be known as the Plantation Office.

The term plantation was commonly applied to great landed estates producing a crop grown primarily for export. The plantations were devoted to such production in the colonial or semi-colonial countries that raised tropical or semi-tropical products such as cotton, sugarcane, rubber, tea, pineapple, banana, etc.

Thus, plantations are essentially tropical institutions. But this is only in regard to location. The capital, machines and profit, all came from or went back to metropolitan countries. It represented an invasion of the tropics by non-tropical people. 'Plantations are intimately bound up with colonial and imperialist conquest and exploitation...the political oligarchy of a plantation country is intimately linked up with foreign interests.'[1]

Indian plantation labourers, particularly in tea plantations, were natives of India but not necessarily people belonging to plantation areas. They were 'forced' and 'indentured' and were kept in strict economic and political servitude.

Plantations in India under British rule, particularly tea plantations, were looked upon as growth centres by the then British administrators and analysts. In the words of George Watt :

> the magnitude of the stake Great Britain...holds in her Indian Tea Plantations will be seen to justify the very strongest statements of the value of India to England ; India has thus its resources developed, its vast expanse of country opened up, and its people enriched by lucrative employment and by the expansion of the cultivated area through the colonization of the tea area. The diversion of British capital to foreign countries cannot be too strongly deprecated both in the interest of the speculator and in the light of the claims of his nationality for support and encouragement. The British investment in Assam tea converted that province into one of the most flourishing and properous in India. A few years

ago the officers who were sent to Assam regarded themselves as condemned to a penal servitude.[2]

According to Watt, tea plantations were the nuclei around which 'a new civilisation' and 'a new commerce' gravitated and their effects were felt from one end of the country to another.

Before the First World War substantial direct areal investments were made in tea, coffee and rubber plantations in the Brahmaputra valley, the Surma valley, the Himalayan foothills in north Bengal and in the Nilgiris.

Unlike tea, many coffee gardens were owned by the local people though their gardens were small and production far inferior to that of the large gardens owned by the Europeans. The average size of each estate in Coorg held by Europeans was 196 acres and by the natives only 8 acres, wrote Watt.[3] In Wynaad in 1868 there were nearly 30 thousand acres under coffee of which 21.4 thousand acres were held by Europeans and about 8.4 thousand acres by the local people.

The entire investment on rubber plantations came from foreigners, particularly the British.

Coal

Coal was being extracted over widely scattered areas in Bengal, Assam, Central Provinces, Nizam's domain and Madras Presidency by the end of the nineteenth century. But the largest number of coalfields lay contiguously in Bengal Presidency forming a compact zone known as the Damodar Valley coalfields. The major coalmines were European owned which raised, at the beginning of the 20th century, more than 4 million tons, those owned by the natives had an output of just about one million tons.

The coal belt of India saw the rapid growth of railways, mushroom growth of settlements, abandoned or ghost settlements left as evidence of the closure of mines, and an unbroken uniformity of cultural landscape—apparent islands of growth amidst a sea of abject poverty. As in the tea gardens, most of the workers in coalfields came from distant places without any connection with the coal belt—except for the back-breaking labour.

These areas were meant to remain isolated, with marginal connections with the Indian hinterland.

Irrigation

At the beginning of this century, investments in irrigation over selected areas in India was almost equal to that in plantation. But the purpose was certainly different, though both involved development of areas. In the case of plantations or mines, the aim was directly to produce a commodity and to that effect to create an areal organization which was, in a way, completely insulated from any interaction with the neighbourhood. In the case of irrigation this insulation could not be created or maintained. Once created, an irrigation canal in the long run was bound to have a profound spread effect on the region through which it flowed.

The First Irrigation Commission in 1901 suggested a number of schemes of which the most important were in western United Provinces (U.P.), in the Madras Presidency and in the Punjab. For parts of Punjab and Sind which are now in Pakistan, the commission had recommended four major schemes such as the lower Bari Doab Canal, schemes on the Sutlej and the Indus rivers, and so on. The U.P. schemes included the Ken Canal, the diversion of Sarda water in the Ganga, etc. In the Madras Presidency projects were recommended on the rivers Cauvery and Krishna.

How concentrated these irrigation investments were can be gauged from the fact that between 1898 and 1914, 50 per cent of the gross irrigation investment in India was in the Punjab alone. Madras also developed a strong base of irrigation infrastructure as a result of large investments during the same period.

Bagchi noted that public investment in the form of irrigation dramatically increased the yield of sugarcane and cotton particularly in the three provinces of Madras, Punjab and Sind. The public irrigation system had rapidly expanded in the above three provinces and irrigated areas constituted the highest precentages of the net sown area in these provinces in that order :

> Public investment in the form of irrigation provided the impetus which could be utilised for producing better varieties of crops ; public investment in the form of research provided the basic information leading to the diffusion of the improved varieties ; the spread of diffusion was far more influenced by the relative profitability of crops in different

> regions than by the mere fact of discovery of imp varieties by the Department of Agriculture.[4]

In fact this diffusion could hardly materialize in the absence of adequate irrigation infrastructure and other facilities going with it, in other parts of India. What, therefore, emerged was enclaves of a different kind, different from plantations or mines, but retaining some of the basic characteristics of pockets of growth unrelated to the economy of neighbouring territories.

II. GROWTH POLES

In post-independence India regional planning exercises have been largely based on the concept of 'growth pole' development both in agriculture and in industry. Isard defines growth poles in the following terms :

> the village, town, or small city at which the strong development impulse occurs is designated as a growth pole, a term frequently used in development literature. This signifies a location's resource base, import substitution possibilities, externalities, exploitation of amenities, and a host of other factors. Such concentration involves developing a great variety of skills in the labour force and of marginal risk-taking abilities among the entrepreneurs of a local elite group. This in turn serves as a propulsive force not only in the growth pole but also at the several other favourably situated villages and towns in a region. The spread effect is closely associated with the fact that industrial growth in other villages and towns can be oriented to the newer and larger markets emerging at the growth pole.... In short as a growth pole develops it can pull the whole or a large part of the region along with it.[5]

Going back to the history of the development of this particular concept, one could point out the various attempts, under conditions of developed capitalism in Europe, to identify the motive force in regional development and use it in regional development strategies. Von Thumen in early 19th century, Christaller of the Central Place Theory fame whose original German book on the subject was published in 1933, or Lösch o supplemented Christaller, developed models of spatial organization in which the motive power was the pulsating central place or the growth pole sending out impulses conducive to regional development. Philosophically, the role of the individual, the banner of capitalism, gets a premium over all other factors. Here the individual is a central place with potentiality. In such a model, development would necessarily be unbalanced; the development occurring in geographical growth points. In fact this model veiws development as necessarily unbalanced.

India in the post-independence period opted for planned development but under a system of private ownership of the means of production, whether in agriculture or in industry. And instead of a strategy of broadside attack on regional disparity or areal inequality left by the British as a legacy, the government undertook a programme which has now become well known as the 'Area Development Programme' of the Fourth Plan. The essence of the programme lay in selecting target areas and groups of people for pumping inputs of development in order to achieve what is called the 'place prosperity' perspective. This is what Isard describes as the 'growth pole' model in which selected centres receiving intensive inputs will set in motion forces that will induce development in their respective hinterlands.

Agriculture

That the area development programme has led to growth of disparity in the country can now hardly be disputed by any scholar. In a major work on this subject, Bhalla and Alag show how in the period between 1962-65 and 1970-73 areal disparity has been accentuated through the 'green revolution'. This is how they characterize the trend of growth in Indian agriculture during the period of their review :

> That a major concentration of modern inputs has taken place is clearly brought out by the fact that with slightly greater than one-fifth of the area, top productivity 69 districts consume about the same proportion of fertilisers as 130 mid-productivity districts with nearly 50 per cent of the area. In fact, there is more disparity in terms of use of tractors and tubewells. . .[1] "almost half the area contributing nearly half the output in India is still in the average productivity

category and as much as about one third of the area producing nearly one-sixth of the output still belongs to the very low productive category.[6]

Even within the 'green revolution' areas, the revolution itself has meant polarization of wealth which capitalist farming is bound to generate. One could mention here that the growth centre strategy fitted well with the strategy of development of efficiently run large capitalist farms.

Industries

If we consider areal distribution of per capita income, average daily employment of factory workers, per capita industrial consumption of electricity and per capita scheduled commercial bank credit, we have eight states which stand below the national average in all the four items, namely, Andhra, Assam, Bihar, Jammu and Kashmir, Madhya Pradesh, Nagaland, Rajasthan and Uttar Pradesh. Two other states, namely, Himachal Pradesh and Orissa, had only marginal advantage in one item each.

The daily employment of factory workers in India in 1971 ranged between 20.8 per thousand in Maharashtra and 2.4 per thousand in Jammu and Kashmir. There were five states which could not even attain the level of 5 factory workers per thousand of population, and 4 states were marginally above 5.

The Monopolies Enquiry Commission of 1965 had noted that 533 large factories were owned by the 75 leading industrial houses in the country of which over 54 per cent were located in only three states : West Bengal, Maharashtra and Tamil Nadu.

Though all planning documents emphasized the need for allocation of economic activity by taking into accout the spatial dimension, and although the serious problems of regional resource constraint were also taken note of, the locational policy of the Government of India degenerated into a veritable scramble for each industrial target by most states.

According to Bhagwati, "Planning for regional balance in India has been at best weak and at worst negligent and negligible. This is demonstrated by reference to industrial location and pricing policies."[7]

The three most important factors that determined the actual decision in regard to industrial location, had nothing to do with the removal of regional disparity, but were, (i) 'export orientation' of industries and locations, (ii) political pulls and pressures by state governments, and (iii) overall failure seriously to work out the space dimension of industrial target.

The 'growth centres' in the industrial development programme could hardly perform their duties. It has been estimated, for example, that the growth of the Rourkela Steel Plant corresponds with a major decline in Orissa's absolute level of diversification. In fact if diversification is taken as an index of industrial development, in 1965 roughly a mid-point between 1951 and 1981, Maharashtra, Madras and Mysore were the three most industrially diversified regions of India while Bihar and Orissa did not show any sign of industrial diversification around major industrial projects located there. At the bottom of the ladder lay four states of India which included Kerala, Assam, Tripura, Jammu and Kashmir. If we consider the present time, more states will join the lowest category, particularly those states in north-eastern India created after 1965.

This 'considerable stability' in the relative levels of industrial diversification over space in India after a fairly long period of planning should indicate the inability of the growth pole strategy effectively to intervene in the situation of disparity existing in the country.

CONCLUSION

In terms of areal organization, the growth poles of today, therefore, do have some resemblance to the enclaves of yesterday. The termination of colonial regime brought about a significant change in so far as it altered the roles of subordination and domination and of the 'enclaves'. But the basic ideology persists, the ideology that views individuals endowed with potentials as the motive force of development, of sending out impulses to be felt all over, as Watt had commented about the tea plantations. The net result has been deepening of regional imbalance between the 'developed' and the 'backward' regions.

Writing about the picture of disparity in urban development all over the developing world, three Soviet authors recently commented as follows :

> Urbanisation in the developing countries accentuates pluralism (including spatial pluralism) of their socio-economic development. Many of its parameters in large and major centres exceed many times over similar indicators for other parts of a country. Such towns become the focal points of attraction of foreign capital and also seats of local monopolies, closely connected with multinational companies. In view of the close ties with the industrially developed countries such centres must often be regarded as a sort of 'imperialist enclaves' and 'support bases of 'internal colonialism' (that is, of subjugation of backward areas by several leading centres.[8]

Urbanization reflects the entire process of economic development within a country. There may be variations of the picture drawn by the Soviet authors from country to country. The Indian scene may not appear as bleak as those found in some other countries. But what we are concerned here is about the basic results flowing out of models put into practice in the name of development. The moot question is, how distant is a 'growth pole' from an 'enclave', at least from the point of view of spatial organization ?

FOOTNOTES

1. E.R.A. Seligman (Ed.), *Encyclopaedia of Social Sciences*, U.S.A., Vol. 11-12, 1935, pp. 149-153.
2. G. Watt, *A Dictionary of Economic Products of India*, Reprint Edition, New Delhi, Vol. 6, Part 3, 1972, p. 423.
3. *Ibid.*, p. 474.
4. A.K. Bagchi, *Private Investment in India, 1900-1939*, Cambridge, 1972, p. 11.
5. W. Isard, *Introduction to Regional Science*, U.S.A., 1975, p. 382.
6. G.S. Bhalla and Y.K. Alag, *Levels of Growth of Agricultural Output in India* (mimeographed), J.N.U., New Delhi, 1975.
7. J.N. Bhagwati, "International and Regional Development", *Economic Development in South Asia*, 1970, pp. 542-543.
8. V.M. Gomhan, *et al.*, "The Economic Geographical Aspects of World Urbanisation and its Specific Features in Developing Countries", *Urbanisation in Developing Countries*, Hyderabad, 1976, p. 43.

NINETEEN

O.H.K. SPATE

THE ROAD TO WAIGANI
Siting the University of Papua New Guinea

IN MY 50 years of university life, I have had many fine students, not a few of whom became more than that—they were dear friends. Both as student and as friend, Mohammad Anas was outstanding by his qualities of mind and heart. I appreciated especially the delicacy and skill with which he used the English language, but beyond this intellectual link we had shared experiences both sad and joyful. To my high regard for him as a geographer there was added a true and deep affection for him as a man. Grieved as I was to learn of his death, I am grateful that at least I have been invited to share in the tribute paid by his friends and colleagues to his memory.

Seeking some theme whereby I might most appropriately make my offering, it occurred to me that it might be found in some aspect of the geography of New Guinea, that country of magnificent landscape and most colourful people, so startling in its sharp juxtaposition of the Stone Age and the Electronic Age, and the subject of Anas's own doctoral thesis, an able, thoughtful, and remarkably comprehensive study.[1]

My topic might be classified as a case-study in non-quantitative geographical analysis applied to a problem in the economics of Academe, which was also a delicate problem in political geography owing to regional or local or even tribal rivalries : the choice of a site for a new university, or rather the university, in a country still under colonial rule (until 1975), far from homogeneous in almost every aspect of social life, and with striking disparities in development.

The eastern half of the great island of New Guinea, with New Britain and other islands, was under Australian rule and covered about 475,000 km^2 (Fig. 19 1). Papua, the southern half of this moiety, became British in 1884 and was transferred to the young Commonwealth of Australia in 1906, while the north and the Bismarck Archipelago was occupied—along the coasts—by the Germans until the 1914-18 War, after which it was administered by Australia first as the Mandated and then as the United Nations Trust Territory of New Guinea. Both sections were under the same Ministry at Canberra, and latterly the distinction had become more nominal than real ; but these differing histories are still reflected in the clumsy double-barrelled style 'Papua New Guinea'.

The environment is tropical and generally difficult except for shifting cultivators. The country consists for the most part of swampy floodplains, or very rugged and deeply dissected mountains covered with dense rain forest or secondry growth flecked by the gardens and fallows of the swidden or slash-and-burn farmers. The axis of the island is a great cordillera rising to 4,900 metres which carries the broad grassy vales and median plateaux referred to as 'the Highlands'. These central ranges are flanked on the south by some of the wildest country in the world, a belt of limestone sinkholes and tangled jagged ridges. South of this belt are the swampy delta of the Fly River and the dry scrubby Oriomo Plateau in western Papua. North of the mountain core is the long trough of the Markham, Ramu and Sepik valleys with large areas of tall grass (probably the result of long-continued burning for shifting cultivation) and in the great floodplains vast swampy 'mosquito reserves'—the story goes that Sepik mosquitoes do not attack aircraft *in flight* : North again are complex mountains and, mostly in the offshore islands, recent and still active volcanoes, sources of disastrous eruptions but also, especially in New Britain's Gazelle Peninsula, of some of the best soils. The lowlands are always hot and, except around Port Moresby, humid ; but the Highlands at 1,200—2,400 metres have a most invigorating climate, a factor of some importance in our enquiry.

Under Australian rule Papua New Guinea had progressed economically, but until the great copper discoveries on Bougain-

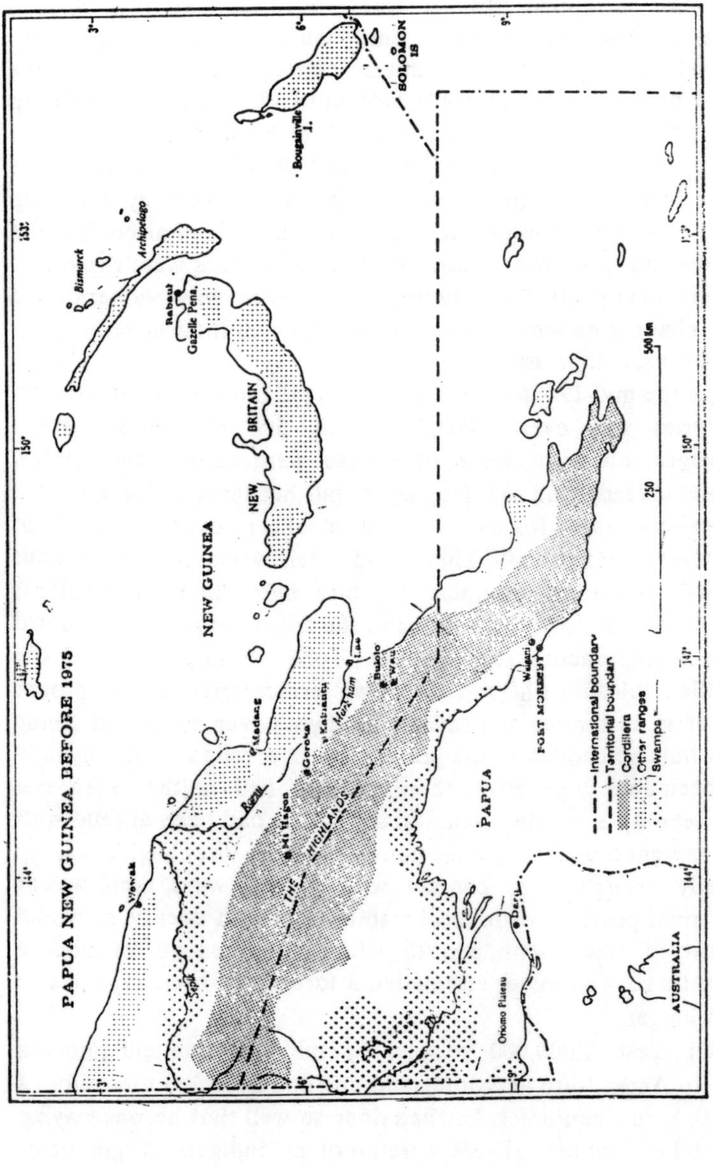

FIG. 19.1 : *Papua New Guinea before 1975*

ville Island in 1965, neither very fast nor very far. These were many coconut plantations around the coasts and a little rubber. The best developed littoral area was in the Gazelle Peninsula, the area of longest substantial European contact where indigenes as well as expatriates produced copra and cocoa. In the Highlands, not seen by white man until 1933 and not opened up before the 1941-45 War, indigenes had a substantial share in coffee production. Around Wau and Bulolo was an important goldfield which had been developed before the War by airlifting heavy dredging equipment—the first such use of aircraft in the world—and the fine stands of Araucaria pine were exploited. But practically all these commodities were very vulnerable to price changes on world markets, and the sum total of production was not very impressive.

In the mid-1960s the indigenous population of the Australian territories was over 2,000,000, with 500 or more distinct languages, although some of course were spoken by very few people indeed, and 37 languages (all but three in three related groups) accounted for over 760,000 speakers, over one-third of the total population.[2] There had been very few indigenous political structures grouping together more than a few villages and 4,000 or 5,000 people, and the tribes ranged from shrewd petty entrepreneurs such as the Tolai cocoa-growers of the Gazelle Peninsula who ran their own fermentaries, to tiny groups in country so remote and savage that flying over them and seeing a few huts embedded in the jungle, one marvelled that anybody had been able to penetrate there. A few had neither seen nor been seen by a white man. Many tribes had been at feud with their neighbours for generations ; and in the Highlands, so recently brought into contact with modern ways, there was at once great pride in virile local traditions and an uneasy envy and distrust of the lowlanders, to whom the 'benefits of modern civilisation', in however attenuated a form, had been available so much longer.

Of these 2,000,000 people, by mid-1965 one had gained a B.Sc. in Agriculture at Sydney and one had qualified there for a Pass B.A. in Economics, but had done so well that he was staying on to take Honours.[3] There were no other indigenous graduates at all. Before World War II education had been left virtually entirely to Christian Missions and as late as 1963 no fewer than 179,000 pupils—81 per cent of the total—were in their hands; but Administration schools accounted for 1,640 out of 2,732 secondary students and for nearly all those receiving any serious technical training.[3] The base for tertiary education was thus narrow indeed and yet—even without pressures from the United Nations which were not lacking—it was clear that, whatever the political future of Papua New Guinea, any type of development whatsoever would need large numbers of properly and sometimes highly trained indigenes for the task of running even a subordinate modern government and its agencies, however modest its scale. If nothing were done there would be the potential for social disaster. The task would be immense, something like transforming pagan Anglo-Saxon England into a twentieth century state in a matter of two decades.

The then Minister for Territories, Mr Paul Hasluck (late Governor-General of Australia), was a man of intellect with a sense of mission, one able to grasp the dimensions of the problem. In 1962 he announced that there would be some institution of university type in Papua New Guinea by 1966. By a sad coincidence two senior public servants assigned to planning it died suddenly, and Hasluck passed the problem to the Australian National University which had a good deal of Papua New Guinean expertise, especially in its Research School of Pacific Studies where Anas had been a Research Scholar. It had indeed rather too much expertise, for the working parties it set up spoke with many and incoherent tongues. They included so many convinced experts that they canvassed, exhaustively and exhaustingly, so many pet projects and variant solutions that at the end of one particularly wearying session I remarked that we would never get anywhere this way, and that the minister ought to appoint a commission, say of three people. I had no idea that he would do just that, and still less that I would be one of the three along with Sir George Currie who had been Vice-Chancellor of two universities, and Dr John Gunther formerly Director of Public Health in the Territory and an administrator noted for a quite unbureaucratic forthrightness.

This Currie Commission took evidence from over 400 people at 22 places in Papua New Guinea and also as far afield as Fiji and New Zealand, and in 12 months produced its report, a very comprehensive survey of the educational system of the Territory and its needs as they then appeared.[4] It would of course be unbecoming were I to say publicly that it is a very good report, though (equally of course) this disclaimer would hardly deceive anybody. At any rate it is a basic document in the history of education in Papua New Guinea.

The commission recommended the establishment of an institute of higher education and a completely autonomous university. The case for complete autonomy was simply that, as has been said, on any political projection increasing numbers of Papuans and New Guineans would have to be equipped to carry out the various activities of a modern society, and beyond this, that the training necessary to this end could best be given in fully independent institutions, so that methods, approaches, contents could be adopted to the peculiar conditions and real needs of the country and not to curricula devised for a completely different universe. I for one had ever before me the question I had once seen in an examination for high school finals in Fiji : 'Give the causes and consequences of the Second French Republic, 1848-51'—so set because the secondary school system of Fiji was geared to entrance to New Zealand universities. Though even so, after the Second Empire and the Third, Fourth, Fifth (and Fifth-and-a-half ?) Republics, one may doubt whether this was a very intelligent question even in France. At all events we were determined not to give any loophole for such nonsenses in Papua New Guinea.

The alternative which was probably favoured initially by most of those who had given their minds to the problem would have been a university college attached to an Australian university, or perhaps—just for extra complications—a consortium of Australian institutions of higher learning, to avoid inter-university jealousies. This had indeed been the most usual common element in most discussions. But this would have meant subordination to organizations and methods which had grown up in the exceedingly different society, culture and economy of Australia with its vastly greater diffusion of 'informal education' in modern ways through all varieties of media. Even within Australia such a relationship was liable to friction, as indeed was shown by the history of Canberra University College, originally attached to the University of Melbourne but later joined in a forced marriage of (governmental rather than academic) convenience with the Australian National University. Not without some stress, this particular marriage has in fact worked out quite happily, certainly more happily than if the spouses had remained separated by over 600 kilometres. Even within Australia it was almost inevitable that such a situation would give rise to friction and to feelings in the subordinate partner of alternate neglect and interference, or indeed both at once, and how much greater would be the likely incidence of such frictions and frustrations if the partners were separated by thousands of kilometres and a vast difference in the polities !

Unfortunately, by the time the report was published Hasluck had been promoted to Minister for Foreign Affairs and replaced at Territories by one who could perhaps fairly be described as a bush or *jangli* politician, with little knowledge of, nor apparently a great deal of concern for, the problems of Papua New Guinea. It took an energetic publicity campaign with the then topical warning of 'a second Congo ?' to induce him to act. However, an Interim Council was formed in October 1965. Under the vigorous leadership of Professor P.H. Karmel of Adelaide as its chairman and John Gunther as vice-chancellor, a very remarkable effort of improvization led to the enrolment of forty-two men and six women students for a preliminary or qualifying year in 1966. Classes were held in the 'tin sheds' of the Port Moresby showground, as were council meetings. Proper first degree work began in March 1967.

Autonomy did not mean a friendless isolation. Australian academics took the infant University to their hearts. The quality of staff offering was excellent, and—one advantage of starting with a clean slate—so too was the motivation of the students who felt they were 'in on the ground floor' of a good thing. To these two factors was owing very much of the initial success of the University of Papua and New Guinea, which almost at once began to make its mark in the cultural life of the country—not least by fostering a remarkable improvement in the quality of public discussions and in encouraging Papuans and New Guineans to speak out, even if in an alien tongue. Although there were no formal organizational ties between it and the Australian National University, the presence of the latter's New Guinea Research Unit on the Papua New Guinea campus led to a most fruitful co-operation in an annual Waigani Seminar on the problems not just of the Australian territories but of Melanesia as a whole, where the offerings of distinguished overseas speakers were balanced by lively indigenous contributions. And traditional Melanesian aesthetic and spiritual qualities found new and dynamic expression in literature and art.

Given this (very sketchy) background, we can approach our geographical question : Where to put the University ? One answer was immediate and obvious but not helpful. Expatriates were often doubtful about the necessity or desirability of a university for indigenes (though many were surprisingly liberal), but since such a concentration of consumers would be good for trade, if there was to be one each local business centre wanted it *here*, and put up the most ingenious arguments for it. All except Daru, and that may be because the commission didn't visit Daru. The businessmen of Rabaul thought that the capital, Port Moresby, was not sufficiently centrally situated. They admitted that Rabaul was not central to Papua New Guinea, but then it was central to the south-west Pacific...At this point Nature took a hand and applied the closure. The town is inside a flooded caldera and has several active volcanoes within sight. The Commission was perched on tubular steel chairs on a dais, which began to quake quite markedly. . . . To give due credit, the town spokesmen saw the point and metaphorically blushed.

More seriously, the technical problem of geography involved in fixing upon the optimum site was immensely complicated by political considerations, some of them highly emotive. Economically, the Trust Territory heavily outweighed Papua, especially in the indigenous sector,[5] and Rabaul, though with only 11,000 people against Port Moresby's 42,000 (1966), was more concentrated and had perhaps more of an urban air. Its *umland* was far more productive than Port Moresby's. The latter of course acted as the 'head-link', the main entry through which external economic, political and cultural influences were canalized and disseminated,[6] but its functioning as such was impeded by the mountain barriers shutting it of from the Trust Territory. But at bottom it may have been more significantly impeded by longstanding antagonism, especially among expatriates, to the seat of a government often regarded by entrepreneurs as too restrictive of commercial enterprise, or exploitation. Even within the bureaucracy there were relics of the old rivalry between Papua and the Trust Territory, which led one official in Lae to remark that the university would be put in Port Moresby over his dead body. (He was still very much alive as a valuable member of council after the university was set up at Waigani on the outskirts of Port Moresby.)

As for indigenous opinion, the three Highlands districts had a third of the total population on an eighth of the area, and on this count alone many or most New Guineans, as distinct from Papuans, might well have favoured a location at one of the chief centres of the region, Goroka or Mount Hagen or Kainantu. Moreover, the Highlanders had recently been introduced to the delights of cash cropping and modernization in general. They were flexing their muscles and prone to assert their weight owing to their numbers. On the other hand, the Highlands were some years behind most of the littoral areas in schools and other social developments, so that they would not be likely to produce large numbers of students qualified for higher education for some time to come, which would weaken the case based on centrality to population density.

Climate was also a strong argument for a Highlands site, and there is no doubt that, especially for expatriates, conditions at about 1,750 metres are much more conducive to sustained energetic work. However, nearly two-thirds of the people lived in hot and humid lowlands, and less than 10 per cent of cash economic activity was carried on above 1,500 metres, although this fraction was increasing. A Highlands site might mean less costly air-conditioning but would not obviate it completely : some laboratories and libraries would still need it. In any case the added costs of a lowland site in respect of this one item of expenditure would be more than offset by savings in transportation costs over a wide range of goods and services. Desirable as the agreeable amenity of a Highlands site might be, it could be bought at too high a price both in the direct capital cost of installation and in running expenditure.

This is owing in the first place to the obvious transport difficulties in a country where until the 1950s it was nearly everywhere a case of 'fly or walk', as it still is in many areas. Goroka, the only Highlands settlement that could very seriously be considered at the time of the enquiry, had a population of about 1,500 and was by an inadequate road, some 260 kilometres from the sea at Lae. The infrastructure for any large installation simply did not exist in the Highlands, and to create it with such inadequate transport would be intolerably expensive. Nigerian experience is to the point : the siting of its first university at Ibadan rather than the port of Lagos added very considerably to the running costs, not to mention endless troubles over such matters as customs clearances, forwarding priorities, and so on ; and Ibadan is only about 160 kilometres from Lagos by road *and rail.*[7]

There were other inland candidates—the gold-mining and timber centre of Wau-Bulolo, Kainantu, Mount Hagen on the advancing coffee-planting frontier. The last two had even less of an infrastructure than Goroka, while Wau-Bulolo is excentric to the belt of denser population and has the climatic disadvantage of lying in an intermont valley at only 700-750 metres. It thus misses out on both of the only really significant factors favouring the Highlands, population centrality and climate.

The Highland sites may then be ruled out. As for coastal sites, Wewak, excentric and with hardly any infrastructure, was not a real contender. Madang was, with a very active airport and important Mission educational institutions, but its climate was as bad as could be. Rabaul was the regional centre for the very progressive Gazelle Peninsula. It probably had more in the way of facilities and institutions that could be considered ancillary or relevant to a university than any other centre except Port Moresby. It had the 'feel' of a real if tiny town, but the location was excentric and the climate poor (not to mention the constant smell of sulphuretted hydrogen !), while the vulcanicity and seismicity presented an unacceptable risk. It is one thing to accept these hazards if one has no option, quite another to move deliberately into them. There remained Lae which could perhaps be considered the most central place, the transport node for the whole country with port, road and airport. But again the climate is poor. It has the strengths and weaknesses of a compromise site, and indeed when differences of opinion on the relationship of the university and the Institute of Higher Technical Education became acute, the compromise solution adopted was to shift the latter (it was as yet unbuilt, which made things easier) from Waigani to Lae, where in due course it was promoted in 1973 into the Papua New Guinea University of Technology.[8]

The main arguments against Port Moresby were its climate (despite the fact that it alone of coastal towns has a dry cool season), that it lacked centrality either geographical or demographic, that it was detached from the most dynamic foci of both indigenous and expatriate economic activity, and that all sorts of governmental activities and facilities were already far too much centralized there. So the university should be elsewhere, as a move towards decentralization if nothing else. Much of the opposition to Port Moresby was emotive prejudice of a type widespread in colonial situations. As we have seen, distrust of the capital was a long-standing tradition in the Trust Territory. This feeling occasionally went so far as to allege that the capital ought to be, and probably would be, moved to the Highlands, central in both area and population. Once well established, however, capitals are not so lightly thrown about—the investment, material and sometimes intangible is simply too great, and shifts of the capital are usually marked by some deep political crisis or a radical change in geopolitical balance or aims. The move from Petrograd (now Leningrad) to Moscow illustrates both points. In the Mandated Territory of New Guinea, Rabaul, excentric and virtually sitting inside a volcano, was the administrative centre simply by the historic accident of inheritance from the Germans. But despite official plans to move to the much more suitable Lae, it retained its position, seemingly from mere capital inertia.

It was also argued that any university should be well away from the capital so that it would be shielded by distance from interference by politicians or bureaucrats. Very oddly this argument came often enough from decentralizers who complained of the neglect of outlying area. It is true enough that, as practically the only pool of wide-ranging as distinct from professionally specialized expertise, the university's staff might be called upon for much enquiring and advising for the government. One might indeed argue that that should be one of its functions if it is not to be an elitist ivory tower. But while there might be fewer such calls if the university were distant from the seat of government, individual tasks might be more disruptive since they would involve more absence from campus duties.

As for more purely political interference, the argument is two-edged. The nearer the university is to the seats of the mighty, the more likely it would be to receive early warning of anything brewing, and the more readily it could take counter-measures through informal contacts with the government or its opponents. On a more material plane, in the early years of a big institution making unusual demands on resources in a poor and poorly developed country, there will be many matters calling for quick administrative decisions such as customs clearances, priorities in supply and communications, extraordinary urgent authorizations, seconding of administrative staff, and so on, all of which could be greatly hampered and made more costly by distance.

Perhaps the strangest argument against placing the university at Port Moresby was that of 'social stratification', meaning that race relations and the colour bar were worse there than elsewhere. This was doubtful in itself but if true would be a function of the greater mass and diversification of the expatriate population there which (as we shall see) is in another aspect a distinctly positive factor. Moreover even were it true, that seems in itself rather an argument for having the university there as a solvent, or better a dissolvent, of prejudice. After all universities have historically had the role of acting upon society and even governments, quite as much as being acted upon. Both *a priori* and as a matter of empirical experience in many countries, a self-respecting university (and universities are nothing if not respectful of themselves !) is more likely to weaken racist attitudes than to be contaminated by them, and it is foolishly defeatist to think otherwise. Of course the social as well as the physical climate of a Highlands site might have been more agreeable for sensitive souls who could flatter themselves on their racial tolerance at mixed tea parties, far away from the heat and dust of the real world of the marketplace and affairs, but that is not doing the job.

One positive advantage of Port Moresby was the immediate availability of a suitable site at 'June Valley' in Waigani, a site adjacent to the most advanced existing educational institution, the Administrative College for the training of indigenous minor officials, and to a Teachers' College. Beyond this rather incidental advantage and the logistical advantages already discussed of proximity to the seat of government, was a great but on the whole indubitably positive intangible : the general intellectual climate of the place. Intellectual snobs would have repudiated the words 'intellectual climate' in this context, and truly it was far from ideal any more than the European society of my first post in the University of Rangoon was ideal. Yet while we expatriate (the word was not then in general usage) members of staff spoke tenderly of how nice it would be to have the university in the delightful little hill station of Taunggyi, we did not really believe it. It would have been too nice, too tight, too boring. I think that while I was on the commission this was a subconscious memory in the back of my mind. Rangoon was hardly a cosmopolitan intellectual centre, but it was the best thing that we and Burma had. Not of course for one moment that the values of Taunggyi and Mandalay, the traditions of Burmese civilisation, were invalid and to be discarded, but to make one's way in the unfortunately inescapable modern world of the marketplace and the multinational, it was essential to have some wider contact with that world than was attainable up-country.

Very far from ideal as it was then, Port Moresby did have the largest, most articulated and most diversified non-indigenous community, even if largely a governmental one, and it had much the largest concentration of professional people—not very large, but the nucleus at least of a society, not just a handful of odd individuals with an interest in cultural affairs. It was also in a way the sounding-board for the whole country, the seat of government, the place where things happened even if they only happened by grace of the minister in Canberra (the university was a factor in changing that !). Papua New Guinea was a country of tiny political groupings, many of them knowing hardly anything of the world beyond a few adjacent valleys. Vast dim potencies, Australia, America, loomed heavily against the old way of life, but were too distant, not just in space but conceptually, to be grasped as realities, though the people were sharp enough in assessing the local agents of these distant entities. For the most part the country people were totally unsophisticated in modern ways, though often well acquainted with such fruits of modern technology as aircraft.

It was vitally important not only that students should be brought together from all corners of the country so as to understand themselves not just as Chimbu or Tolai or Motu or any other tribal 'wantok' (='one-talk') but as Papuan New Guineans, but also that they should be brought together in the locus where there was the maximum possible contact with 'affairs' in the broadest sense. University communities very often tend to live too much to themselves, and this tendency could only be accentuated if the university were placed in some tiny settlement or backwater trading town, no matter how agreeable a home in itself. An indigenous student could have no hope of grasping the nature of the 'outside world', so powerful and so mysterious to the lad or lass from some tiny village or remote Mission school, if he or she met nobody but missionaries, planters and polite, college dons. It is significant that younger people, whatever part of Papua New Guinea they hailed from, tended to favour Port Moresby for their university-to-be. For some without doubt this was simply the attraction of the 'bright lights', jazz and rock-'n-roll, but some specifically stated that their reason was that Port Moresby was the centre where the action was, where it was most feasible to study the running of the country and its wider relationships. A sound instinct.

A number of minor factors could be reckoned as favouring Port Moresby. It had much the widest range of sporting facilities and competitions, the sphere in which black and white most readily intermingled, and an important factor in building up *esprit de corps*. It had the greatest number of people who could benefit by part-time studies ; access and accommodation for overseas visitors were most readily available there. But to my mind, and I feel assured that this is also what would have appealed to most to the fine humanistic temper of Mohommad Anas, beyond these material factors of cost, infrastructure, accessibility, the most important factor in the case for Port Moresby was that indicated in the preceding paragraph.

Port Moresby was the place where the university could most effectively be worked in the interest of the whole country, where dialogue and discussion were most likely to be informed and wide-ranging. To quote directly from the Currie Commission Report itself :

> The real social issues, in the study and resolution of which the University should have a major role, will be in the marketplace, in the developing towns. Particularly in, an

> emerging society, it would be a dereliction of duty, and one with adverse affects on the University itself, were this not realised. It is for the University to bring wider and more objective views to bear on public issues, and it cannot well do this from an aloof location. On the other hand, a university community itself needs criticism and debate to keep it on its intellectual toes ; in an isolated locality, it could too easily become a mutual admiration society in an ivory tower.[9]

Since those brave days of the sixties, the University of Papua New Guinea has had its vicissitudes. In the seventies, it had its full share or even more of student unrest. Indeed, very early in its existence the students fulfilled that essential role of a colonial university and launched a demonstration, and later on some student movements were not quite non-violent nor devoid of political motivation and effect, though the effect was sometimes negative. There has been criticism of the university as fostering an élite, but what society has ever subsisted let alone advanced without an élite ? It has been called to order by the government of independent Papua New Guinea, and not always perhaps without warrant, for a too doctrinaire and academic approach, neglectful of the pressing needs of a new and spirited but struggling state.

It has certainly evolved into something very different from the ideal blueprint of the Currie Commission. We thought our planning bold and indeed by the standards of the day it was. But nobody in 1965 thought that independence would come to Papua New Guinea by 1975 as it did. The original concept was caught up by the acceleration of events and one might almost say that it became a back number with the rapid advance of a pragmatic nationalism. Yet it laid sound foundations for the changed structure of today.

From the start the university was a factor of great significance in the cultural and political life of the country, and it has served that country well. The smoothness of the transition from a backward colony into an independent state surprised even ardent well-wishers. To a great extent this was owing basically to the good sense and the feeling for consensus of so many Melanesian people. Yet even with those assets the transition would assuredly have been far more difficult, if not impossible, without the cadres that the university trained and put into the field in what seems, looking back, to be an incredibly short time. The University of Papua and New Guinea has done much, and of set purpose, to foster the essential sense of nationhood. It has provided a forum for wide and informed discussion of national affairs and a hearth for artistic and literary achievement.

I am proud of my part in all this, and proud too to dedicate these recollections to the memory of Mohammad Anas, who also knew and loved the country and its people.

FOOTNOTES

1. Mohammad Anas, 'Australian New Guinea : A Geographical Survey', Australian National University, 1958.
2. S.A. Wurm, "Language and Literacy", in (Ph.D. Thesis, *New Guinea on the Threshold*, ed. E.K. Fisk (Canberra : ANU Press, 1966), pp. 135-48.
3. O.H.K. Spate, 'Education and its Problems', in *New Guinea on the Threshold*, pp. 117-34.
4. *Report of the Commission on Higher Education in Papua New Guinea*, Canberra : Dept. of Territories, 1964.
5. H.C. Brookfield and D. Hart, *Melanesia : A Geographical Interpretation of an Island World*, London : Methuen, 1971, pp. 296-302.
6. O.H.K. Spate, 'Factors in the Development of Capital Cities', *Geographical Review* 32 (1942) : 622-31.
7. K. Mellanby, *The Birth of Nigeria's University* (London : Methuen, 1958), p. 43 ; incidentally, there is not and never has been, a mile of locomotive railway in Papua New Guinea.
8. I. Howie-Willis, *A thousand graduates : Conflict in university development in Papua New Guinea, 1961-76* (Canberra : ANU Press, 1980), 72-81, 135-9 ; see this book for later developments.
9. *Report of the Commission on Higher Education*, 179-80.

SECTION FIVE

AGRICULTURAL GEOGRAPHY AND RESOURCE MANAGEMENT

TWENTY

SHER SINGH DHILLON, RATTAN LAL SHARMA AND
JASBIR SINGH

A CLASSIFICATION OF INDIAN AGRICULTURE INTO MAJOR SYSTEMS

INTRODUCTION

UNLIKE mining, manufacturing and service industries, agriculture requires a much larger area of arable land for each person engaged in cultivation. It implies that agriculture is more widely distributed over the earth's surface than other economic activities. The diversities in agriculture result chiefly from two factors : first, the varied opportunities offered by surface configuration climate-soil complex which is variously restrictive—the more so in the underdeveloped areas ; and second, the socio-economic, infrastructural controls which stand at markedly different levels of material culture, commanding various skills and technical know-how and seeking different objectives (Singh 1974). Therefore, agriculture includes a wide variety of agricultural activities and man adopts his agricultural practices to spatially differing environmental conditions both physical and cultural. Obviously, the production of crops and raising of livestock cover most of the global terrestrial space and these are most easily observable in the field.

The primary objective of this paper is to classify Indian agriculture into major agricultural systems that may help in the description, analysis and synthesis of different agricultural distributions at all levels. Thus classification is to be considered as a prerequisite for agricultural planning as it offers a helpful approach that steers the investigator to make a thorough study of the problems of agriculture in the regions of plenty and poverty. The composite circumstances that contribute to the existing problems facing agricultural activities need to be appreciated critically in a spatio-temporal perspective. As a consequence, it seems worthwhile to attempt a subdivision of India into major systems of agriculture. The concept of major agricultural systems as used here means an areal unit of the surface of the earth which possesses throughout its extent similar basic agricultural characteristics or functioning forms of agriculture.

The present study is based on the agricultural data that are officially documented in abundance in various departments with a fair amount of efficiency. With these in hand an attempt has been made (i) to identify the major agricultural systems possessing their own distinguishable inherent characteristics on the land surface, and (ii) to show their effective distribution on a map of India. The delimitations have been checked and confirmed with the help of statistical measurements. Both the empirical and quantitative approaches are used to solve the problem of classifying Indian agriculture into major systems.

BASES OF CLASSIFICATION

In this exercise the overwhelming varieties of agriculture in India have been classified into major forms of agriculture. Various observable functioning forms of agriculture as suggested by Whittlesey (1936) have been used for this purpose, the most important ones being (i) types of cropping, such as fine foodgrains, coarse foodgrains, oilseeds, and olericulture, horticulture, industrial, beverage and forage crops, (ii) animal husbandry, (iii) farming techniques and implements employed such as wooden ploughs, iron ploughs, tractor horse power, and intensity of irrigation, (iv) dependence on labour to perform farm operations, (v) purpose of production, (vi) capital investment in farming enterprises, and (vii) the level of agricultural productivity.

While classifying Indian agriculture the types of cropping or animal husbandry, or combination of both, consider only the predominant type of crops raised or animals husbanded, for example, 'fine foodgrains' (such as rice and wheat), or 'coarse foodgrains' (such as *bajra*, *jowar*, *ragi*, barley, maize, pulses and gram), cultivated as the dominant crops, or in rotation with oilseeds and forages. Where either fine or coarse foodgrains or both are the major types of cropping, the farming systems are designated as 'fine foodgrains' or 'coarse foodgrains' or 'fine and coarse foodgrains' farming systems, respectively. Here the cropping techniques refer to non-irrigated, partly irrigated, moderately irrigated, or irrigated farming. Besides, it includes the tools with which the fields are tilled such as the density of wooden and iron ploughs or tractor horse power per 100 hectares of cultivated area. The purpose of production means the manner in which farm production is utilized. Consumption of the farm products by the farmer and his family is the basis of 'subsistence agriculture', whereas consumption of the farm products by outsiders involves exchange and is termed as 'commercial agriculture'. Anyhow most subsistence farmers have some crops that are sold, but this is not cash-cropping. In this exercise the purpose of farming has been inferred with the help of regional variations in levels of agricultural productivity. Lastly, the idea of the extent of use of chemical fertilizers and high-yielding varieties can be expressed with reference to the intensity of irrigation and the cultivation of high-yielding and high value wheat and rice crops. These guidelines are used to analyse more sharply the basic elements of the classification, and 13 basic systems of agricultural occupance of the arable land are identified in India. These are believed to be the primary systems in vogue today and Figure 20.1 gives the distribution of these systems. The more important characteristics of functioning forms of agricultural occupance of the major systems of agriculture and the spatial spread of each are set out in Table 20.1.

MAJOR SYSTEMS OF AGRICULTURE

The major systems of agriculture as summarized in Table 20.1 present not only the distribution of agricultural activities but also the factors that create particular forms of agriculture in particular areas. These factors may help in understanding the natural environmental conditions that influence the successful production of particular crops, and also the economic conditions (institutional and infrastructural) that favour or restrict production of these crops and the techniques employed. In addition, the 'Political Ecology' which may over-ride environmental or economic condi-

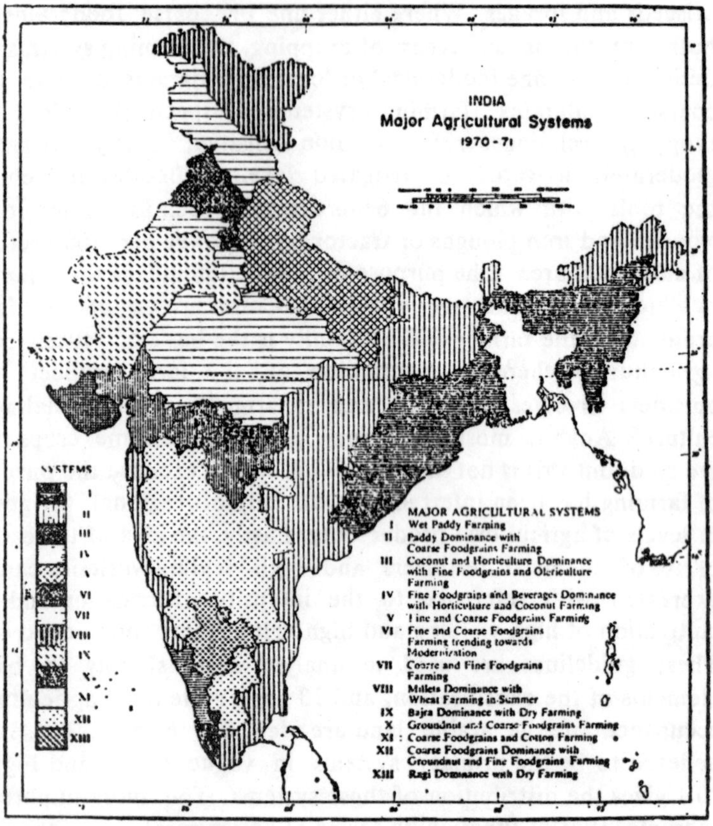

FIG. 20.1 : *India, major agricultural system, 1970-71*

tions to ensure the production or non-production of the crops must be taken into account. Also, the cultural pressures that may inhibit or stimulate farming decisions by the farmers must also be considered. The political and cultural pressures need to be considered at the time of detailed investigation of individual major farming systems at meso or micro level.

The production of foodgrains constitutes a significant and even a major proportion of the farming systems in India. On this basis the basic agricultural systems may be grouped under four broad categories :

1. The rice region extending from Assam Valley through

TABLE 20.1 : **Functioning forms of agricultural**

Sr. No.	Major agricultural systems	Spatial spread	Crop combination	Livestock association
(1)	(2)	(3)	(4)	(5)
I	Wet paddy farming	Mizo-Manipur hills ; Tripura plains ; Meghalya plateau and Assam hills and valley ; West Bengal plain and basin ; Balasore plain ; Mahanadi delta and basin ; Garhjal hills ; Ranchi plateau ; Chatrapur plain ; eastern ghats of Orissa, and northern coastal plain of Andhra Pradesh.	Fine foodgrains* (69.39)**	Cattle dominant
II	Paddy dominance with coarse foodgrains farming	North-eastern Himalaya ; Bihar plains ; Hazaribagh plateau ; Baghelkhand plateau ; Chatishgarh basin ; eastern section of Satpura range ; Dandakaranya ; Chanda plain ; Godavari basin ; eastern ghats of Andhra Pradesh ; Godavari-Krishna delta ; southern coastal plain of Andhra Pradesh ; northern coastal plain of Tamil Nadu ; Karnataka (Mysore) coastal plains ; Konkan (Maharashtra coastal plains), and western slopes of western ghats.	Fine foodgrains (49.75) and coarse foodgrains (21.79)	Cattle dominant, but cattle-buffalo at places in coastal plains

occupance of the major systems of agriculture in India

Percentage of net area sown recropped	Percentage of net area sown irrigated	Cultivators per 100 hectares of cultivated area	Wooden ploughs per 100 hectares of cultivated area	Iron ploughs per 100 hectares of cultivated Area	Tractor horse power per 100 hectares of cultivated area	Level of agricultural productivity
(6)	(7)	(8)	(9)	(10)	(11)	(12)
22.25 Low	15.69 Low	50-65 High	Over 35 High	Under 3 Low	Nil	Medium
21.15 Low	28.43 Medium	50-65 High	25-35 Medium	Under 3 Low	Nil	Low in upland areas and high to very high in coastal plains

TABLE 20.1 : (*Contd.*)

(1)	(2)	(3)	(4)	(5)
III	Coconut and horticulture dominance with fine foodgrains and olericulture farming	Southern plains of Kerala, and south Kerala hills.	Oilseeds (26.75) Horticulture (20.09) Fine foodgrains (17.99) Olericulture* (17.82)	Cattle dominant
IV	Fine foodgrains and beverages dominance with horticulture and coconut farming	North Kerala hills ; northern plains of Kerala, and the extreme southern part of western ghats.	Fine foodgrains (34.31) Beverage (19.18) Horticulture (14.28) Oil-seeds (12.19)	Cattle dominant but cattle-buffalo at places in coastal plains
V	Fine and coarse foodgrains farming	Uttar Pradesh Himalaya ; western and eastern plains of Uttar Pradesh, uplands of Uttar Pradesh.	Fine foodgrains (41.91) Coarse foodgrains (36.46)	Cattle dominant but cattle-buffalo at places in intensive irrigation agriculture
VI	Fine and coarse foodgrains farming trending towards modernization	Punjab plains and North-eastern and eastern Haryana plains.	Fine foodgrains (45.39) Coarse foodgrains (31.07)	Cattle-buffalo dominant but buffalo-cattle at places in areas of transformed agriculture

(6)	(7)	(8)	(9)	(10)	(11)	(12)
33.39 Medium	Negligible Very low	50-65 High	25-35 Medium	Under 3 Low	Nil	Very high
22.32 Low	3.54 Very low	50-65 High	25-35 Medium	Under 3 Low	Nil	High to very high
38.58 Medium	36.94 Medium	Over 65 Very high	25-35 Medium	6-10 High	Negligible	Medium to high
38.25 Medium	54.47 High	35-50 Medium	Under 25 Low	Over 10 Very high	6-15	High to very high

TABLE 20.1 : (*Contd.*)

(1)	(2)	(3)	(4)	(5)
VII	Coarse and fine foodgrains farming	South Kashmir Himalaya ; Himachal Pradesh Himalaya ; western, south-western and southern parts of Haryana plains ; the extreme north of Rajasthan bagar ; Chambal basin ; Aravalli ranges ; east Rajasthan uplands ; Malwa plateau ; Bhind-Marena plains ; Vindhyan scraplands, and central part of Satpura range.	Coarse foodgrains (48.81) Fine foodgrains (28.31)	Cattle-buffalo ; cattle, and cattle-sheep-goat
VIII	Millets dominance with wheat farming	North Kashmir Himalaya.	Coarse foodgrains (75.73) Fine foodgrains (18.23)	Sheep-cattle
IX	Bajra dominance with dry farming	Marusthali ; Rajasthan bagar ; Kutch peninsula, and the extreme northern part of Gujarat plains.	Coarse foodgrains (72.77)	Cattle-camel-sheep-goat-buffalo
X	Groundnut and coarse foodgrains farming	Kathiawar peninsula.	Oilseeds (45.58) Coarse foodgrains (35.32)	Cattle-buffalo
XI	Coarse foodgrains and cotton farming	Gujarat plains save its extreme northern part ; Tapti basin ; Narwada valley ; eastern plateau of Maharashtra, and western part of Satpura range.	Coarse foodgrains (48.02) Industrial crops (28.67)	Cattle and cattle-buffalo

(6)	(7)	(8)	(9)	(10)	(11)	(12)
18.38 Low	20.64 Low	50-65 High	25-35 Medium	Under 3 Low	Negligible	Very low to medium
6.37 Very low	100.00 Very high	35-50 Medium	Under 25 Low	Under 3 Low	Nil	Very low
5.69 Very low	9.83 Very low	35-50 Medium	Under 25 Low	Under 3 Low	Nil	Medium
4.68 Very low	10.25 Low	35-50 Medium	Under 25 Low	Under 3 Low	Nil	Medium
4.70 Very low	5.84 Very low	35-50 Medium	Under 25 Low	Under 3 Low	Nil	Low

(1)	(2)	(3)	(4)	(5)
XII	Coarse foodgrains dominance with groundnut and fine foodgrains farming	Eastern slope of western ghats ; western plateau of Maharashtra ; Karnataka low plateau ; Telangana plateau ; Krishna basin ; Tamil Nadu uplands and Cauvery basin.	Coarse foodgrains (41.48) Oilseeds (16.06) Fine foodgrains (15.29)	Cattle, and cattle-buffalo
XIII	Ragi dominance with dry farming	Karnataka high plateau.	Coarse foodgrains (63.85) Fine foodgrains (12.18) Oilseeds (9.76)	Cattle, and cattle-buffalo-goat/sheep

%Refer to Figure 20.1.

*Fine foodgrains include rice and wheat ; coarse foodgrains comprise bajra, jowar, ragi, maize, gram, and barley ; groundnut, coconut etc, ; industrial crops cover cotton, jute, mesta, sugarcane etc. ; beverages include coffee and

**Figures in brackets are percentage of the total cropped area.

TABLE 20.1 : (*Contd.*)

(6)	(7)	(8)	(9)	(10)	(11)	(12)
8.20 Very low	19.91 Low	35-50 Medium	Under 25 Low	3-6 Medium	Nil	Very low
12.47 Low	18.08 Low	50-65 High	25-35 Medium	3-6 Medium	Nil	Medium

*Olericulture covers vegetables, potatoes, etc. ; horticulture includes tree crops and fruit trees ; oilseeds comprise tea.

West Bengal Basin to include a very large part of the north-eastern and the south-eastern India, with a narrow strip along the western coast.

2. The wheat region occupying most of north-west India practising irrigation agriculture.

3. The millet region comprising mainly the dry Deccan Plateau in the centre of the Indian Peninsula and the arid parts of Rajasthan and Haryana—a dry farming region.

4. The temperate Himalayan region including Kashmir and Himachal Pradesh where olericulture and horticulture crops are as important as cereal crops such as maize and rice.

The agriculture of the broad categories mentioned as well as of the major agricultural systems show an enormous diversity owing to differences of altitude, rainfall and cultural ecology. As such it is possible to have further subdivisions with the help of statistical and field surveys at meso as well as micro levels. These subdivisions will be useful for the formulation of intelligent policies for effective agricultural planning and development.

Wet paddy farming and paddy dominant with coarse food-grains farming are the simplest agricultural systems in India. Over most of the plains (coastal, flood or alluvial), basins, etc. wet paddy cultivation is dominant. Elsewhere in upland districts it occurs in combination with the cropping of coarse foodgrains, particularly pulses. The relative contribution of cattle every-where, but buffalo at places, is mainly as a biological force for ploughing, puddling, transportation, etc. Although, the intensity of irrigation is low or medium, yet sufficiently high and reliable rainfall supplements the artificial irrigation for paddy culture. As the density of cultivators is high, the farming is family based ; labour input is, therefore, high and farming is intensive. Besides, the farm implements used are traditional as is evident from the medium to high density of wooden ploughs, low density of iron ploughs and the absence of tractorization. With the exception of the coastal plains the level of agricultural productivity is generally low to medium. It confirms the fact that the peasantry has not yet adopted modern practices of farming such as the intensive use of biochemical techniques. On the whole these systems may be identified with the 'intensive subsistence tillage with paddy dominance' of Whittlesey.

Coconut, horticulture (dry fruits and condiments), olericulture (vegetables), beverage (coffee) and wet paddy farming as an important phase of the agricultural business occurs only in the wetter section of the southern part of the western coast where soil moisture deficit is negligible. It exhibits the most extreme type of tropical agriculture. On the whole this agricultural system comprises a very small proportion of the geographical area of the nation. Paddy is a wet crop on coastal alluvium, and on uplands and hilly laterite and red loams coconut, dry fruits, condiments, vegetables and coffee cover the extensive stretches. As the density of cultivators is high, so is the labour input. Consequently, the level of agricultural productivity per unit area is between high and very high but is low per capita. Thereupon, farming is intensive and subsistence.

The third group of agricultural systems comprises the areas of fine and coarse foodgrains farming. Except where transformed by an exceptional human element, high intensity of irrigation, high density of iron-based appliances, tractorization and intensive use of bio-chemical techniques, elsewhere it is intensive and subsistence because of very high cultivators' density and traditional inputs. The farmer is near-subsistence and semi-commercial if placed in world perspective but in the Indian context it is commercial. Besides, milch stock is also husbanded in the Punjab-Haryana alluvial plain and Ganga-Yamuna Doab where irrigation agriculture is popular. Still there is scope, with smaller gestation period, for a large number of schemes such as minor irrigation, digging of wells and sinking of tubewells. These schemes would provide immediate relief to the farmers and insulate them from the vagaries of nature, rainfall in particular.

Other groups of agricultural systems neatly illustrate the contrasts arising from a single basic difference in the crop association, a difference owing to dry climate and scarcity of irrigation facilities. The latter alone forces the cultivators to grow coarse foodgrains and oilseeds, groundnut in particular, which are drought-resistant as well as drought-escaping crops. Therefore, the methods of dry farming—a subsistence system—are in practice. In this system where the land lies outside the wet rice or irrigated rice and wheat area, crops such as wheat and rice are either ruled out or hold a negligible proportion of the cropped area primarily owing to intense soil moisture deficit. Besides, a good deal of land is out of agricultural service during part of the year owing to low or very low intensity of irrigation. As a result the percentage of area sown more than once is low or very low. The abiotic and biotic handicaps influence the agricultural systems in subtle but profound ways. For instance, the rice and wheat dominance is substituted by several coarse foodgrains such as *bajra*, *jowar*, *ragi*, maize and gram. These grains are not very productive ones ; they are low value and low-yielding crops. Therefore, the level of agricultural productivity per unit area as well as per capita is low to very low, hence the subsistence level of farming systems. At places the improved institutional ecology (irrigation) and fertile soils have encouraged the production of wheat and cotton. On the whole these systems are identified with 'intensive subsistence tillage without paddy dominance' of Whittlesey.

CONCLUSION

While describing various systems of agricultural production an attempt has been made adequately to illustrate the principles involved in the location of agricultural activities. The ecological and socio-economic factors significantly influence the location and intensity of agricultural systems. Ecological influents in many instances restrict human effort in utilizing land for agricultural activities. Such influents are, however, treated as important elements in the land use dynamics because such conditions determine the range of enterprises that are practicable in a given region, depending on the relative degree of suitability of the regional environment for the particular form of land utilization. On the other hand, the role of socio-economic factors is active because such elements determine as to which of the feasible enterprises the farmer will choose and the input-intensity with which he farms.

On the whole, India is grappling to overcome a deficit between supplies and national subsistence requirements of foodgrains. Successful solutions to these problems may perhaps lie in the relatively more efficient use of available resources by the farming communities throughout the country. The farmers of some parts of India have already moved much ahead and achieved comparatively high levels of productivity. Any further increase in productivity is expected to come from industry-type organizations with increased scale of agricultural activities, high degree of mechanization and higher levels of organization. At the same time, the non-agricultural sectors may also be called upon so as to provide increasing support to the agricultural sector because of ever-increasing and alarming pressure of population on the agricultural land resources, at places even exceeding the 'critical density' of population.

Generally, in most systems of agriculture changes come slowly ; but with the adoption of modern methods in many parts of India changes have occurred more rapidly. In the dry farming regions of India little change has occurred in the continuity of agricultural traditions. Even simple methods are not available to increase productivity per person, per unit of land or per unit of capital invested. But in the worst affected communities agriculture inevitably demands the involvement of the state, both directly or indirectly. Although some efforts have been made to bring about some changes in farming methods, yet increase in population at an unprecedented rate in the last 30 years, with no parallel increase in non-agricultural employment to relieve the increasing rural crowding, has given rise simultaneously to the problem of subdivision and fragmentation, underemployment, low standard of living and malnutrition. These are serious setbacks for agricultural transformation. Unfortunately, farmers in certain parts of India neither have capital nor access to the new inputs which could transform their agricultural activities. They are, therefore, forced by circumstances to resort to the traditional methods in order to increase output, which are not only inadequate but also inefficient.

REFERENCES

Singh, Jasbir, *An agricultural atlas of India : a geographical analysis* (Kurukshetra : Vishal Publications, 1974), p. 13.

Whittlesey, Derwent, S., 'Major agricultural regions of the earth', *Annals of the Association of American Geographers* 26 (1936) : 199-240.

TWENTY ONE

JERZY KOSTROWICKI

THE TRANSFORMATION FROM THE TRADITIONAL TO THE MARKET-ORIENTED OR SOCIALIZED AGRICULTURES AS SEEN FROM THE TYPES OF AGRICULTURE MAP OF EUROPE

THE types of agriculture Map of Europe is a result of the co-operation of a number of European geographers, initiated by the International Geograpical Union Commission of Agricultural Typology in 1964-76 (Kostrowicki 1979). The map is also based on the concept and method elaborated by that commission (Kostrowicki 1966 ; Kostrowicki and Tyszkiewicz 1970, 1979, 1983 ; Reeds 1973 ; Vanzetti 1972, 1975 ; Rakitnikov, Kuzina and Pyshov 1976). Based on that concept and method the first version of the typology of world agriculture was presented for discussion in 1972 (Kostrowicki 1972, 1973), then tested by a number of regional studies, discussed and finally accepted as a basis of further work on types of agriculture in individual countries (Kostrowicki 1980a).

The aim of the commission also was to draw, based on the accepted concept, method and techniques, the types of agriculture maps of individual continents and then the types of agriculture map of the world. For financial reasons the whole work of the commission was considerably delayed. In spite of that, types of agriculture maps were made for a number of countries (Dongmo 1973, 1975 ; Gorbunova, Komleva and Skishkina 1979 ; B.P. Panda 1979 ; Rikkinen 1983 ; Scott 1975, 1983 ; V.R. Singh 1979 ; Stola 1975 ; Szczesny 1979 ; Troughton 1979 ; Tachudi 1983 ; Tyszkiewcz 1975, 1976). This ambitious task could not be accomplished within the 12-year span of the commission's activity.

A number of typological studies of agriculture were also carried out after the commission had been discontinued (Bonnamour 1980; J. Felizola and Diniz 1980; Galczynska 1984; Hill 1982, 1983; Sharma 1983a, b; J. Singh 1983; Stola 1983; Szezesny 1981, a, b; Szyrmer 1984; Tyszkiewicz 1980, 1982). Realising that the best way to convince others that the task could still be accomplished was to show results, it was decided to start working on the Types of Agriculture Map of Europe. To do so, first the statistical data for all countries of Europe had to be collected, processed and typified.

The first draft of the Types of Agriculture Map of Europe, on the scale of 1 : 2.5 million with a number of gaps and inconsistencies, was completed in 1980 and shown to the 24th International Geographical Congress in Tokyo. The second draft was presented to the International Cartographic Conference held in Warsaw in July 1982 (Kostrowicki 1982). The present paper is based on the third and final draft of the map.

To have a better understanding of the map, first the concept and method of agricultural typology has to be briefly summarized (Kostrowicki 1977, 1980a).

According to that concept, the identification of types of agriculture is based on a set of selected variables representing social, operational, productional and structural attributes of agriculture. External conditions, whether natural or otherwise, in which agriculture develops are not included in the typology. They are, however, used to explain why a particular type of agriculture has developed in a given time and place.

The selected variables are all presented in a measurable way by indices showing percentage, proportion, density, size, etc. of individual attributes. As such indices are expressed in various measures, to make them comparable they have to be normalized. This has been done by reducing the value of those indices to five classes representing world ranges of given attributes. The figures representing those classes are arranged in codes. The codes for over 2,000 investigated cases, whether agricultural holdings or such aggregates as agricultural regions or administrative units, are grouped based on their similarity into the types of three orders for which the model codes have been established.

With such model codes, every code representing agriculture of an investigated unit can be compared. If such a code does not differ from any of the model codes by more than 30 per cent of the total variance for the types of the 1st order, 20 per cent of the 2nd order and ten per cent of the 3rd order, they can be identified as being of the type represented by such a model code. In case the difference is more, and there are a number of cases deviating from one or more model codes in the same direction, they can be used for describing a new type of agriculture.

Based on the research conducted to date in various countries of the world, six types of the 1st order, over 20 types of the 2nd order and more than 100 types of the 3rd order have been identified (Kostrowicki 1980a). While it is doubtful whether many more types of the 1st order can still be identified, it is certain that more types of the 2nd order and particularly those of the 3rd order can be detected particularly in countries little investigated to date. But even based on the material collected for the Types of Agriculture Map of Europe a few additional types of the 2nd order and many types of the 3rd order have recently been identified and described.

The following six world types of the 1st order : type E. Traditional Extensive Agriculture ; type L. Traditional Large-scale Agriculture (Latifundia) ; type T. Traditional Intensive (Peasant) Agriculture ; type M. Market Oriented Agriculture ; type S. Socialized Agriculture ; and type A. Highly Specialized Livestock Breeding,* established in the way presented by Kostrowicki (1984) and differing from each other by more than 30 per cent of the total variance, that is, by more than 11 out of the total 27 variables ($27 \times 4 = 108$ $0{,}3 = 32{,}4$) could in different proportion still be identified in Europe, though it is not possible to mark them all on the present scale of the map.

For example, extensive traditional agriculture which in its most intensive form of current fallow, subsistence or semi-subsistent agriculture was dominant in Europe in the Middle Ages and was still being practised in many countries till the 19th century or later, survived the inter-war period only in more remote or underdeveloped areas (Boesch 1966). Even after World War II it was still being practised in south-eastern Europe. It

* The names of the types reflect first of all their most characteristic attributes out of 27 ones, that have been used for their identification.

disappeared with the collectivization of agriculture in Rumania and Bulgaria, but could still be found until the early sixties in north-eastern Poland (Biegajo, 1964) and exists in some parts of southern Yugoslavia even now. Elsewhere this type was replaced either by traditional intensive agriculture or socialized agriculture. On the provisional Types of Agriculture Map, type 'T' appears at present only in Yugoslavia and Poland along with more intensive Traditional Agriculture.

The specialized livestock breeding which in its intensive form is represented by highly industrialized livestock breeding and which is quite common both in western and eastern Europe, does not appear on the map because the basic territorial units of research are too large. On the other hand, in its extensive form it appears in the extreme northern part of Europe, in Norway, Sweden, Finland and the northernmost parts of the U.S.S.R. as reindeer grazing, as well as in the south-eastern margins of the European part of the U.S.S.R. as other forms of livestock grazing.

As a result, three types of the 1st order are now dominant in Europe ; traditional intensive agriculture, market oriented agriculture and socialized agriculture.

Characterized by rather heavy inputs of labour and animal power, low capital inputs and low commercialization, traditional intensive agriculture appears at present both in some less developed capitalist countries and in some socialist countries where private agriculture is still in existence. It is particularly true of Poland and Yugoslavia where private agriculture still covers about 80 per cent of the total agricultural land, and to a much more limited extent of Rumania and Czechoslovakia where it survives only in the mountains. During the last 30 years the development of private agriculture in these countries has been rather slow, following vicissitudes in agricultural policies of the states. Nevertheless certain progress has been achieved.

Traditional intensive agriculture is still quite common in southern Europe from Greece through southern Italy to Spain and Portugal.

It appears on the types of Agriculture Map of Europe in several types of the 2nd and the 3rd order mainly as labour intensive, traditional, mixed agriculture with livestock raising prevalent (Tmk) in southern Poland, parts of Yugoslavia and the Rumanian mountains as well as in northern Spain ; as mixed agriculture in which crop and animal production are almost equal in proportion (Tmk) in central Poland and north-western Yugoslavia; and as mixed agriculture with crop production dominant (Tmg, Tmc, Tmf) in Greece, southern Yugoslavia, southern Italy, Spain and Portugal, etc.

In various transitional forms between traditional and market oriented agriculture (Mmz, Mmg) it has also been identified in some less developed, mainly in the mountainous areas of various countries such as the Massif Central in France, the Austrian Alps, the Scandinavian countries and elsewhere.

From the typological studies that have been made in some of these countries such as Austria, Finland, Norway (Stola 1975; Szczesny 1979), it can be concluded that these transitional forms are gradually disappearing and are being substituted by similar forms of market oriented mixed agriculture with livestock raising dominant (Mmw, Mma). At the same time in southern and eastern Europe the area of traditional agriculture is being reduced and substituted by various types of market oriented agriculture (Mmr, Mmz, Mmh, Mmp, Mmf, Mln). With growing capital inputs and decreasing inputs of labour and animal power, more capital intensive, more productive and more commercial types of agriculture are expanding. These are still mostly mixed types of agriculture, either with livestock raising dominant (Mma) such as in most countries of northern Europe; fully mixed agriculture (Mmm) in western Poland, central Germany, Denmark, southern Sweden and Norway, southern-eastern England, French Brittany etc.; and mixed agriculture with crop growing prevalent (Mmr, Mmf) in central France, parts of Italy, Spain and Portugal. The most intensive types of mixed agriculture (Mmi, Mmv) are characteristic of Netherlands, northern Belgium (Flanders) and parts of northern Italy (Lombardy). At the same time in parts of central France and elsewhere more specialized large-scale farming (Mlm, Mlc) is appearing. However, because of the scale of the basic units of study, growing specialization of west European agriculture could not be observed on the map except in the south —in southern France, parts of Italy and southern Spain—where larger and larger areas are being covered by more labour intensive market oriented agriculture, more or less specializing in fruit or vine growing (Mif).

In Spain considerable areas, particularly in the central parts of the country, are occupied by large-scale, rather extensive (fallow), more or less market-oriented agriculture of the transitional character between the old traditional latifundia type and the market oriented large-scale agriculture type (Mel). Simiiar types of agriculture have also been identified in Eastern Portugal. As a result of the land reform following the so-called 'Carnation' Revolution of 1974, in southern parts of Portugal some forms of socialized agriculture had been established at that time as collective farms organized by peasants who spontaneously occupied former landed estates. In spite of their different social character, the operational, productive and structural attributes of these farms do not differ much from the surrounding extensive large-scale private farming. In the present day socialist countries of Europe the transitional paths from traditional agriculture to the socialized one have been different.

In the U.S.S.R., before the Revolution of 1917, traditional agriculture dominated the whole empire with extensive traditional agriculture practised not only in the southern semi-arid and arid lands, but to a considerable extent also in northern Russia. Along with peasant villages, liberated from serfdom only in the middle of the 19th century, there existed numerous landed estates that, in spite of their rather traditional agriculture, supplied most of the agricultural products to the home market and for export. The collectivization of agriculture in the thirties entirely transformed this picture.

The small plots of land belonging to the peasant villages were merged to form collective farms, while former great landed estates were converted into state farms. Later some of the collective farms were merged with or transformed into state farms. Gradually the differences between collective and state farms decreased to form one type of agriculture of the 1st order—socialized agriculture. This is essentially large scale, highly mechanized, commercial agriculture with low labour inputs, varying intensity and productivity and various orientations. Based on these differences, that reflect both different natural, climatic or soil conditions, degree of development, distances from the market etc. several types of the 2nd and 3rd order have been identified.

In the semi-arid or arid south-east, various types of extensive, highly commercial agriculture specialized in grain crops (Scc), or grain crops with subsidiary livestock raising (Scm) have been identified, as well as some intensive agriculture specialized in cotton growing on irrigated land (Sac).

In the central and northern parts of the U.S.S.R. mixed agriculture (Sm) predominates, characterized by decreasing productivity and growing proportion of animal products towards the north (Sma) and increasing productivity and growing proportion of crop products (Smu), towards the south, particularly in south-west Ukraine.

Around big cities this mixed type of agriculture is substituted by a more intensive type of mixed agriculture in close proximity with socialized horticulture (Sh), which, however, could not be shown on the map owing to its scale.

More labour intensive types (Shf, Ssc) of agriculture specialized in fruit, grapes or tea production have been identified in some parts of the Caucasus and Crimea.

The transition from traditional to socialized agriculture was different in the socialist countries of east-central Europe. In the inter-war period, except eastern Germany and western parts of Czechoslovakia and Poland, where market oriented agriculture has already been developed, commercial production was mainly supplied by the landed estates that were particularly numerous in Hungary and Poland. In spite of the fact that all forms of villein services of the peasants were abolished in these countries by the end of the 18th or the middle of the 19th century at the latest, some of the landed estates particularly in the east still practised traditional agriculture, not differing much from feudal latifundia. Only in eastern Germany, western Poland and western Hungary had they been transformed into large-scale market oriented farms characterized by lower labour but higher capital inputs, higher labour productivity and commercialization. Most of the peasant farms of these regions, however, continued to practise traditional, intensive agriculture.

A few years after the post-World War II land reforms which either distributed the landed estates among peasants or converted them into state farms, agriculture in most countries of east-central Europe was collectivized, following the Soviet model. Only in Poland and Yugoslavia does socialized agriculture not cover much more than 20 per cent of agricultural land.

The development of socialized agriculture was fastest in the

German Democratic Republic (GDR), Hungary and Bulgaria, and most retarded in Rumania where the incipient forms of socialized agriculture (Se), typical of the transitional period after collectivization, could still be identified.

The development of socialized agriculture in these countries is mostly owing to the heavy capital inputs that resulted in growing intensity and productivity without much change in agricultural orientation. At present the most intensive and productive mostly mixed agriculture with livestock raising prevalent (Smd) is dominating in GDR and covers considerable areas in Czechoslovakia, Hungary and western Poland. In Bulgaria a quite intensive mixed agriculture with crop growing prevalent (Smu) is developing, but this is much less common in Rumania.

A more specialized agriculture mainly in fruit trees and vine growing (Ss or Sh) has also developed in some parts of Bulgaria, Ramania and Hungary, which, however, because of the scale seldom could be marked on the map. Many Yugoslavian socialized farms, usually small, have also specialized in fruit or grape production. A labour intensive and highly productive agriculture specializing in tobacco growing combined with extensive livestock grazing (Sgt) has been reported from the mountains of southern Bulgaria.

In view of the changing pattern of agricultural development in Europe, the question about future trends may naturally arise. Some answers could be obtained by the use of typological methods. This would require a study of past trends as reflected in the distribution of various types of agriculture during the last two or three decades. Projection of such tendencies to the future would show what could happen if the past tendencies and the rate of development are continued. The plan for future development could then be formulated in such a way that the unfavourable tendencies of the past are eliminated and favourable ones strengthened. Comparative studies of agricultural development in countries or regions that have already achieved a higher level under comparable external conditions, could be very useful in chalking out future developmental programmes.

In fact such studies have already been made for Poland (Kostrowicki 1975, 1976a). The conclusions drawn from that study and extrapolation of the past tendencies in the development of private farming have been compared with the results of similar studies relating to Denmark, northern part of the German Federal Republic, Belgium and Austria. The comparison reveals that the set of variables characterizing, for example, Belgium agriculture in 1950 were substantially similar to those characterizing Polish private agriculture in 1970. It could, therefore, be concluded that Poland is about 20 years behind Belgium in agricultural development (Stola 1983). Similarly, the agricultural development gap between southern Poland and Austria is about 15 years (Szczesny 1979).

Eliminating those aspects of the external environment that could not be reproduced, conclusions were drawn as to the suitability of foreign models for application to the problems of spatial planning of agricultural development in Poland.

Based on such assumptions and a careful study of external conditions of agricultural development such as natural conditions, location in relation to a market etc. models for the future, desirable types of agriculture for a given area could than be proposed and the process of transition from the present types to the desirable future types be outlined. Such plans would only be worthwhile as academic exercises unless they were implemented. For this, however, geographers could not be made responsible unless they became planners themselves.

REFERENCES

Biegjo, W., 'The Ways of transition from the three I-I field system to the modern farming as currently observed in Poland's underdeveloped region of Biaystok', *Geographia Polonica* 2 (1964) : 153-58.

Boesch, H. and M. Bronhofer, 'The decline of the Dreizelgen-system in North-eastern', in *Comptes Rendus du XVIII-e Congres International de Geographie*, Vol. 4, Rio de Janeiro, 1956, pp. 21-25.

Bonnamour, J. and Ch. Gillette, *Les types d'agriculture en France, 1970. Essai methodologique* (Paris : Centre National des Recherches Scientifiques, 1980).

Dongmo, J., 'Typologie de l'agriculture camerounaise. d'application a l'Afrique Noire de la methode . . . du professeur Kstrowicki', *Annales de la Faculte des Letters et Sciences*, Universite de Yaounda 5 (1973) : 19-40.

Dongmo, J., Typologie de l'agriculture camerounaise', in *Agricultural Typology and Land Utilization*, ed., C. Vanzetti, Verona : 1975, pp. 115-45.

Felizola, Diniz, J.A., Tipos de agricultura em sergipe, *Cadernos Sergipanos de Geografia*, no. 11, 1980.

Gorbunova, L.I. ; Komleva, M.V. and Skishkina, L.V., 'Agricultural typology of the USSR', *Geographia Polonica* 40 (1975) : 83-92.

Grigg, D.B., 'Agricultural regions of the World, Review and reflections'. *Economic Geography* 45 (1969) : 95-132.

Hill, R.D., *Agriculture in the Malaysian Region*, Budapest (1982).

Hill, R.D., 'The Malaysian Region and the World typology of agriculture'. *Geographia Polonica* 46 (1983) : 21-48.

Kostrowicki, J., Tipologia geografica de la agriculture mundial : Principos y Metodos (Union. Geogr. International Conferencia Regional Latino-americana. Mexico DF. Vol. 2, (1966) pp. 793-807.

Kostrowicki, J., 'A preliminary attempt at a typology of world agriculture', in *International Geography*, Vol. 2, eds. W.P. Adams, and F.M. Helleinter (Montreal : 1972) : pp. 1097-1100.

Kostrowicki, J., 'The typology of world agriculture. A preliminary scheme', in *Agricultural Typology and Land Use*, ed., L. Reeds (Hamilton, Ontario : 1973), pp. 2-52.

Kostrowicki, J., 'The Typology of World Agriculture. Principles, Methods and Model Types'. Mimeographed, Warsaw, 1974. Reprinted in *Agricultural Typology and Land Utilization*, ed., C. Vanzetti (Verona : 1975), pp. 429-79.

Kostrowicki, J., 'An attempt to apply typological methods for forecasting and/or programming further changes in spatial organization of agriculture', in *Agricultural Typology and Land Utilization*, ed., C. Vanzetti (Verona : 1975).

Kostrowicki, J., 'World Types of Agriculture', Warsaw (mimeographed), p. 49.

Kostrowicki, J., 'Agricultural typology as a tool in planning spatial organization of agriculture.' *Geoforum* 7 (1976a) 245-50.

Kostrowicki, J., 'Agricultural Typology. Concept and Method'. *Agricultural Systems* 2 (1977) : 33-45.

Kostrowicki, J., 'Twelve years' activity of the IGU Commission on Agricultural Typology', *Geographia Polonica* 40 (1979) : 235-53.

Kostrowicki, J., 'A hierarchy of world types of agriculture', *Geographia Polonica* 43 (1980) : 125-62, Reprinted in *Perspectives in Agricultural Geography*, Vol. 1, ed., Noor Mohammad (New Delhi : Concept Publishing Company, 1981), pp. 165-208.

Kostrowicki, J., Geografia dell agricultura. Ambiente, societa, sistemi, dell agricoltura. Milano, 711 p. F. Angeli editore.

Kostrowicki, J., 'The Types of Agriculture Map of Europe', *Geographia Polonica* 48 (1982) : 79-92.

Kostrowicki, J., 'Types of Agriculture of Europe. A Preliminary Outline'. *Geographia Polonica* 50 (1984) (in print).

Kostrowicki, J. and Tyszkiewicz, W., 'Essays on Agricultural Typology and Land Utilization', *Geographia Polonica* 19 (1970).

Kostrowicki, J. and Tyszkiewicz, W., 'Agricultural Typology', *Geographia Polonica* 40 (1979).

Kostrowicki, J. and Tyszkiewicz, W., 'Agricultural Typology and Rural Development', *Geographia Polonicu* 46 (1983).

Panda, B.P., 'Agricultural types in Madhya Pradesh', *Geographia Polonica* 40 (1979) : 133-56.

Rakitnikov, A.N. ; Kuzina, I.M. and Pyshov, W.G., The 'problem of agricultural typology', Commission on Agricultural Typology. XXIII International Geographical Congress. Odessa (1976).

Rikkinen, K. and Vapaaoksa, T., 'The application of world agricultural typology to Finland', *Geographia Polonica* 46 (1983) : 93-106.

Reeds, L.G., *Agricultural typology and land use* (Hamilton, Ontario : 1973).

Scott, P., 'The application of world agricultural typology to Australia', in *Agricultural Typology and Land Utilization*, C. Vanzetti (ed.) (Verona : 1975).

Scott, P., 'The typology of Australian Agriculture', *Geographia Polonica* 46 (1983) : 7-20.

Sharma, B.L., 'A typological analysis of agriculture in the Rajasthan state', *Geographia Polonica* 46 (1983a) : 71-77.

Sharma, B.L., *Agricultural Typology of Rajasthan* (Udaipur : publisher, 1983b).

Singh, J., 'A typology of agriculture : The Indian experience', *Geographia Polonica* 46 (1983) : 49-70.

Singh, V.R., 'Agricultural typology of India', *Geographia Polonica* 40 (1979) : 113-32.

Stola, W., 'Changements dans less types de l'agriculture belge dans les annees 1950-70 in ed. C. Vanzetti.

Stola, W., 'Essai d'application des methodes typologiques a l'etude comparee sur le developement des agricultures belge et polonaise,' *Geographia Polonica* 46 (1983) : 159-74.

Szc esny, R., 'Changing types of Austrian Agriculture 1960-70', *Geographia Polonica* 40 (1979) : 161-70.

Szczesny, R., Transformations of agricultural types in Poland 1970-76 in *Perspectives in Agricultural Geography*, ed., Noor Mohammad (New Delhi : Concept Publishing Company, 1981a), pp. 209-28.

Szczesny, R., 'Typy rolnictwa Szwajcaril (Sum : Agricultural types in Switzerland), *Przeglad Geograficzny* 54 (1982) : 511-32.

Szyrmer, J., 'Essai de la typologie de l'agriculture autogeree algerience', *Geographia Polonica* 50 (in print).

Tschudi, Aa, B., 'Types of agriculture in Norway by the typogram method', *Geographia Polonica* 46 (1983) : 83-91.

Troughton, M., 'Application of the revised scheme for the typology of world agriculture to Canada', *Geographia Polonica* 40 (1979) : 95-112.

Tyszkiewicz, W., 'Types of agriculture in Poland as a sample of the typology of world agriculture', in *Agricultural Typology and Land Utilization*, ed., C. Vanzetti (Verona : 1975) pp. 391-408.

Tyszkiewicz, W., 'Agricultural typology of the Thracian basin. Bulgaria, as a case of the typology of world agriculture', *Geographia Polonica* 40 (1976) : 171-86.

Tyszkiewicz, W., 'Types of agriculture in Macadonia as a sample of the typology of world agriculture', *Geographia Polonica* 43 (1980) : 163-85.

Tyszkiewicz, W., 'Zastoscwante metcd typolcgicznych do badan rolnictwa Szeecji (An application of typological methods to the study of Swedish agriculture)', *Przeglad Geograficzny* 53 (1982) : 551-70.

Vanzetti, C., *Agricultural Typology and Land Utilization* (Verona : 1972), pp. 1-448.

Vanzetti, C., ed. *Agricultural Typology and Land Utilization* (Verona : 1975), pp. 1-498.

TWENTY TWO

HORST G. MENSCHING

THE ROLE OF GEOGRAPHY AND GEOGRAPHERS IN THE FIELD OF RESOURCE MANAGEMENT IN DRYLANDS
Theory and Practice

GEOGRAPHY in its widest sense embraces the complex study of geo-factors (e.g climate, landforms and soils, water balance, vegetation as natural factors and all human factors deriving from the activity of man) and the interaction between them. Geography is concerned with the significance and differentiation of all these factors or components within the relationship between 'Man and the Biosphere' (MAB), in their local and regional distribution.

Modern geography has integrated into its research the investigation of the 'dynamic processes' forming the natural landscapes, that at the same time are used and transformed through human impact in different ways and different periods. Consequently, geography has to take part in the studies and discussions on environmental problems, the solution of which is of such fundamental and paramount importance for our world today. The degradation of ecosystems and natural resources in most climatic zones all over the world today is going on in such a way as to upset the ecological balance and possibly the climatic conditions on the continents.

Geography must be aware of this problem, and recently geographers have done extensive work on it. In particular the severe drought period in the Sahelian zone of Africa between 1969 and 1973 and during the following years right up to today has triggered many studies in this field. This is also reflected in the activities of the International Geographical Union (IGU).

Shortly after the Sahel drought the IGU 'Arid Zone Commission' changed its aims and objectives to a more applied direction of studies and discussions, namely, the study of the world-wide problem of 'desertification'. The IGU Working Group on 'Desertification in and around Arid Lands' was followed in 1980 by the Working Group on 'Resource Management in Drylands'.

This working group's first meeting was held in Egypt in 1982 and discussions then focused on the problem of irrigated land management in traditional and modern areas of the old Nile Valley and the so-called 'New Valley' in the Libyan desert. The 1983 meeting was held in India and focused on a comprehensive study of resource management problems in the Rajasthan dry region. These experiences revealed that the question of what the geographical contribution to the 'management' of natural resources actually means is open to further discussion and needs to be more clearly defined.

The most important issue is that of land use and misuse, the exploitation of natural resources in drylands. Within the scope of this issue it seems, however, that one problem has been seriously neglected. There is a lack of studies illuminating the difficult boundary area between the natural processes and the anthropogenic processes in the same field.

Erosion, for example, is a very common morphological process in all arid and semi-arid landscapes. Human impact through numerous land use practices can, however, tremendously accelerate such processes, if management does not take this into consideration. Yet where is the line between natural and anthropogenic processes to be drawn ? There are few studies on this problem. But they are essential for the important decisions as to what can and must be done against a wrong or not suitably adapted management in dryland land use systems.

Studies on water management are another important field of applied dryland geography. One of the most significant impacts on available resources in drylands derives from water use. Therefore, geographical research must be concerned with the detailed investigation of water resources and water use. Fossil water supplies, for example, are natural deposits that are eventually depleted, if their exploitation exceeds their recharge. Geographers are called upon to investigate the connection between available surface and groundwater supplies and the meth ods of their use, especially since the employment of modern technology in water extraction and distribution is connected with many dangers of misuse.

More attention must be paid to all human activities influencing the ecosystem within the man-biosphere relationship, for example, through all kinds of degradation of the vegetation cover, the soils and the whole water balance, including the increase of evaporation from surface water and soil moisture deriving from anthropogenic influences. Social, economic, political and ethnic, even administrative activities can all contribute to ecological change for better or for worse.

Geographers must concentrate their research on the complex interdependencies between the natural resources and man's activities. This is their field of specialization, completely apart from strictly sociological, economic or political issues. The questions of natural resources management, that is the focus of the present IGU drylands working group, centre around the interactions between the environment and the human impact on it.

The value of such geographical work in practice is considerable. My own experience with development projects in dry areas in western and eastern Africa, as well as south-east Asia has been extensive. In all these development projects geographers are working very successfully. This success is only guaranteed, however, if the aspect of 'integrated' development is kept in mind, and if the implementation of development planning pays attention to the interaction of all geographical factors, natural and human. Therefore, it is absolutely necessary to analyse the processes and the effectiveness of specific factors not by themselves, but by their efficacy in relation to all others within the whole ecosystem.

These tasks assigned to modern geography are closely related to the practical needs of modern development and planning decision making, which to be successful must be based on the sound knowledge of ecosystems and human patterns of interaction with them. This comprehensive knowledge of natural, economical and social factors helps planners make better decisions than is possible from a merely technical standpoint.

Professor Anas, to whom these brief remarks are dedicated in kind memory, especially with respect to our last discussions at the meeting of the IGU Working Group on Resource Management in Drylands in Delhi in March 1983, passionately stood up for the realization of these modern geographical aims. His active participation in the meeting and his lively chairmanship of the subsequent colloquy at the Aligarh Muslim University shortly before his death will remain memorable for all fortunate to be there.

SECTION SIX

PHYSICAL GEOGRAPHY

TWENTY THREE

A.B. MUKERJI

LANDSLIDES AND MAN
Development and Destruction in Western Himalaya

INTRODUCTION

ECOLOGY neutral development processes are in the ultimate analysis destructive to both ecology and society. Destruction is most widespread and intensive, both to natural and cultural landscapes, in a fragile environment. The Himalaya provide us with an exceptionally instructive example of such a tropical mountain environment. The most significant element destroyed, significant for both nature and man, is the slope. Slope damages constitute the most widely recognized and discussed phenomena among the natural hazards encountered by human communities in the Himalaya.

Landslides, avalanches and other types of mass movements are the most widespread and most frequent among the changes in the geomorphic landscape of the hills and mountains and they form the core of geomorphic damages. All geomorphic landscapes can be resolved into slopes, 'the lowest common denominator' as Butzer (Butzer 1976) describes them. All geomorphic damages in the mountains are really slope damages.

The objectives of the paper are twofold, (i) to analyse the regional and temporal patterns of landslides and avalanches, and (ii) to identify the human dimensions.

HYPOTHESIS

The basic hypothesis of the paper is that most geomorphic

damages like many natural hazards are caused, in fragile environments and settled areas, by rapid developmental activities. It follows that these damages become more frequent and concentrated during the times of intense human activities and in the areas where nature is sought to be destructively transformed.

In the western Himalaya (Fig. 23.1), a ravaged ecological setting, the geomorphic damages would reveal a striking increase in number and intensity of destruction during the post-independence era and in the lower and middle elevation zones, both related to development activities. Also, the era experienced the

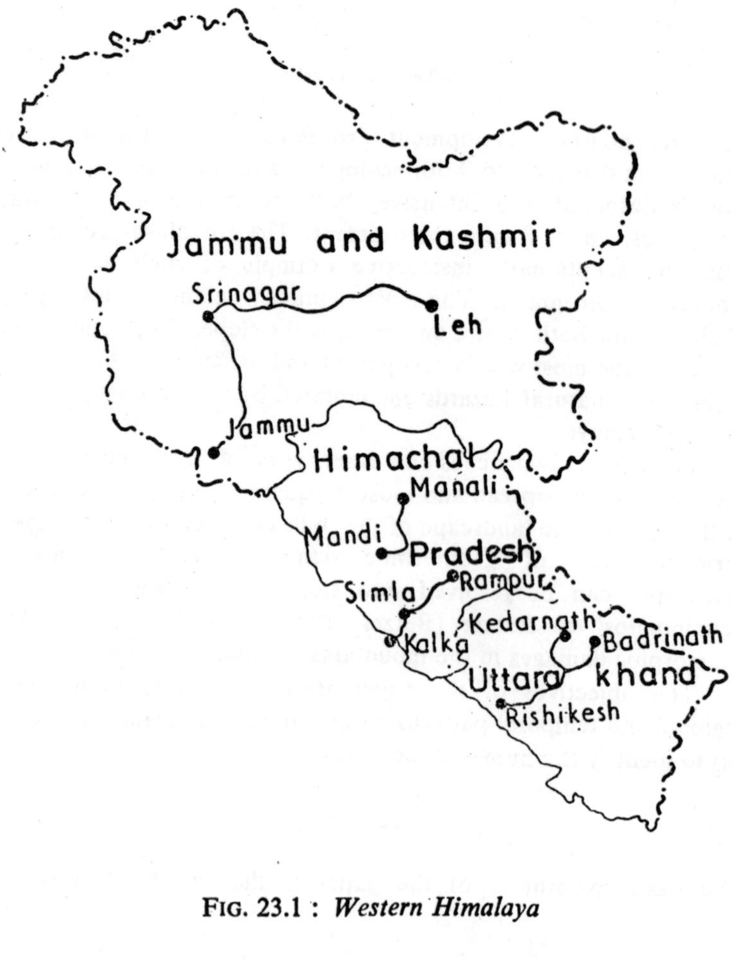

FIG. 23.1 : *Western Himalaya*

effects of the implementation of the five-year development plans. The area receives heavy and high intensity monsoon rainfall and is composed of weak rocks.

METHODOLOGY

From the reports of the residents and observations of the landscape it is known that a larger number of landslides and avalanches have occurred than reported in the national and regional newspapers. However, most of the occurrences along the highways and the main roads and those that have caused large damages to life and property, and some of those located at distances from the highways in the interior and less accessible areas are also reported. These reports cover the period 1957-81, during which six five-year development plans were implemented, and are abstracted from two national, daily newspapers, *the Hindustan Times* and *the Times of India*, and two regional, daily newspapers, *the Tribune* and *Indian Express*. Each day's issue of all the four newspapers were scanned. These reports have provided significant data on the spatial and temporal distribution and causes of landslides and related damages and provide a pointer to the assessment of man's role in generating them.

The landslides and avalanches reported for four national highways, Rishikesh-Badrinath (275 km.), Kalka-Rampur (200 km.), Mandi-Manali (100 km.), and Jammu-Leh (650 km.) have been analysed to illustrate the basic theme of the study (Figs. 23.2 23.3, 23.4 and 23.5).

The roads extend through the following Himalayan zones : Outer Himalaya, Lower Himalaya, Lesser Himalaya, Greater Himalaya, and Trans-Himalaya. The zones are differentiated among themselves in terms of location with reference to the north Indian plains, elevation, lithological formation, climate, vegetation, soil, hydrography and types and patterns of human occupance. The first three roads traverse the first four zones while the Jammu-Leh road cuts through all the five zones.

Lithological formation with its internal micro-structures, elevation zone, and the form, nature and amount of precipitation—the former an internal and the latter two external—emerge as the dominant natural determinants of the site and regional pattern of landslides and avalanches.

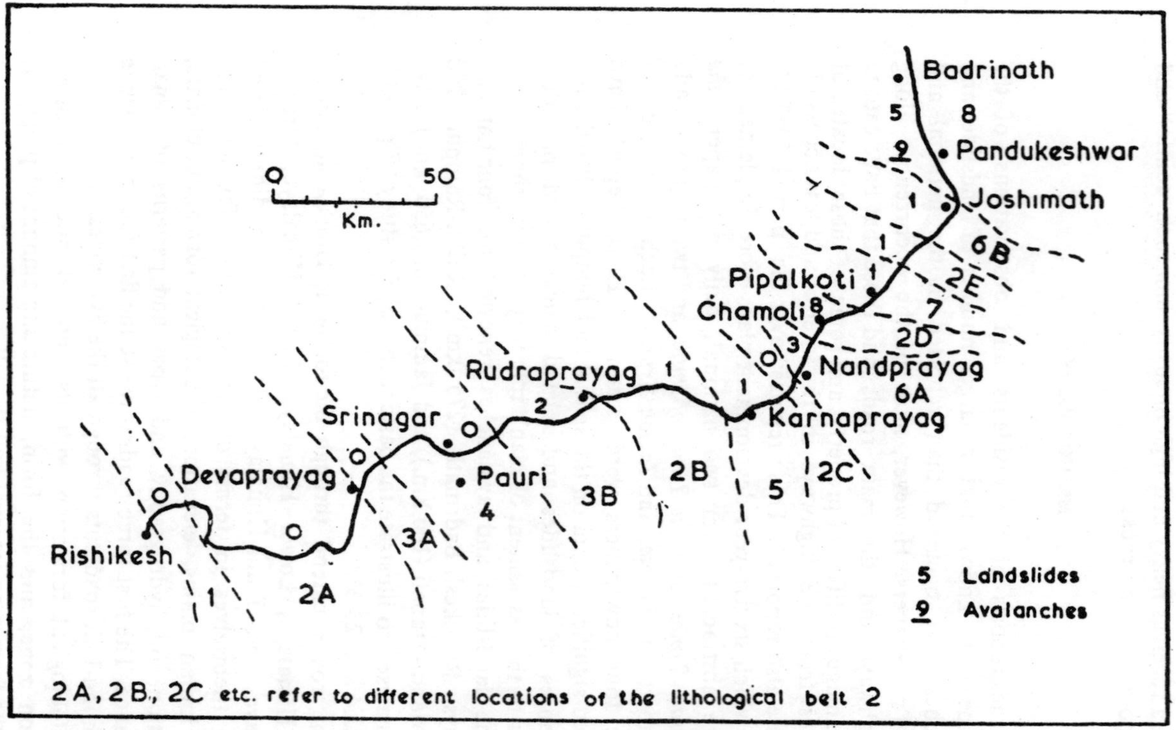

1 Krol (list, shales, sst.)
2 Nagthat (orthoquartzite, shales)
3 Damta (grey-wacke, slumped beds)
4 Chandpur (shales, slates, quartzite [black])
5 Basics (intrusive, extrusive)
6 Almora Group (granite, gnelss)
7 Tejam Group (phyllite, slote, blue color limestone)
8 Joshimath formation (gneiss quartzite)

FIG. 23.2: *Rishikesh—Badrinath highway. lithological zones*

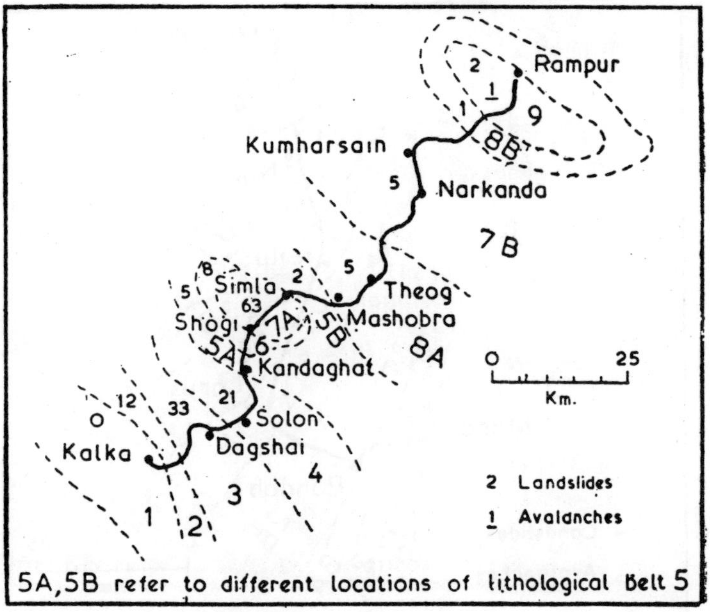

FIG. 23.3 : *Kalka-Rampur highway litholagical zones*

1 **Alluvium**
2 **Lower Siwalik (grey-wack, conglomerate)**
3 **Subathu/Murree (red sh., list)**
4 **Krol (shales, list, sst.)**
5 **Nagthat (orthoquartzite, shales)**
6 **Blaini (siltstone, slates)**
7 **Low Grade (quartzite, slate, phyllonites)**
8 **Basantpur (black sh., basics)**
9 **Basics (Intrusive, extrusive)**

For the purposes of correlation, description and explanation of distribution of landslides. the roads have been divided into their constituent elevation and lithological zones, each experiencing different frequencies of landslides.

Elevation associations

The four highways have their origins and destinations at different elevations and, therefore, the range of height through which they extend is different (Table 23.1). The second and the fourth

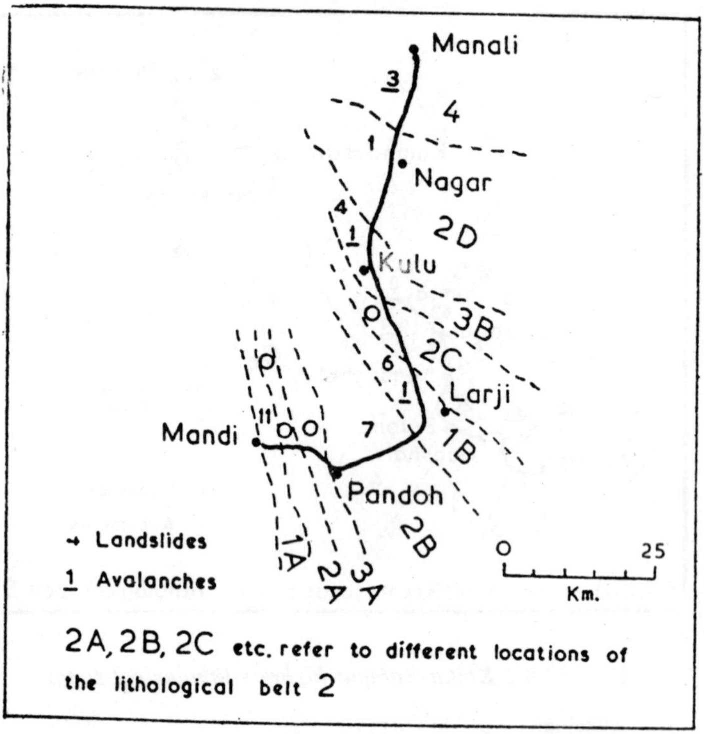

FIG. 23.4 : *Mandi-Manali highway lithological zones*

1. Larji Shali Formation (lime stone, dolomite)
2. Salkhala Formation (schist, quartzite)
3. Manjir Formation (conglomerate)
4. Central Gneiss Formation (gneiss, schist)

roads—the former having a high origin and the latter a high destination—have experienced a larger number of landslides. The Jammu-Leh road which has a much longer traverse in the Central and the Trans-Himalayan zones have recorded the largest number of avalanches. On the other hand, the Mandi-Manali road originating at a high elevation has a higher linear density (Table 23.2). The highest density of 79 is suggestive of the instability and man's destructive role in the natural-ecological system along the Kalka-Rampur road. This highway has been widened to

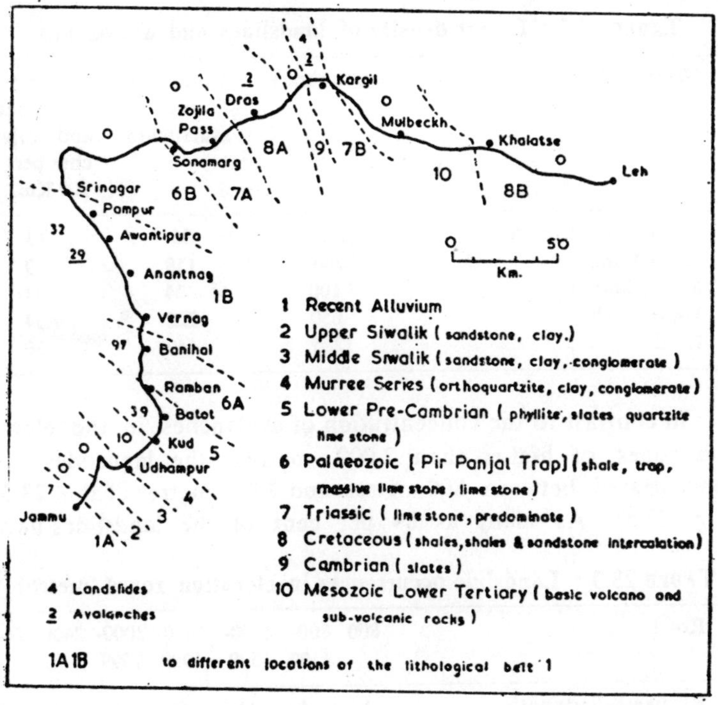

FIG. 23.5 : *Jammu-Leh highway lithological zones*

TABLE 23.1 : **Topographic range of highways and number of landslides and avalanches (Heights in metre)**

	Origin	Height	Destination	Height	Range of height	Landslides and avalanches
1.	Rishikesh	348	Badrinath	3096	2748	35 (26+9)
2.	Kalka	580	Rampur	1007	327	158 (157+1)
3.	Mandi	754	Manali	1926	1172	34 (29+5)
4.	Jammu	366	Leh	3514	3148	222 (189+33)

(The first and second figures in the parenthesis are respectively for landslides and avalanches.)

permit two-way traffic and for most of its length it is two-lane movement line.

TABLE 23.2 : **Linear density of landslides and avalanches**

	Name of the road	Length (km.)	Landslides and avalanches	Number of landslides and avalanches per 100 Km.
1.	Rishikesh-Badrinath	275	35	13
2.	Kalka-Rampur	200	158	79
3.	Mandi-Manali	100	34	34
4.	Jammu-Leh	650	222	34
	Total	1225	449	36

In contrast to the concentration of avalanches in the elevation zones of higher than 2,000 metres, the landslides are concentrated between 800 metres and 2,000 metres (Table 23.3) (Fig. 23.6). As many as 84 per cent of the landslides have

TABLE 23.3 : **Landslide occurrences in elevation zones (metre)**

	Road	800	800-1199	1200-1599	1600-1999	2000-2399	2400	Total
1.	Rishikesh-Badrinath	1	5	15	0	1	4	26
2,	Kalka-Rampur	0	29	66	13	48	1	157
3.	Mandi-Manali	0	18	11	0	0	0	29
4.	Jammu-Leh	3	63	97	15	5	6	189
	Total	4	115	189	28	54	11	401

occurred in the lower elevation zones that constitute the locale of heavy, high intensity monsoon rainfall and of weak, sedimentary formations of the Lower Himalaya. Also, the most devastating effects of human occupance have been experienced here.

There is a striking increase in the occurrence of avalanches with elevation as is clearly indicated by the Jammu-Leh road and the Mandi-Manali segment (Table 23.4). Even on the former road as many as 31 of the total of 33 avalanches have occurred at elevations of more than 1,600 metres, the zone above 2,000 metres having recorded more than 26. Undoubtedly this elevational concentration is related to heavy winter snowfall in the Srinagar—Leh and Kulu-Manali segments where the slopes, typical of the Trans-Himalayan zone, are bare of vegetation and soil and

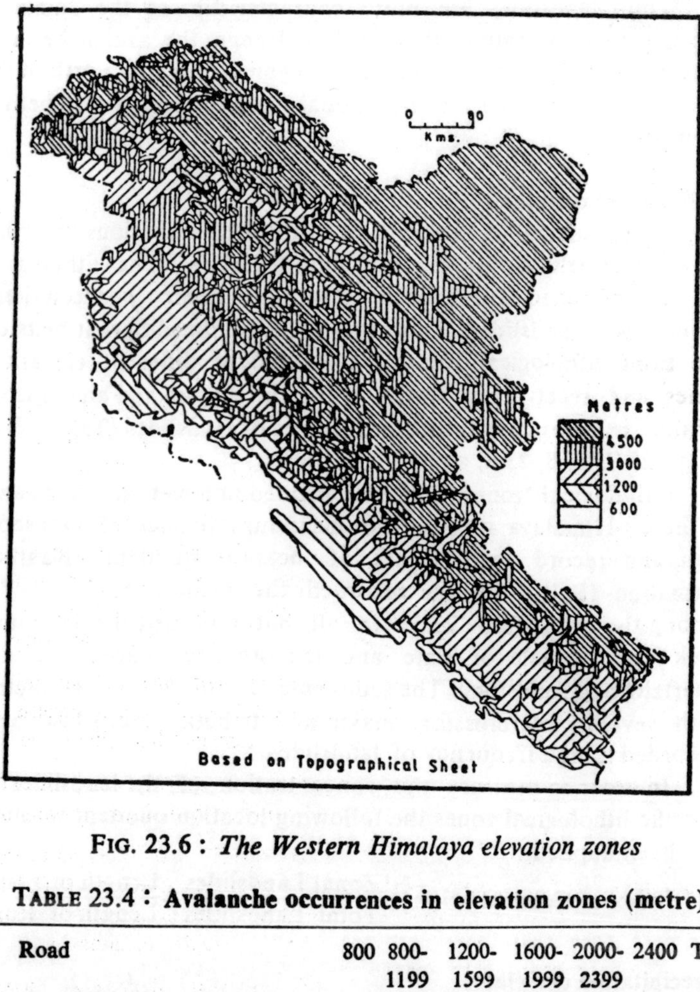

FIG. 23.6 : *The Western Himalaya elevation zones*

TABLE 23.4 : **Avalanche occurrences in elevation zones (metre)**

	Road	800	800-1199	1200-1599	1600-1999	2000-2399	2400	Tota
1.	Rishikesh-Badrinath	0	0	0	0	0	9	9
2.	Kalka-Rampur	0	0	0	1	0	0	1
3.	Mandi-Manali	0	0	2	3	0	0	5
4.	Jammu-Leh	0	1	1	5	11	15	33
	Total	0	1	3	9	11	24	48

intense mechanical weathering produces extensive grus cover. Likewise all the nine avalanches of the Rishikesh-Badrinath road occurred at elevations of more than 2,400 metres.

While landslides are more characteristic of the Jammu-Srinagar section south of the Zanskar Range the avalanches are concentrated in the Srinagar-Leh segment running north of it. The former is related to monsoonal rainfall, the latter to heavy snowfall.

Lithological correlates

There are several repetitions at different elevations of the lithological belts across the highway traverses. The lithologies are distinguished from each other by their characteristic proneness to landslides. This explains why the different belts of the same lithological formation located at different elevation zones and receiving different types, intensities and amounts of rainfall, experience different frequencies of landslides (Table 23.5) (Figs. 23.7, 23.8, 23.9, and 23.10).

Lithological zones that are located at lower elevations and in the cis-Himalaya receive the maximum impact of monsoon rain and record high values of location quotient. Nagthat formation (Belt 2 D) has recorded the highest values of 7.9 among the belts (Table 23.5). In all but one belt the dominant rock type is shale or slate and the other types are limestone, quartzite and sandstone. The sedimentaries and the zones afflicted with several intercrossing, major and tributary faults have also recorded a high frequency of landslides.

In order to measure the concentration of the landslides in specific lithological zones the following location quotient formula has been applied :

$$\text{Index of Concentration} = \frac{\text{Zonal Landslides}}{\text{Total Landslides}} \div \frac{\text{Length of Zone}}{\text{Length of Road}}$$

Precipitation correlates

Although precipitation is highly variable in the region, three basic tendencies relevant to the understanding of the regional patterns of landslides and avalanches must be noted. First, precipitation is mainly in the form of rain south of the Central Himalaya and in the form of snow in the Trans-Himalaya. Second, precipitation in the former region increases with elevation till the peak belt extending over the Central Himalaya is reached. Third, while snow is almost entirely concentrated during winter, rain is heavily concentrated, about 70 to 80 per cent of the

TABLE 23.5 : **Lithological formation located at different elevation zones**

Road	Lithological Zone	Elevation Range (metre)	Zone Length/ Road Length	Zonal Slides/ Road Slides	Location Quotient
1. Rishikesh—Badrinath	Nagthat 2D (Shale, Orthoquartzite)	1200—2000	12/275	8/23	7.9
2. Kalka—Rampur	Subathu—Murree 3 (Shale, Limestone)	1200—1400	30/200	33/157	1.4
3. Kalka—Rampur	Krol 4 (Shale, Limestone, Sandstone)	1400—1600	17/200	21/157	1.6
4. Kalka—Rampur	Low Grade 7A (Shale, Quartzite)	1600—2100	17/200	63/157	5
5. Mandi—Manali	Larji Shali 1A (Limestone, dolomite)	1000—1300	5/100	11/29	7.4
6. Jammu-Leh	Lower Precambrian 5 (Slate, limestone, quartzite, phyllite)	800—1900	45/650	39/189	2.8
7. Jammu-Leh	Palaeozoic 6A (Shale, limestone)	800—2400	57/650	97/189	6.4

2D refers to the number in highway maps

annual total, during the summer monsoon (June through September).

The three tendencies of precipitation explain why the landslides are concentrated in the cis-Himalaya and during the summer monsoon period. As many as 84 per cent (336) of the total (401) reported landslides occurred in the region and about 65 per cent (219) of them occurred during the monsoon. In contrast 35 of the reported 48 avalanches occurred in the Trans-Himalaya and all but one in winter and spring (October through May).

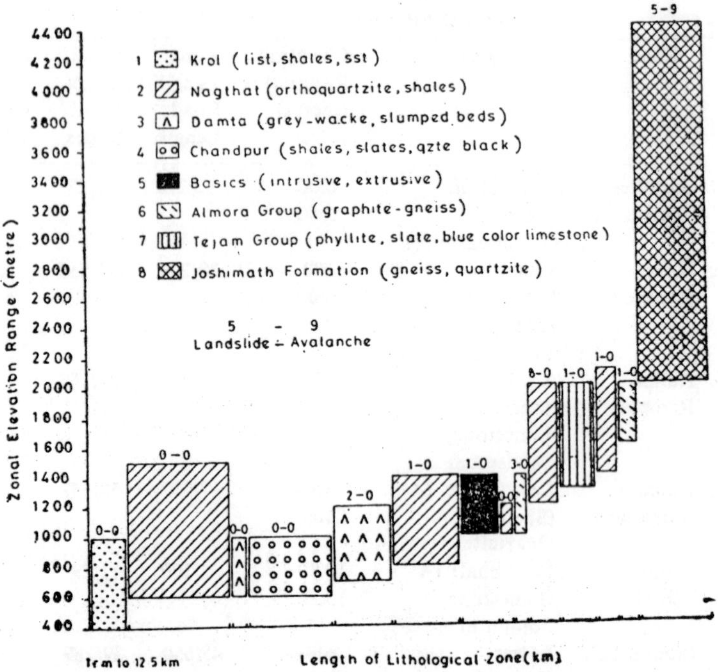

FIG. 23.7 : *Rishikesh-Badrinath highway : Number of landslides and avalanches*

Temporal aspects

Temporal distribution of landslides and avalanches is expressed through two dimensions, variations through the years and variations through the seasons. All the highways have experienced larger number of landslides in the post-1967 period, which coincides with the tenure of the fourth, fifth, and sixth five-year plans, than in the pre-1967 period (Table 23.6). Similarly, of the 33 avalanches that have occurred along the Jammu-Leh highway as many as 29 were concentrated in the post-1967 era. The era was particularly marked by extensive development of roads, towns, agriculture and industry. Large scale removal of forests for creating agricultural, terraced fields became widespread.

During the fourth, fifth, and sixth five-year plans two special programmes, Sensitive Border Area Roads and Hill Region

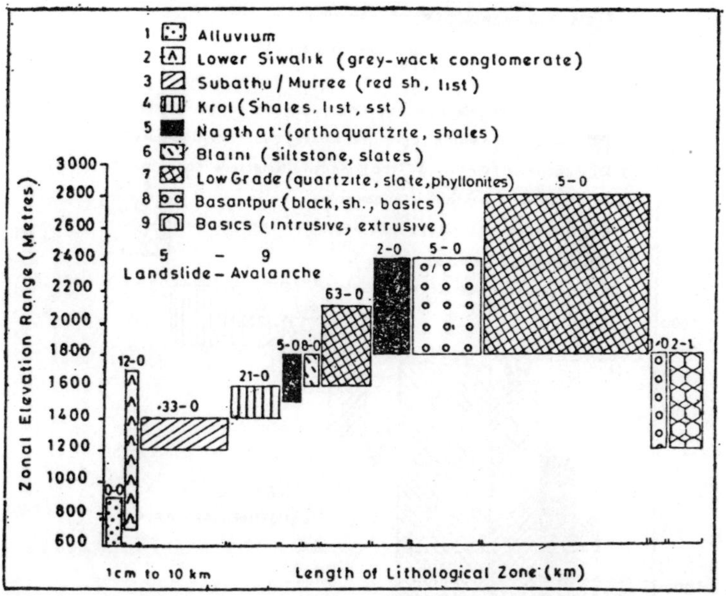

FIG. 23.8 : *Kalka—Rampur highway : Number of landslides and avalanches*

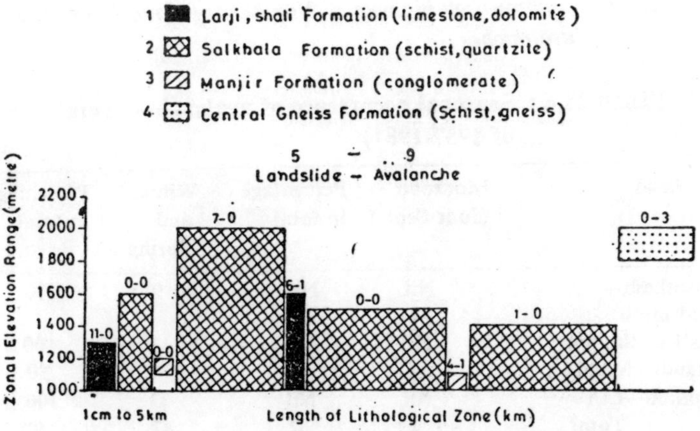

FIG. 23.9 : *Mandi-Manali highway : Number of landslides and avalanches*

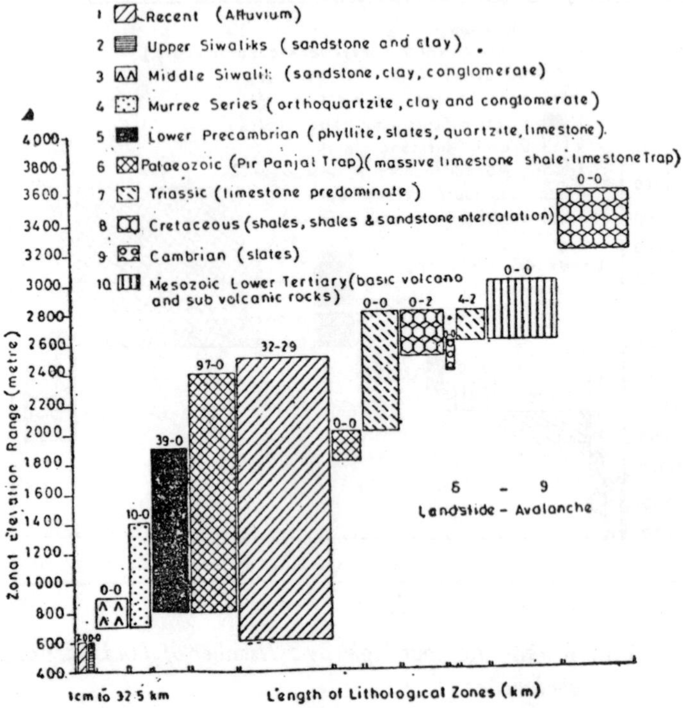

FIG. 23.10 : *Jammu-Leh highway : Number of landslides and avalanches*

TABLE 23.6 : **Seasonal occurrence of avalanches (Total of 1957-1981)**

Road	Monsoon (June-Sept.)	Percentage in total	Winter and Spring	Percentage in total
1. Rishikesh—Badrinath	Nil	Nil	9	100
2. Kalka—Rampur	Nil	Nil	1	100
3. Mandi—Manali	1	20	4	80
4. Jammu—Leh	Nil	Nil	33	100
Total	1	2	47	98

Roads, resulted in a rapid and large expansion of road network, particularly in those parts where the vegetation is sparse, mechanical weathering is intense and rocks are weak. The Middle-, Central-, and Trans-Himalayan zones, in particular, experienced a large number of landslides as a response to these developments.

Seasonality of occurrence of landslides both for the total and for the first three roads is very marked (Table 23.7). The first two roads which are located in the eastern parts of the

TABLE 23.7 : **Seasonal occurrence of landslides (Total of 1957-1981)**

Road	Monsoon (June-Sept.)	Percentage in total	Winter and Spring (Oct.-March)	Percentage in total
1. Rishikesh—Badrinath	24	92.30	2	7.7
2. Kalka—Rampur	134	85.35	23	14.65
3. Mandi—Manali	17	58.62	12	41.35
4. Jammu—Leh	44	23.28	145	76.72
Total	219	54.61	182	45.39

western Himalaya have had more than 85 per cent of the annual totals concentrated in the monsoon season. The characteristically high intensity of rainfall is particularly conducive to deep lubrication in large masses of weathered and crushed rock bodies resulting in massive landslides. Bare hillside slopes with thin soil covers are the immediate victims of landslide processes. In the aggregate of the four highways as well, the monsoonal concentration (219) is higher than the winter occurrences (182).

In contrast to monsoonal concentration of landslides in the eastern parts, the Mandi-Manali and Jammu-Leh roads, central and western locations, respectively, have experienced a higher frequency in the winter and spring seasons. The Jammu-Leh road has had about 77 per cent of the occurrences during the winter. Essentially this is attributed to the larger proportion of rainfall and snowfall occurring during the winter. During July to September the Kashmir Valley section of the road receives only 20 per cent of the annual rainfall (Kaul 1980).

It is interesting that of the 48 avalanches reported for the study area as many as 33 had occurred along the Jammu-Leh road and all of them were concentrated during the winter and spring seasons. This is true for all other roads as well (Table 23.8). Obviously, the relationship here is with heavy snowfall concentrated in the winter. Further, the Jammu-Leh road has the major segment in the Trans-Himalayan zone beyond the Zanskar ranges where most of the precipitation falls in the form of snow and during the winter months.

TABLE 23.8 : **Temporal concentration of landslides**

Road	Pre—1967 (1)	Post—1967 (2)	Percentage of (2) in total
1. Rishikesh—Badrinath	9	17	66
2. Kalka—Rampur	48	109	69
3. Mandi—Manali	7	22	75
4. Jammu—Leh	23	166	88
Total	87	314	78

Human dimensions

While a fairly large and conclusive body of literature exists on the natural conditions and correlates of landslides and avalanches the role of human activities and human occupance of the earth has tended to be ignored. Observations made in the field suggest the following human activity have generated causes of the landslides that have occurred in the area :

a. ill-maintenance of narrow irrigation channels ;
b. removal of forests by uncontrolled felling without a proportional reafforestation, and removal of vegetation cover by grazing and of grass by cutting for fodder ;
c. mining and quarrying ;
d. urban growth through increase in the number of towns and through physical expansion of towns ; and
e. increase in the transportation network.

Irrigation channels (*kuhls*) descending from one to the other terraced field either down the slope or along the contours are widespread in the region. Constructed of mud the channels carry water throughout the year except during severe winters. Breaching of channels because of poor maintenance and freeze expansion is quite common, leading to large seepage and surface run-off. The cumulative effect of repeated breaching ultimately contributes to the generation of landslides. Kienholz *et al.* (1983) and Spate (1967) have emphasized that the most serious form of erosion leading to landslides results from damaged canals and gullying. Since the Middle Himalaya (1,200 m. to 2,000 m.) contains the densest network of channel (*kulh*) irrigation, it is here that the number of landslides related to this factor is the highest.

Forests are being continually felled as a part of large-scale commercial exploitation and there is not much reafforestation. There is widespread grazing by cattle, goat and sheep and the destructive effect is particularly marked in the higher, steep slopes and in the Trans-Himalayan zone.

Mining and quarrying, particularly along the highways, were intensified during the last decade along with the increase in construction activities of urban buildings and roads. Particular mention must be made of extensive quarrying of limestone in the Lower Himalayan zone along some of the major roads, and of slate in the middle zone.

A whole series of new market centres, administrative towns and industrial complexes have emerged along the four highways, with their major cancentration in the Lesser and southern flanks of the Central Himalaya. What accelerate the slope damage are the upward expansion from the road of these construction activities on the slopes that have already been weakened, the construction of multi-storey houses almost to the edge of overhanging cliffs against which the niche of the road has been cut, and extensive lineation of urban built-up belts without any drainage, leading to heavy infiltration in small pockets.

There has been a continuous increase in the number of urban places in the region, 104 in 1961, 126 in 1971, and 170 in 1981. Most of the towns tend to remain concentrated in the Lower and Lesser Himalayan zones and play a destabilising role in their ecologies.

There has been a significant increase in the length of the existing roads and in the construction of new roads. Elaborate networks of border roads constructed for strategic reasons and as a major component of regional development plans also have significantly contributed to an increase in the number of landslides. Particularly significant as a causal factor has been the widening of the roads without the construction of protective measures such as a retaining wall. Further, the absence of bridges on steeply sloping wide gullies which carry enormous and rapid discharge during the monsoon days results in local damming and rapid drainage within the regolith, finally leading to massive landslides.

A contributory factor, not generally recognized, has been the blasting of large bodies of slopes in road construction which leads to internal disintegration of the regolith and rock body, thus promoting landslides. Many sites of landslides are associated with the widening of roads through blasting of rocks.

All the five conditions became acute during the post-independence period and more so in the post-1967 era when the later five-year development plans were being assiduously implemented.

Human casualties and property damage

The casualties and damage have been analysed at two levels and presented in six tables (Tables 23.9, 23.10, 23.11, 23.12, 23.13 and 23.14). Tables 23.9 to 23.12 give details of casualties and damage caused by landslides and avalanches in different lithological and elevation zones. Tables 23.13 and 23.14 provide details of casualties and damage along each road in the pre-1967 and post-1967 periods.

In the first two roads (Tables 23.9 and 23.10) the two formations worst affected are Nagthat and Joshimath, both located in elevation zones of 800 metres to 2,400 metres. The casualties highlight human life lost and the damage reflects a wide spectrum of property, not limited to roads. The worst hit are the terraced fields which involve the efforts of several generations. Effects of continuous urban building are indicated in the avalanche damage in Joshimath and its surroundings.

Larji Shali formation of limestone and dolomite is associated with extensive damage in the elevation zone 1,000 metres to 1,400 metres along the Mandi-Manali highway, interesting is the obstruction to the construction of Pandoh dam (Table 23.11). Palaeozoic shales and slates and recent alluvium at elevations of

TABLE 23.9 : **Rishikesh-Badrinath highway**

Index No.	Lithological Zones	Zonal Elevation Range (metre)	Landslide Casualties		Avalanche Casualties		Landslide Damage	Avalanche Damage
			Killed	Injured	Killed	Injured		
1	2	3	4	5	6	7	8	9
1	Krol	400-1000	0	0	0	0	—	—
2A	Magthat	600-1500	0	0	0	0	—	—
3A	Damta	600-1000	0	0	0	0	—	—
4	Chandpur	600-1000	0	0	0	0	—	—
3B	Damta	700-1200	0	0	0	0	—	—
2B	Nagthat	800-1400	34	0	0	0	4 big houses and 8 shops were buried.	—
5	Basioo	1000-1400	0	0	0	0	—	—
2C	Nagthat	1000-1200	0	0	0	0	—	—
6A	Almora Group	1000-1400	0	0	0	0	In the Mandakini valley 17 km long stretch of Nandprayag ghat, motor road had been badly damaged. Several houses collapsed and damaged.	—

TABLE 23.9 : (*Contd.*)

1	2	3	4	5	6	7	8	9
2D	Nagthat	1200-2000	146 25 cattle	7	0	0	Buried the whole of the small mountainous village named Daduwa. All the terraced fields of the villages have disappeared. Total loss of crops, property cattle is Rs 1 lakh. Destroyed 8 houses, 8 goshalas, 19 houses, 23 cattlesheds and grains valued Rs 40,000 damaged.	—
7	Tejam Group	1300-2000	0	0	0	0	—	—
2E	Nagthat	1400-2100	0	0	0	0	14 villages are sinking.	—
6B	Almora Group	1600-2000	0	0	0	0	—	—

8	Joshimath	2000-4400	128	3	24	4	Half of the Nepali house, kitchen building of Andhra dharmsala site, power house, temple establishment building, Paliwal's Gujrati Niwas, Vaisha-nava Bhavan, Sant Niwas and some private houses damaged. Some buildings of Kali Kamli Wala, the outhouse of Andhra building of Badrinath temple damaged. All the houses in the market area of the 10,30C0 feet town at the base of the Nilkantha damaged. Birla house was partly damaged. Also destroyed the civil police post, barracks and all the three schools.	—

TABLE 23.10 : **Kalka-Rampur highway**

Index No.	Lithological Zones	Zonal Elevation Range (metres)	Casualties Killed	Casualties Injured	Landslide Damage	Avalanche Damage
1	Alluvium	600-900	0	0	—	—
2	Lower Siwalik	700-1700	0	0	—	—
3	Subathu/Murree	1200-1400	0	0	A sudden landslide pushed down a running track along with parapet. The truck fell about 200 feet below.	
4	Krol	1400-1600	0	0	—	—
5A	Nagthat	1500-1800	1	0	—	—
6	Blaini	1600-1800	0	0	—	—
7A	Low Grade	1600-2100	9	2	The landslide has resulted in the slipping down of the circular road below the Rivoli cinema and damaged a building close by. Disrupted electricity and water supply of Subhash Nagar area. Damaged a saw mill. Houses have sunk about 30 cm. while	

					a number of others had developed cracks owing to landslides. Telegraph and telephone lines have been disrupted at several points.	
5A	Nagthat	1800-2400	10	2	—	—
8A	Basantpur	1800-2400	0	0	—	—
7B	Low Grade	1800-2800	4	8	—	—
8B	Basantpur	1200-1800	0	0	—	—
9	Basics	1200-1800	8	0	—	—

TABLE 23.11 : Mandi-Manali highway

Index No.	Lithological Zones	Zonal Elevation Range (metres)	Landslide Casualties		Avalanche Casualties		Landslide Damage	Avalanche Damage
			Killed	Injured	Killed	Injured		
1A	Larji Shali Fm	1000-1300	7	3	0	0	Breach in water mains.	—
2A	Salkhala Fm	1000-1600	0	0	0	0	—	—
3A	Manjir Fm	1100-1200	0	0	0	0	—	—
2B	Salkhala Fm	1000-2000	6	11	0	0	A motor truck rolled down and was wrecked 'A major landslide on the top of one of the 2 support hills of the proposed dam . . . upset the construction' (site at Pandoh).	—
1B	Larji Shali Fm	1000-1600	2	4	0	0	Roads have been breached at several points.	—
2C	Salkhala Fm	1000-1500	0	0	0	0	—	—
3B	Manjir Fm	1000-1200	0	0	5	5	The landslide has disrupted water supply.	—
2D	Salkhala Fm	1000-1400	0	0	0	0	—	—
4	Central Gneiss Fm	1800-2000	0	0	4	5	—	—

400 metres to 2,400 metres are associated with damage and deaths, both indicating the suddenness and major impact of movements along the Jammu-Leh highway (Table 23.12).

Both landslide and avalanche damage and casualties have been more numerous in the post-1967 than in the pre-1967 era (Tables 23.13 and 23.14). The Rishikesh-Badrinath highway has recorded the most gruesome destruction of the burial of a whole village and the killing of as many as 173 persons. Both landslides and avalanches have recorded the largest number of deaths along the Jammu-Leh highway traversing through limestone and shale formations.

The avalanches and landslides that occur at higher elevations and in the Trans-Himalayan zone pose the greatest difficulty to the problems of their management. They are relatively inaccessible and are located at great distances from the local and regional centres of administration and have poor or practically no communication facilities with the main settled areas.

CONCLUSION

Developmental activities accompanies by landslides have two dimensions, spatial and temporal. Accordingly, landslides also have these two distributional components. Spatially, the landslides and the human casualties and property damage caused by them are concentrated in the cis-Himalaya. Temporally, they were concentrated in the post-1967 period of the post-independence era.

Slate, shale, limestone, and orthoquartzite, particularly in the zone of crushing along the faults, comprise the lithological factor. The steep slopes, mainly a function of the original structural dips, provide a conducive topographic setting. High intensity and high amount of rainfall and snowfall complete the list of natural causal factors of landslides.

Removal of forests, development of urban mesh, and elaboration of transportation network have been the principal human correlates.

Man, through imprudently operating the processes of development, causes extensive damage to the ecological resources and distortions in the arrangement of ecological settings. The damage and the distortions retard the development processes.

TABLE 23.12 : Jammu-Leh highway

Index No.	Lithological Zone	Zonal Elevation Range (metres)	Landslide Casualties		Avalanche Casualties		Landslide Damage	Avalanche Damage
			Killed	Injured	Killed	Injured		
1A	Recent	400-600	25	34	22	—	Destroyed a labour shed near the site of a Hydel project at Salal.	—
2	Upper Siwalik	400-600	0	0	0	0	—	—
3	Middle Siwalik	700-900	0	0	0	0	—	—
4	Murree Series	700-1400	0	0	0	0	Road was badly damaged.	—
5	Lower Precambrian	800-1900	0	0	0	0	—	—
6A	Paleozoic	800-2400	93	0	0	0	A truck was caught in the landslide.	—
1B	Recent	600-2500	25	0	132	35	—	Destroyed a house. Some huts were also destroyed. Power supply disrupted. 'Blackout' was

									caused by disruption of electric lines and suspension of the working of the Ganderbal power-house whose feeding channel choked by avalanche. Telecommunications also disrupted.
6B	Paleozoic	1800-2000	0	0	0	0	—		—
7A	Triassic	2000-2800	0	0	0	0	—		—
8A	Cretaceous	2500-2800	0	0	0	0	—		—
9	Cambrian	2400-2600	0	0	0	0	—		—
7B	Triassic	2600-2800	0	0	0	0	—		—
							Srinagar-Leh suffered damage.	Road extensive	One of the huts was completely washed away.
10	Mesozoic	2600-3000	0	0	0	0	—		—
8B	Cretaceous	3200-3600	0	0	0	0	—		—

TABLE 23.13 : **Landslide damage**

Highway	Casualties				Property Damage	
	Pre-1967		Post-1967		Pre-1967	Post-1967
	Killed	Injured	Killed	Injured		
Rishikesh—Badrinath	135	—	173	10	Buried the whole of the small mountainous village named Daduwa. All the terrace fields of the village have disappeared. Total loss to crops, property, cattle Rs one lakh.	Road had been badly damaged. House collapsed. Destroyed 8 houses, 8 ghoshalas, 19 houses and 23 cattle sheds. Grains valued Rs 40,000 damaged. 14 villages between Pipalkoti and Joshimath are sinking.
Kalka—Rampur	13	12	19	0	Resulted in the slipping down of the circular road below the Rivoli cinema. Number of buildings have been damaged. Damaged telegraph and telephone communication from and to Simla. Damaged a saw mill.	Some houses have sunk about 30 cm while a number of other had developed cracks owing to landslides. Telegraph and telephone lines have been disrupted at several points.

Mandi—Manali	2	4	13	14	A motor truck rolled down and was wrecked. Mandi-Kulu road has been breached at several points.	Breach in water mains. Disrupted the construction of dam at Pandoh site. Disrupted water supply.
Jammu—Leh	25	0	118	34	—	Destroyed a labour shed. Road was badly damaged. Destroyed one truck. Extensive damage to Srinagar-Leh Road.

TABLE 23.14 : **Avalanche damage**

Highway	Casualties				Property Damage	
	Pre-1967		Post-1967		Pre-1967	Post-1967
	Killed	Injured	Killed	Injured		
Rishikesh—Badrinath	8	—	16	4	Half of a Nepali house, Kitchen building of Andhra dharmsala site. Power-house, temple establishment building, Paliwal's Gujrati Niwas, Sant Niwas and private houses damaged. Some buildings of Kalikamliwala, the outhouse of Andhra building of Badrinath temple damaged. All the houses in the market are damaged. Birla house was partly damaged.	Avalanches also destroyed the civil police post, barracks and all the three schools.

Kalka—Rampur	0	0	0	0	
Mandi—Manali	9	10	0	0	
Jammu—Leh	15	0	139	38	Some huts were destroyed. Power supply disrupted. 'Blackout' was caused by disruption of electric lines and suspension of the Ganderbal power-house. One of the huts was completely washed away.

Development thus implemented neutral to ecology becomes counter-productive to both man and nature.

There is hardly any part in the Himalaya which is not standing at the threshold of destruction because of the conspiracy being continually hatched against it by the combined role of lithology, topography and precipitation among the natural factors, and their relationship with the patterns and activities of human occupance. It is not only the natural but human factors as well which trigger off geomorphic damage and once they are initiated they set out on an unending chain of self-destruction.

REFERENCES

Butzer, Karl W., *Geomorphology from the Earth*, New York : Harper (1976), p. 79.

Keinholz, H. et. al., 'Mountain Hazards Mapping in Nepal's Middle Mountains with Maps of Land Use and Geomorphic Damages (Kathmandu Kakani Area), *Mountain Research and Development* 3 (1983) ; 195-220.

Kaul, A.K., 'Monsoon Component of Rainfall in Kashmir Valley', *Vayu Mandal* : (1980) 22-23.

Spate, O.H.K., *India and Pakistan* (London : Methuen & Co. 1980).

TWENTY FOUR

N. AHMAD AND S. RAIS

HIMALAYAN GLACIATION
Some Misconceptions

A SCIENTIFIC study of the Himalayan glaciers was started towards the end of the last century and most of the early workers were those who had studied and worked in the Alpine areas. These early workers not only compared the Himalayan glaciers with those of the Alpine region, they also correlated the Himalayan glacial events with those of the Alps and judged their behaviour with the same criteria in mind.

The Himalaya are the only mountains in the world to have developed in the tropical zone in an east to west direction. Nearness of the Himalaya to the equator has influenced the development and movement of their glaciers. Though some-apparent similarities may be seen between these tropical Himalayan glaciers and the temperate Alpine glaciers, observed data do not suggest that the glaciers of these areas are in any way comparable. Some of the conclusions drawn by early workers that do not conform to observed facts are as follows :

1. The Snowline in the eastern Himalaya is lower than in the western Himalaya.

2 The Himalaya have experienced four or five glacial advances which can be correlated with the four phases of Alpine glaciation.

3. The Himalayan glaciation started with the general lowering of the snowline in the northern hemisphere, the first glacial advance in the Himalaya being contemporaneous with the Gunz glaciation of the Alpine series.

Charlsworth (1957) and Wadia (1957) have reported that the snowline in the eastern Himalaya is lower than that in the western part. They have come to this conclusion on the basis of the reports of the visitors who were present at the time of some summer snowfall. The other basis for this presumption is that the eastern Himalaya have the highest ranges and that the annual moisture precipitation in the east is more than that in the west. Therefore, the annual accumulation of snow and ice in the east must be greater and the snowline lower.

The best criterion for fixing the snowline level is not the observed lowest level of individual snowfall, it is the altitude at which local cirques occur. The cirque level data show that except in the western Himalaya (Pakistan and Afghanistan) the cirque line gradually rises from the west towards the east (Ahmad and Mayweski 1981).

Cirque levels in Himalaya

Location	*Upper Cirque Level*	*Lower Cirque Level*
Western Himalaya	4900-1500m	4600-4800m
Kashmir valley	3600-3800m	3200-3400m
Central Himalaya	3900-4100m	3300-3400m
Eastern Himalaya	4100-4200m	3800-4000m

The gradual rise of the snowline level towards the east is owing mainly to the differences in the occurrence of annual precipitation and latitudinal position of different parts of the Himalaya (Kashmir 35°N, Garhwal 30°N and Sikkim 27°N). In the eastern Himalaya most of the precipitation takes place during summer while in the western Himalaya it mostly occurs during winter. Winter precipitation is mostly in the form of snow that collects and adds to the thickness of glacial ice while the summer precipitation is in the form of rainfall that induces melting. The effect of this difference in the timing and type of precipitation is that in the western Himalaya net annual accumulation of snow is more than that in the east. This effect is responsible for the level of the snowline being higher in the east and lower in the west. In Afghanistan and Pakistan the snowline is higher because by the time the summer monsoon reaches there it is very weak and the volume of winter snowfall is not enough to maintain a lower snowline. The eastern Himalaya being located in lower latitudes must also have a higher snowline than the western Himalaya which fall in higher latitudes.

De Terra and Patterson (1939) have reported the occurrence of a gravel bed with vertebrate fossil bones in the Sombur quarry (Kashmir valley). This bed rests unconformably on Triassic limestone and is overlain by upper Karewa varves. They are of the opinion that the bone bed separates the upper Karewa deposits from the lower Karewa and that the latter are pre-glacial. From the bone bed they have reported *Elephus hysudricus*, which has been recorded from the Tatrot and Pinjor beds of the upper Siwaliks deposits (of late Pliocene and early Pliestocene age). Except for the occurrence of this bone bed no other fossil bearing beds below the base of glacial deposits have so far been reported.

Ahmad (1982) has found an extension of the Sombur bone bed about a kilometre west of Sombur village. Outside the village this bone bed occurs conformably in between Karewa varves. In this locality the exposed part of the underlying varves are more than 30 m thick. From the stratigraphic position of the bone bed in this new locality, it is inferred that it had been laid during first glacial or the first inter-glacial period. In any case it was deposited after the start of the Himalayan glaciation. This bed also indicates that by the end of the Tatrot or by the beginning of the Pinjor period the glaciers had descended to lowest levels.

Wadia (1957), Gupta (1975) and Krishnan (1982) are of the opinion that the Himalaya attained their present height in the mid-Miocene period. During this phase of the uplift most of the Himalayan peaks had nearly attained their present heights. Possibly it was during this time that most of the Himalayan peaks also went above the regional snowline and the development of the glaciers began. However, it was only by the end of the Tatrot period that the intensity of the glaciation reached its maximum and the terminal end of the glaciers had reached below the 1,600 m level.

The large basin-like upper level cirques of the Himalaya suggest that these had developed over a long period of glacial activity. The absence of any organic fossils other than a few bones may be because the time gap between the uplift of these ranges above the regional snowline and the advance of the glaciers was not long enough for the high altitude flora to migrate and for the fauna to adapt to low temperature and establish themselves on the slopes of these ranges. Moreover, whatever early plant life may have been there, was scraped off by the advancing glaciers and it is possible that the lignite deposits in Lower Karewa may have come from this source.

Gupta (1976) has reported that terminal moraines of the first glacial advance were laid below 1982.5 m and that of the fourth advance between 2,450-3,050 m levels. Wadia (1957) has reported terminal moraines of the first advance at 1,620 m, of the second advance at 1,800 m, of the third advance at 2,100 m and that of the last advance between 2,740-3,650 m level. Ahmad and Mayweski (1981) have observed that in the Himalaya terminal moraines were laid at six levels. Three of these were laid by transverse glaciers that terminated below 2,740 m level and the other three by longitudinal glaciers at 2,740 m, 3,048 m, and 3,650 m levels. However, in every valley all the three terminal moraines are not preserved. For instance in the east Liddar valley the 3,048 m moraine has been eroded and in the west Liddar valley the 3,650 m moraine is absent. In Milam valley in Garhwal and Suru valley in Ladakh the 2,740 m moraine is absent.

De Terra and Patterson (1939), Wadia (1957), Gupta (1976) and Krishnan (1982) have suggested that during the Pleistocene period the Himalaya experienced four glacial episodes and these were equivalent to the four Alpine glacial episodes.

CONCLUSION

The International Geological Congress in its 1948 session concluded that the Pliocene-Pleistocene boundary be placed between the Tatrot and Pinjor beds. Sahani and Khan (1946) and Prasad (1964) have concluded from the available faunal evidences that the Pinjor beds should be placed at the base of the Pleistocene and Tatrot beds near the top of the Pliocene period. From the stratigraphic position of the Tatrot and Pinjor beds it is now established that the Himalayan glaciation started in the Tatrot period. The Himalayan glacial episodes could not be equivalent in time to those of Alpine glacial episodes because the former started in the upper Siwalik Tatrot stage (De Terra and Patterson 1939; Wadia 1957; Krishnan 1982) of the Upper Pliocene. In the Alps region even the pre-Gunz high level Alpine glaciation, Donau, started in the Lower Pleistocene. All persons working in the Himalaya have identified four or five glacial advances on the basis of the number of terminal moraines they recorded in different valleys. Unfortunately in most of the Himalayan valleys all the three high level terminal moraines are not preserved and sometimes recessional moraines have been misidentified as re-advance moraines. Moreover, longitudinal and transverse glacier moraines were also not separately identified.

In the Himalaya all the longitudinal glaciers terminated above 2,740 m altitude and transverse ones below that level. The three terminal moraines lying above that must, therefore, have been laid during the three advances of longitudinal glaciers and the three terminal moraines below that altitude must have been deposited during the three advances of transverse glaciers. These two sets of three terminal moraines each, therefore, represent three Himalayan glacial advances and not four or five advances as earlier believed. Those three advances as identified by Ahmad (1982) are as follows :

Pahalgam stage	First glacial Maximum advances	advanced to	2740 m level
Aru-Martoli stage	Second glacial Maximum advances	advanced to	3048 m level
Milam-Sheshnag stage	Third glacial Maximum advances	advanced to	3650 m level

REFERENCES

Ahmad, N. and Mayewski, P.A., 'Himalayan Glaciation : An unresolved issue', *The Geographer* 28 (1981).

Ahmad, N., 'Glaciation in the Himalayas', in *Symp. Quart., Episodes in India, Neotectonism, Eustasy and Palaeoclimates* (Abst.), Baroda : M.S. University, 1982, p. 28.

Charlsworth, J.K., *The Quarternary Era*, 2 Vols. (London : Edward Arnold Pubs. Ltd., 1957).

Gupta, V.J., *Indian Cenozoic Stratigraphy* (Delhi : Hind Pub. Corpn., 1976).

Krishnan, M.S., *Geology of India and Burma*, 6th eds. (Delhi : CBS Pubs. & Distributors, 1982), p. 536.

Terra, De H. and Paterson, T.T., *Studies on the Ice Age in India and Associated Human Culture* (Washington D.C. : Carnegie Inst., 1939), p. 345.

Wadia, D.N., *Geology of India*, 3rd edn. (New York : Macmillan & Co. 1957), p. 535.

SECTION SEVEN

ENVIRONMENTAL POLLUTION AND MANAGEMENT

TWENTY FIVE

ALLEN G. NOBLE

THE GEOGRAPHY OF NOISE IN THE SOUTH ASIAN CITY
A Case Study of Kandy, Sri Lanka

FEW geographers have yet become involved in analyses of the patterns of urban noise. However, problems of noise pollution are likely to increase as the urban population grows. The components of urban noise are somewhat different in western and Asian cities. This study examines, albeit in a preliminary fashion, the geography of noise pollution as well as its temporal variation in Kandy, a city of about 100,000 population in central Sri Lanka. This is the first such study for a south Asian city. How typical Kandy is of other south Asian cities is difficult to say, but a distinctive profile of noise exists in this city, quite different from that identified in several studies of American cities. The variance results because noise sources are not the same as in American cities and the daily rhythm of activity is also different.

The observations upon which the present study is based were made in April 1980. Of necessity the readings were limited, consequently only the broadest noise patterns could be identified and the conclusions drawn must remain tentative. Nevertheless this study represents the first attempt to identify a noise profile which may typify south Asian cities.

METHODOLOGY

Six observation sites were selected rather arbitrarily, but they were with a view to choosing locations representative of the central and hence noisiest parts of Kandy (Fig. 25.1). Site One is

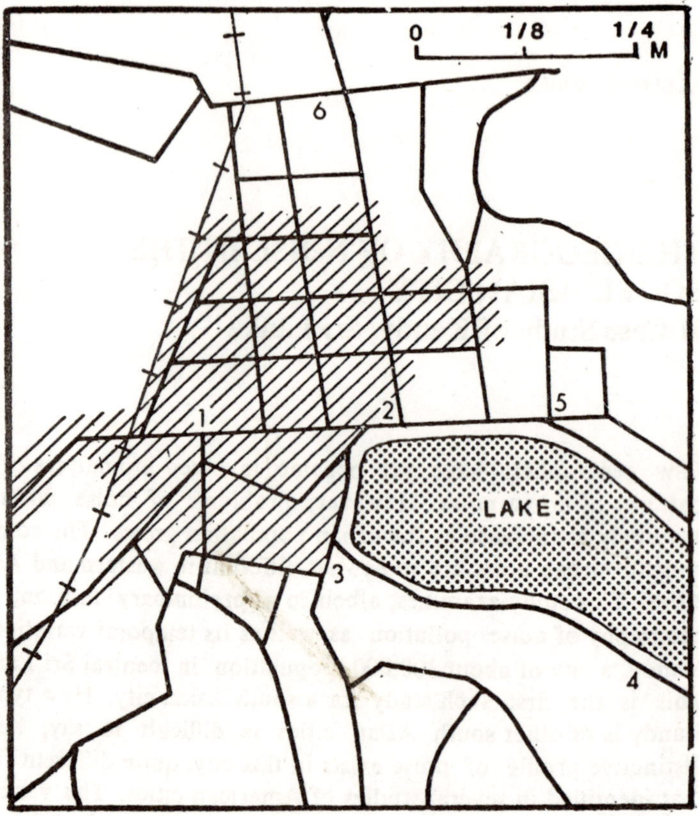

FIG. 25.1 : *Observation sites*

adjacent to the Clock Tower near the centre of the busiest bazaar section of the city. Site Two is the open, elevated verandah of the Queen's Hotel at the opposite end of the central business district main street. Site Three, termed Lake Corner in this study, marks the south-eastern limit of the central business district and the junction of several types of land use. Site Four, called Temple, lies adjacent to a small Buddhist temple on the south shore of Kandy lake in a neighbourhood that is partly residential and partly institutional and commercial. Across the lake is Site Five located in the garden before the Buddhist Temple of the Tooth, one of the major Buddhist pilgrimage centres in Sri Lanka.

The final site is at the northern extremity of the closely built-up, central part of Kandy where residential and commercial land uses are mixed. Each of these six sites reflects different kinds of urban activity and land uses. Together they provide a cross section of daytime noise levels within the central city.

Readings were taken on five non-consecutive weekdays during the month of April 1980. None of the days had any unusual activities associated with them and are thought to be representative of much longer periods. The weather was clear with no high winds or rain enhanced pavement noise. The range of readings given in Figures 25.2 to 25.7 is an average of all observations taken over the five days. In addition the highest individual readings throughout the day are also given, since they are the most noticeable and most annoying of sounds. These readings are quite comparable to the L-90 readings of standard American acoustical studies.[1] By the same token, the midpoint of the noise range given in the graphs of the Kandy sites is comparable to the L-50 level statistic.[2]

ANALYSIS

Readings at each of the observation sites produce profiles that are quite disparate. That at Site One, Clock Tower (Fig. 25.2), records very high noise levels in the early morning, much higher than at other sites. Both the central market, with a large surrounding bazaar, and the central bus terminal are adjacent to the Clock Tower. Both are places of great activity in the early morning and the high noise levels are a result of the combination of these activities. The secondary increase in noise between 10 A.M. and 11 A.M. reflects the rather late hour of opening of many of the shops and small businesses in the Kandy central business district. A midday lull in downtown activity shows up in declining noise levels from about 11 A.M. to about 3 P.M. followed by the afternoon rush hour.

The noise profile at Site Two, Queen's Hotel (Fig. 25.3), is very much different from that of Clock Tower. Noise readings generally increase in volume through the morning but decrease sharply after noon. However, by 3 O'clock noise levels equivalent to those of the late morning have been regained, but almost immediately sounds begin to decline. The biggest source of noise

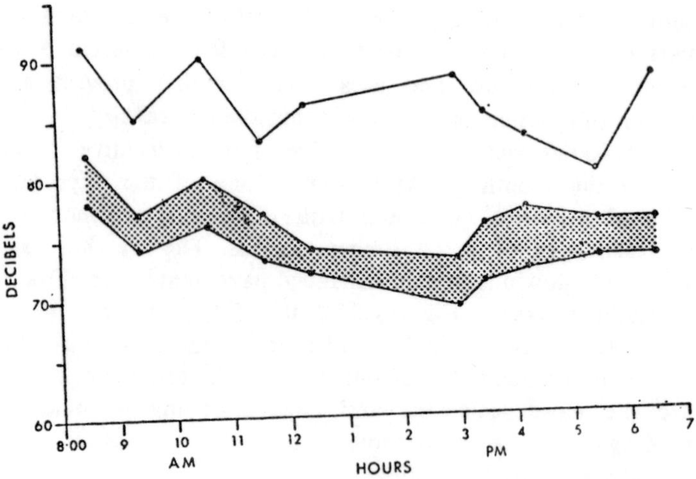

FIG. 25.2 : *Clock tower*

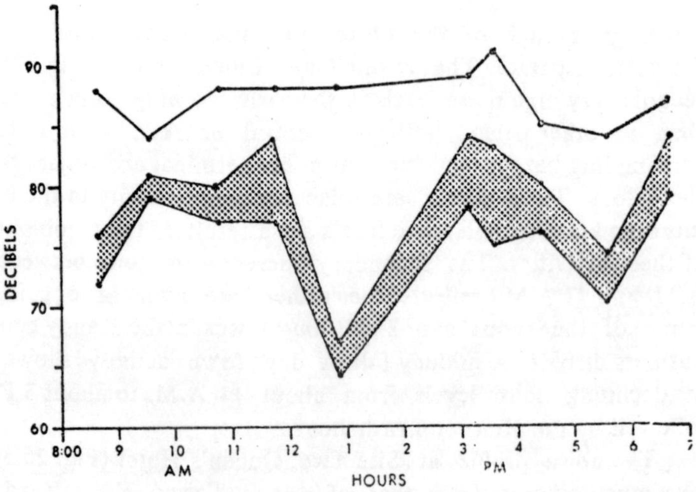

FIG. 25.3 : *Queens hotel*

at this site is that of vehicles, especially lorries and buses, but the late morning and mid-afternoon peaks do not reflect rush hour traffic conditions ; rather they are a combination of vehicular and pedestrian noise. A final peculiarity of noise at this site is the sharp increase experienced between 6 P.M. and 7 P.M. the result of large numbers of roosting birds in banyan trees opposite to the hotel. The noise levels at Queen's Hotel at noon, 3 P.M. and 7 P.M. are as high as those of early morning at Clock Tower and make this location the noisiest in Kandy, almost at a point sufficient to cause hearing damage with sustained exposure.[3] Part of the reason for the intensity of noise here is that the roadway is depressed about five feet below the flanking sidewalks which act as sounding boards.

A quite similar pattern of noise is experienced at site Three, Lake Corner, the major difference being the wider range over which noise fluctuates at each hour (Fig. 25.4). At Lake Corner noise increases up to noon, followed by a rapid decline and

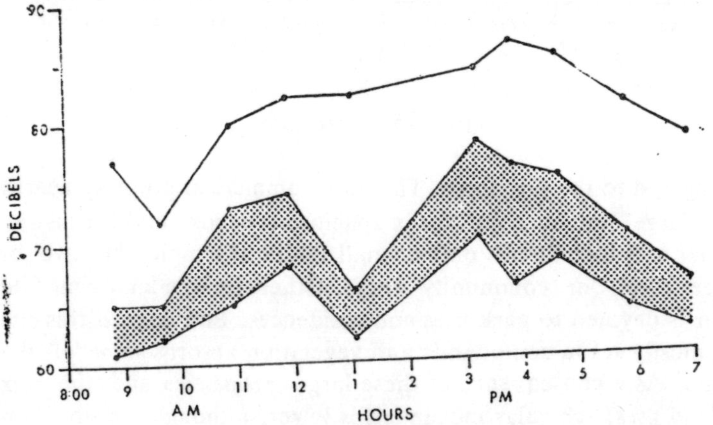

FIG. 25.4 : *Lake corner*

slower recovery. By 3 P.M. earlier noise levels have been attained. Indeed they are even higher than the noon level. These higher readings appear to be related to much higher pedestrian movement and sounds in the afternoon than in the morning. Lake Corner is a shady location adjacent both to the lake and an attractive hillside park. Sounds of pedestrians form a major component of noise at this site. From 3 O'clock onwards, the level of noise declines steadily to nightfall.

The sounds recorded at Site Four, Temple, are generally much lower than at the first three locations (Fig. 25.5). This is in large part a reflection of the different land uses at this site as

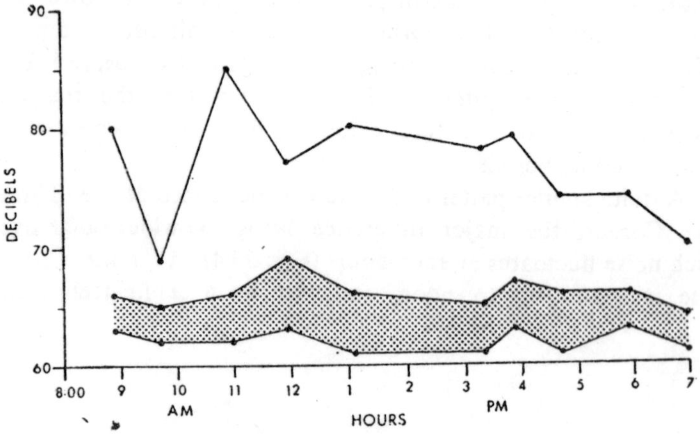

FIG. 25.5 : *Temple*

compared to the first three. The only commercial property nearby is a large tourist hotel set in spacious grounds, and the institutional uses include not only a small Buddhist temple, but two or three social or community clubs. Otherwise the land near Site Four is devoted to park uses and residences. Buildings at this site are mostly set in compounds and vegetation absorbs some of the sound. As a consequence of these larger properties and the mix of land uses, vehicular movement is lower, although the site is on a major arterial road. The peaks of very loud noise are almost always the sounds of poorly muffled buses. Generally speaking, the range of noise at Site Four is quite uniform, although the noontime and late afternoon peaks can be discerned. The afternoon noise peak is at 4 P.M. later than that at Sites Two and Three reflecting both the more peripheral location of Site Four and the more central location of the other sites. Noise levels migrate outward from the central business district as pedestrians and vehicles move toward residential areas. However, since the central bus depot is at Clock Tower, noise levels remain higher at the later hour there.

Recording Site Five, termed Garden, is situated in the garden or open space which separates the Temple of the Tooth from the central business district (Fig. 25.6). The temple is one of the main Buddhist pilgrimage centres in Sri Lanka. The profile of

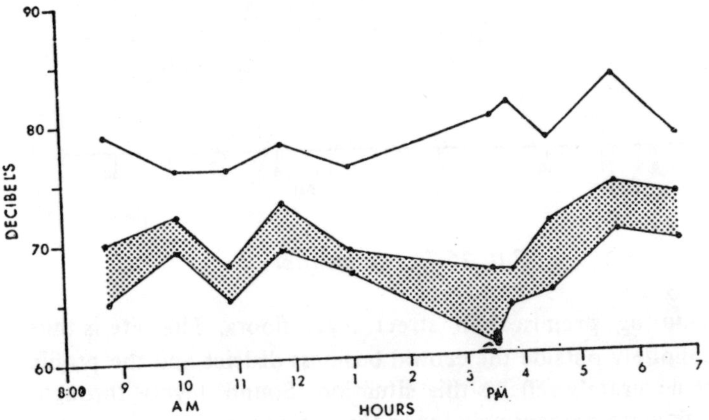

FIG. 25.6 : *Garden*

noise at Garden is quite unlike that at other sites. The greatest number of worshippers gather at the temple around daybreak and again at sunset. However, only sunset noise level has been recorded in this study ; sunrise levels would be approximately equal to those of sunset. In any event, the magnitude of these sounds is considerably below those of the central business district sites. Such reduced level is largely the result of two conditions. First, because the readings are taken in the centre of the garden, the distance to a roadway, about 20 feet, is greater than at all the other sites. Hence both pedestrian and vehicular sounds are lower. Second, the park contains trees, shrubs and other vegetation whose surfaces absorb sound, although this effect is not consciously aimed at as in Chinese cities.[4]

The final Site Six, Residential, was chosen to represent the transition between commercial and high density residential land uses (Fig. 25.7). Most of the properties in the vicinity are multi-family residential buildings with a scattering of shops and small

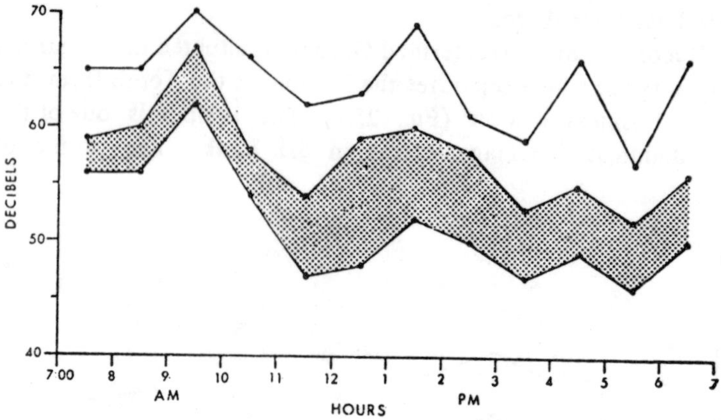

FIG. 25.7 : *Residential*

manufacturing premises on street level floors. The site is thus quite definitely outside the central business district and the profile of noise accurately reflects this situation. Sound levels throughout the day are perceptively lower than at other recording sites. Even maximum sounds are lower and usually represent a poorly muffled bus or truck. The mid-morning noise peak is related to the late opening of shops and increased vehicle and pedestrian traffic to these places. From 11.30 A.M. onwards this neighbourhood is noisy, firstly a consequence of school children, and later of workers returning for the midday meal at home. By 2.30 P.M. residents begin to return to the neighbourhood.

THE KANDY NOISE PROFILE COMPARED WITH OTHER CITIES

An average or generalized noise profile for Kandy has been constructed in Fig. 25.8 by taking the averaged midpoints of the readings at the six individual sites for each hour of the day. For practical purposes this profile may be considered the L-50 level for the community. Generally speaking the morning hours are noisier than afternoon, with major change in sound level between noon and 1 P.M.

In Fig. 25.8 the generalized noise profile for Kandy is compared with similar profiles obtained in two studies of specific

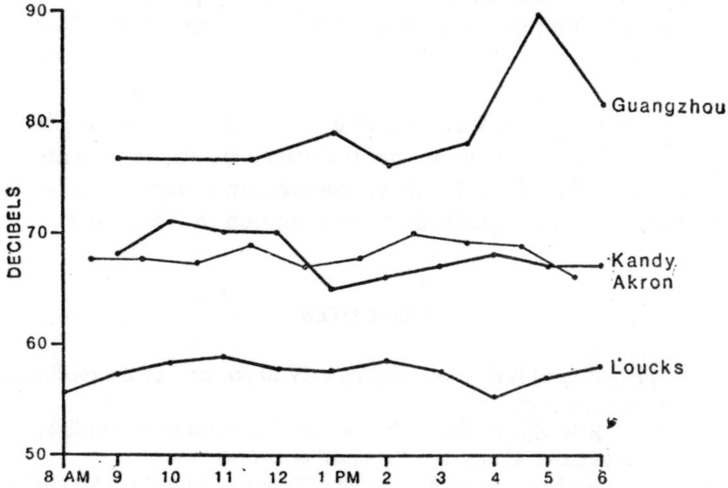

FIG. 25.8 : *Noise profiles of spactic cities* : *A comparison*

cities[5] and an L-50 profile by Loucks, in which noises occurring in many cities were averaged.[6] In all these studies, because recordings began too late in the day, the morning rush hour noise is not included, but an afternoon period of increased activity and noise does show up. In both the Loucks and Akron studies, a midday quiet period occurs. In Kandy the quiet period is somewhat delayed and in Guang-zhou (Canton), China, increased noise is encountered, probably the result of an influx of people to the combined commercial-residential neighbourhood where the single recording station in that study was situated.

In both the Loucks and Akron studies, sound peaks in mid-afternoon and then declines, most sharply in the Loucks study. In Kandy, the afternoon noise peak is at 4 P.M. and in Guang-zhou unusually loud noise levels are experienced at about 5 P.M.

CONCLUSION

The geographical variation of noise in Kandy appears to be closely related to land use. The daily profile of noise at each recording site varies, reflecting both land use of the surrounding district and the relative location of the site in Kandy's urban structure. The overall level of noise in Kandy appears to be quite comparable to that in a middle-sized American city, both of which are somewhat noisier than average as given in the Loucks study. However, Kandy is significantly quieter throughout the day than a residential-commercial section of a large city in southern China. All these conclusions must be tentative, based as they are on fragmentary data. They do offer, however, an intriguing point of departure for future studies of noise pollution in Asian cities.

FOOTNOTES

1. The L-90 level is that exceeded by only 10 per cent of the readings at a recording site.
2. The L-50 level is that exceeded by just 50 per cent of the readings at a recording site. Hence it is the mean or average level.
3. Allen G. Noble and R. Harnapp, 'Toward a Model for Monitoring Community Noise', *The Environment : Chinese and American Perspectives*, in Laurence J.C. Ma and Allen G. Noble (New York : Methuen, 1981), p. 286.
4. Allen G. Nobel, 'Noise Pollution in Selected Chinese and American Cities', *Geo-Journal* 4 (1980) : 573-75.
5. Noble, Noise Pollution, for Guang-zhou data ; Vern R. Harnapp and Allen G. Noble, 'Time Variations in Urban Traffic Noise : A Case Study in Akron, Ohio', *Sound and Vibration* 16 (1982) : 20-22.
6. Daniel P. Loucks, 'Noise : The Fourth Pollutant', *Engineering Cornell Quarterly* 15 (1980) : 9.

TWENTY SIX

S.L. KAYASTHA AND V.K. KUMRA

QUALITY OF HUMAN ENVIRONMENT
A Case Study of Kanpur City

THE environmental quality of cities in the country is steadily deteriorating. The standard of cleanliness, accommodation, transportation and civic amenities has fallen, while noise levels and dust concentration in the air have considerably increased in the past 20 years, and our cities are now becoming less and less suitable places to live in. This imbalance of the ecosystem is mainly attributed to overcrowding, uncontrolled growth of city population, lack of resources and basic facilities, lack of clean potable drinking water and sewage disposal systems, high traffic density and foul air. Yet our metropolises continue to attract ever-increasing numbers of people from the surrounding rural areas as well as from other parts of the country.

An explosive increase in population and industrialization have together wrought havoc with the urban environment. In the social context this has been reflected in two forms : more people are ill-housed, malnourished and underprivileged than ever before, while on the other hand a continuous degradation of environment has been noticed. India had a total urban population of nearly 108.8 million or 19.87 per cent in 1971 ; more than the population of most countries such as the U.K. Canada, Japan or France. It has been noticed that there has been a steady shift towards urbanization during the last 45 years. During this very period the total urban concentration rose from 12.5 per cent to 25 per cent of the total population of the country. The four largest cities, Calcutta, Bombay, Madras and Delhi continue to experience heavy growth. It is estimated that by the end of the 20th century about one-third of the population of the country will reside in urban areas. Along with these cities there are also other metropolitan cities such as Ahmedabad, Kanpur, Hyderabad and Bangalore where population congestion is rapidly increasing. The population of Kanpur is expected to rise from nearly 1.70 million (1981 census) to 3.0 million by the end of this century.

These cities are growing according to their master-plans in order to curb or eliminate irrational development. The impending city population explosion has posed a difficult problem for planners and administrators, with the colossal demands for potable water, adequate sanitation and sewage systems, proper disposal of solid waste, housing and slum clearance, not to mention the control of affluents from power plants, industrial establishments and diesel and petrol-driven vehicles which are continuously polluting the urban air as well as water bodies.

The present study relating to one of the worst-hit urban industrial environments of the country has been attempted in the above context. Kanpur, the biggest industrial city of Uttar Pradesh, highlights the fact that it is one of the dirtiest cities of India. The startling realities of the extent of environmental pollution in the city should well be an eye opener for persons who yet subscribe to the illusory notion that 'Indian cities have no such problems'.

A mass of old industrial establishments, hundreds of squatters and unauthorized colonies lack even the basic requirements of sanitation. There has been an unprecedented growth in migration from rural areas. A city with an area of about 300 km^2 is simply not ready to handle such a heavy influx of population. About 1.5 million people (1975) live in the city and the disparity between the rural population and those living in the urban agglomeration is alarming. The central part of the city, covering mostly old settlements, has a density of more than 60,000 persons/km^2. One of the important consequences of this rapid and unchecked population increase has been an acute housing shortage. Density of houses is found to be maximum in the central part of the city (100-250 houses/acre covering Moolganj, Rotigodown, Bakarmandi, Chamanganj and Birhara Road localities of the city. About 500,000 persons in the city live in slums. Unauthorized construction and slums have cropped up all over the city giving it a squalid look. According to the Master Plan the present rate of growth would require more than 200,000 dwelling units. Steep differences in rents is another factor for disparity in the density of population in Kanpur city. On the one hand the city today can boast of a number of sky scrapers in posh colonies with high rentals, while on the other hand many a family has to be content with a single room with common sanitation facilities.

The quality of environment of Kanpur is a result of several factors such as air, water and noise pollution, and unsatisfactory garbage disposal and sewerage systems.

AIR POLLUTION

The foul air in Kanpur city may force one to compare it with the conditions prevailing in London, Los Angles or Manchester. Dense smoke and toxic gases, covering the city especially during the winter months, have resulted in a dramatic rise in the occurrence of smog and an increasing incidence of all kinds of respiratory diseases. Plants have also been observed to be affected by dust particles emitted from industries and thermal power plants.

Several localities of the city have already started showing ominous signs of smoky effluence. Unlike Calcutta, Bombay and Delhi, air pollution in Kanpur has not been systematically surveyed, but a few studies made by the Indian Institute of Technology (I.I.T.) Kanpur reveal that air pollution in Kanpur has assumed serious proportions. The number of automobiles in Kanpur is increasing at a staggering rate of 8.5 per cent per year. In addition to these, there are many more vehicles that enter the corporation limits and go back from the city or pass through the highways. Faulty air exhausts are a regular feature of the city buses and interstate transport buses and lorries that belch out thick black smoke. Owing to the immense quantities of fuel exhaust, toxic pollutants such as carbonmonoxide, hydrocarbons, nitrogen oxides, organic vapours, lead and smoke are being added to the urban atmosphere. In a study conducted by one of the authors, it was observed that at certain congested crossings of the city such as Bara Chauraha, Parade and Nanarao Park crossing, the emission of Carbon dioxide varied from 1,185 to 2,085.0 g in a day (Fig. 26.1). The maximum concentrations of lead was found to be 8.3 $\mu g/m^3$ at Parade crossing and the minimum at Rawatpur crossing (0.95 $\mu g/m^3$). In some parts of the city, concentration of lead in air exceeds even the WHO air quality standards for lead level criterion of 2 $\mu g/m^3$. Some of the pollutants have proved to be mutagenic and carcinogenic. Increased accident liability which is dependent among other factors on visibility, vehicular density and noise is another direct consequence.

Kanpur is one of the biggest industrial cities of the country. Various types of industries, located in and around the city, discharge huge quantities of smoke, particulates and gaseous pollutants. The city annually consumes about 1.25 million tonnes of coal, which causes severe air pollution problems in the city. For a geographical analysis of air pollution intensity it is essential to know the spatial distribution of various pollutants in the city. For this purpose data have been collected from different localities of the city. Table 26.1 shows the extent of air pollution in certain localities of the city.

Table 26.1 reveals that the concentration of sulphur dioxide was found to be higher in the civil lines than in Anwarganj and Swaroopnagar because the civil lines is nearer to the old thermal power plant and the Elgin Mill which produce huge quantities of sulphur dioxide. But the concentration of dust particulates was maximum in Chamanganj locality (600 $\mu g/m^3$) because it is a very congested locality of the city. It was also recorded that the average monthly concentration of ammonia during different periods of the day varied from 20 $\mu g/m^3$ to 200 $\mu g/m^3$ from one locality to another. In addition to the huge quantities of smoke, continuous vibrations cause serious nervous disorders. With the conversion of the once secluded railway station into a junction, Kanpur Central Railway Station is yet another noisy colony in the making.

Another factor causing serious atmospheric pollution is the smoke emitted from countless domestic hearths. The consumption of domestic coal alone is about 10,000 tonnes a month, in addition to the use of dried animal dung and firewood. Railway engines also create serious smoke problems near the Railway

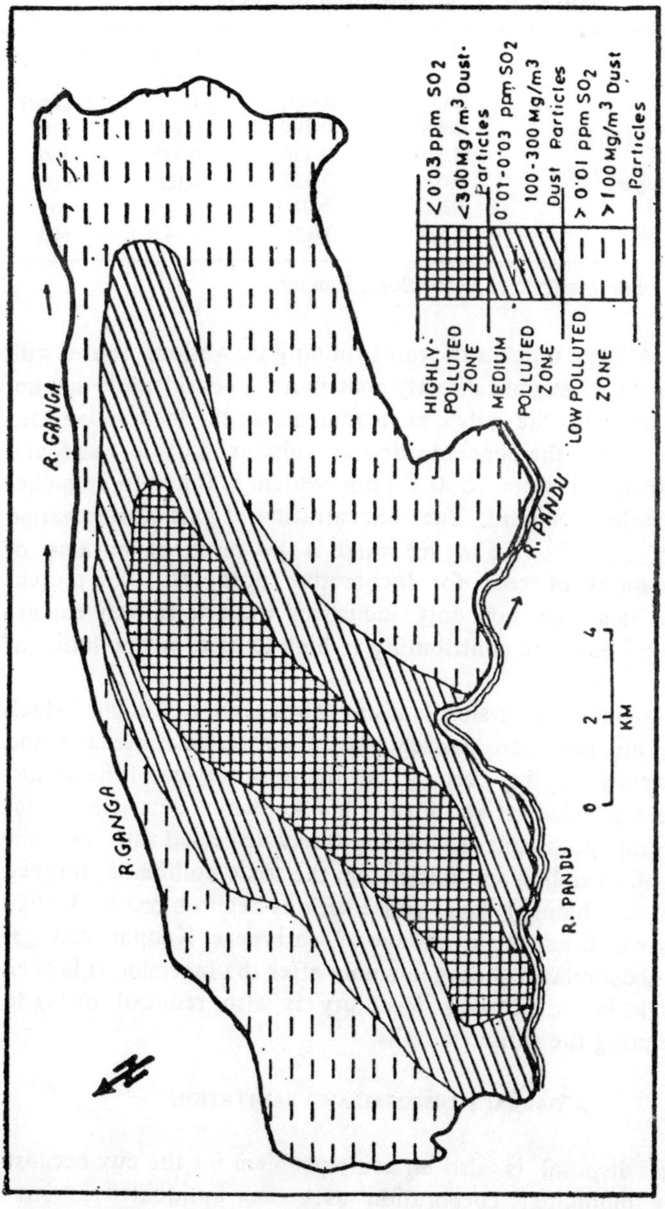

FIG. 26.1 : *Air pollution in Kanpur*

TABLE 26.1 : **Air pollution in Kanpur city**

Localities	So_2	No_2	Ozone	Particulates
	(ppm)	(ppm)	(ppm)	($\mu g/m^3$)
Civil lines	0.400	0.300	0.040	120
Anwarganj	0.110	0.130	0.135	400
Swaroopnagar	0.025	0.020	0.028	60
Vishnupuri	0.011	0.029	—	65
Chamanganj	0.038	0.050	—	600

Source : Indian Institute of Technology, Kanpur.

station. A large number of trains running on various routes with steam or diesel engines directly contribute a considerable amount of pullutants to the city's environment. In the locomotive yard of Kanpur city, the concentration of sulphur dioxide was found to vary from 0.1 ppm to 0.3 ppm which is found to be higher than the tolerance limit. The concentration of particulates varied from 3.53 mg/m^3 to 7.6 mg/m^3 which is the result of burning of huge amounts of coal for locomotive purposes. The present study reveals that not only industries but also railway engines and automobiles are contributing to high level of air pollution of the city.

Temperature inversion is also experienced in Kanpur which adds to air pollution particularly in winter. In winter the land cools rapidly in the evening but the upper layers of the atmosphere retain their heat and form a blanket which acts as a barrier and prevents pollutants from being freely dispersed into the air. Instead of ascending and dispersing, the toxic pollutants trapped beneath the blanket layer cause serious health hazards. Unlike Bombay which has the advantage of sea breeze, Kanpur can get rid of accumulated heavy air only after the inversion is broken by sunlight in the morning. Visibility is also reduced owing to smoke during the winter months.

A GARBAGE DISPOSAL AND SANITATION

Garbage disposal is also an acute problem for the city because the city municipal corporation uses an antiquated garbage removal system. Dumps of garbage situated at every street corner in the city remain uncleared for days together for want of an efficient system of garbage removal. This provides an open invitation to stray animals and birds, and constitutes an ideal breeding ground for flies which ultimately cause a hazard to human health. In the old and congested parts of the city areas such as Bakarmandi, Chamanganj, Baseerganj, Fazalganj, Rotimohalla, etc. it is a common sight to see pigs feeding on the filth and spreading the refuse on roads, ably assisted garbage by pickers looking for paper, plastic and whatever else they can find to sell. The stink from the open trucks carrying garbage persists in the areas through which they pass as they invariably spill some of the rubbish as they go along. Owing to the lack of adequate composting facilities through incineration, the refuse is generally dumped in the low-lying areas along the Grand Trunk road and Jhansi road on the outskirts of the city. Very recently a compost plant has been installed at Panki with a capacity of 200 tonnes of garbage every day, but it is not working to full capacity.

In addition, the sanitary facilities in the city can be safely categorized as inadequate and unhygenic. Water-logging owing to open drains can be seen in the old parts of the city. Choked sewers and uncleaned public latrines and urinals are a very common feature in various localities of the city. Some of the *ahatas* in the main city have open drains that carry waste water even from latrines. A canal passes through the main city which is the source of bad odour and causes severe pollution problems for the city.

WATER POLLUTION

The problem of water pollution is becoming a serious one for the city administration. In this industrial capital of Uttar Pradesh, more than 40 per cent of the water supply is contaminated. The residents of Kanpur thus suffer from diseases such as typhoid, dysentery, cholera, infectious hepatatis, gastroenteritis, etc. The Ganga, which is the main source of water supply to the city, is becoming only an effluent discharge stream carrying filth and poisonous industrial wastes. The main sources of water pollution are : discharge of sewage water, industrial waste water and surface run-off water containing pesticides and insecticide residuals in river and other water bodies of the city.

The Ganga receives the city's sewage and lethal amounts of toxic effluents and at the same time is the source of drinking water for the citizens of the city. About 70 per cent of the waste water of the city flows into the Ganga at various places, polluting the only source of water supply to Kanpur. It is because of this that one can never be sure of the quality of drinking water one gets in the city.

Every day about 40 million gallons of waste water are discharged into the Ganga through sewers. An analysis of sewage water showed that concentration of B.O.D. (Bio-chemical Oxygen Demand) in winter was 270 mg/l, while in summer it was only 240 mg/l, which shows that in summer waste water was bacteriologically more polluted than in winter. Concentration of B.O.D. per capita per day was found to be 31.7-39.0 g/l in Kanpur city sewage water. In addition to this, concentrations of sulphides, chloride and toxic heavy metals were also found to be high.

There are only three sewage pumping stations in the city but owing to the frequent breakdown of power and the inefficiency of the Kanpur Municipal Corporation they are non-functional during a major part of the day. The only sewage treatment plant at Jajmau is also not working to full capacity.

Along with sewage water, a number of outfalls carrying effluents of textile and tanning industries also join the river. The study of river water characteristics showed that at certain locations, the concentration of B.O.D. was 33.4 mg/l (at Bengali Ghat) which is above the tolerance limit. The concentration of chloride varies from 9.33 mg/l at Bithoorghat to 17.37 mg/l at Guptar Ghat. The mean value of dissolved oxygen (D.O.) was observed as 3.05 mg/l at Guptar Ghat and 5.5 mg/l at Bithoorghat. The reason is that at Bithoorghat no outfall joins the river and the water is suitable for drinking purposes. But after Bithoorghat, several nallahs join the river carrying raw effluents from industries and which cause a high level of pollution in the river Ganga (Fig. 26.2).

Another source of waste disposal is the river Pandu which receives huge amounts of waste water from the southern part of the city. The main source of pollution to the river is the Indian Explosives Limited and a few other small industries. An analysis of Pandu's water showed that it was highly contaminated with

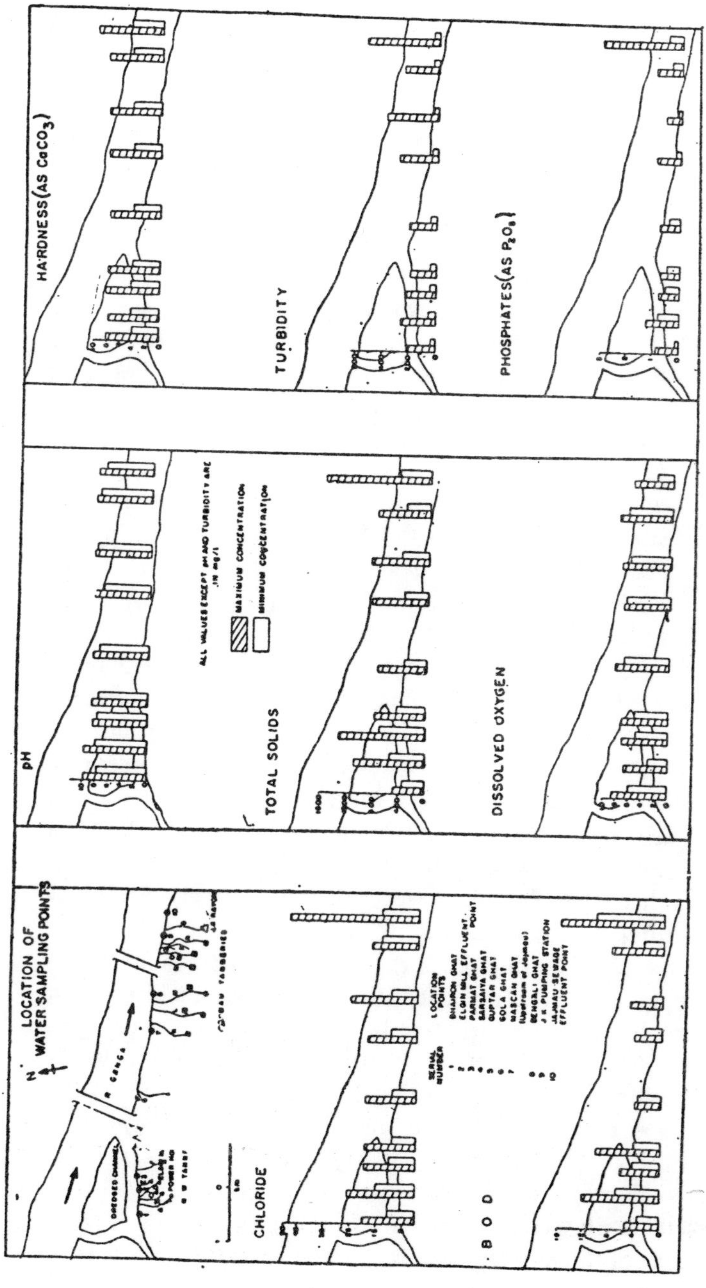

FIG. 26.2 : *Level of pollution in River Ganga*

urea, sulphides and ammonia nitrogen, resulting in a serious hazard to aquatic life. Sulphide concentration of 2.0 ppm to 5.0 ppm is more pronounced at the railway station area, Sersai village and waste water discharge joining point of Govindnagar.

A recent survey of Kanpur wells showed a very high proportion of intestinal bacteria and parasites. In addition to this, the concentration of fluoride was found to be higher than 1.5 mg/l in the water of most of the wells. More than 1.00 mg/l concentration of fluoride has resulted in the incidence of fluorosis among village children varying from 40 per cent to 75 per cent from one village to another. Taking into consideration the tolerance limits prescribed by the Uttar Pradesh Water Pollution Control and Prevention Board for total hardness, alkalinity, chlorides, sulphides, nitrates, bi-carbonates and flourides, water of nearly 65 per cent of the samples was found to be sub-standard. Most of the old villages in Kanpur lack potable drinking water facilities. For the Kalyanpur, Chakeri, Jajmau and Rawatpur settlement localities contaminated water is the root cause of most of the residents' ailments. A study revealed that scabies, boils, conjunctivitis, dysentery, malaria, typhoid, and in some areas jaundice were on the increase. In 1975 about 41 persons died of cholera. The incidence of jaundice increased from 2,380 in 1973 to 4,750 in 1977. This shows that because of the lack of potable drinking water facilities in certain areas of the city, the incidence of disease is increasing.

Kanpur's total requirement of water is estimated to be 307.5 million litres daily. The demand is likely to be about 500 million litres by 1991. However, the actual supply is far less than the total demand and out of this only 70 per cent may be considered as safe. On the outskirts of the city there is no piped water supply and in those localities where this facility exists, people have to queue up for hours for a bucketful of water. On the other hand a few colonies get round the clock water supply which is mostly used for watering gardens. This shows the misuse of water resources of the city in certain high-class residential localities.

The intolerable whispers

As a result of high noise stress, peace and tranquility have become more or less non-existent thus causing severe mental tension. In highly congested areas such as Parade, Mall Road, Gumti No. 5, Rawatpur and Birhana road the dinning noise caused by racing trucks, buses, cars, silencerless motor bikes and scooters makes life extremely uncomfortable. In addition to this the Kanpur citizen is also exposed to some additional noises made by the factory *kirtans* (devotional group singing) in the night which greatly disturbs sleep.

It is essential to evaluate the pattern of noise pollution in the city. Noise levels may substantially vary by location and time as well as by density of population. If the density of population is high, the level of noise pollution will also be high. Table 26.2 shows that the level of noise varies in different localities.

Kanpur is next only to a few cities of the country as far as noise level is concerned. Most of the traffic congested roads do not limit themselves to the International Noise Standards, which suggest a minimum distance of 15 ft. from the roads for residential dwellings. At some places in the old city and the Mall, roads pass so close to the houses that one could simply come out of the house and step right on to the footboard of a passing bus.

It has been observed that at a few roads crossings, the noise level varies from 80-88 decibels, which is intolerable to human beings. The accoustical division of the National Physical Laboratory, New Delhi measured an average noise level of over 90 decibels, seldom falling below 60-65 decibels. Bombay is also one of the noisiest cities in the world with a peak average of 75 decibels. Though the noise level of Kanpur is not as high as that of Delhi, Bombay, and Rio de Janeiro (the noisiest city of the world), there is every possibility that as the traffic density and industrial agglomerations increase, they will cause acute problems after one or two decades.

Many pathological effects of noise have been recognized. The effects of noise may vary from mild annoyance to irreversible changes and permanent hearing loss In addition to this, noise interferes in communication often causing accidents. It is possible to infer how 4.9 million people in the U.S.A. suffer from hearing damage—largely caused by noise jobs and environment. Data relating to diseases produced by noise are not available for Kanpur or any other Indian metropolitan city as noise research in the country has just begun. However, Kanpur doctors do

TABLE 26.2 : **The range of outdoor noise for different residential localities**

Residential Locality	Typical Range decibel (A)	Average decibel (A)
Quiet suburban residential	36-40	38
Normal suburban residential	41-45	43
Urban residential	46-50	48
Noisy urban residential	51-56	53
Very noisy urban residential	56-60	58

report many cases of central nervous system disorders, neurosis headaches, high blood pressure and loss of memory among the residents because of the increasing nuisance of noise. The Medical College, Kanpur has registered an alarming increase of over 6 per cent in the incidence of deafness, particularly among the young people, during the past three years. In addition to this, industrial workers working in machine rooms have reported many cases of deafness to the hospitals.

Noise of high level affects the heart and decreases supply of oxygen to the brain. Interference with digestive processes has also been reported owing to problems caused by noise interference in communication (R.F. Christman, U.S.A.).

TABLE 26.3 : **The thresholds of noise pollution for human health**

Levels	Decibels
Threshold of audibility	0
Pain threshold	120
Extreme limit of human tolerance	140
Prolonged exposure causing insanity	140
Prolonged exposure causing burning of skin	150

Thus the analyses carried out reveral that Kanpur's environmental quality is deteriorating day by day. In order to preserve the environmental quality of the city and strive for a better environment, comprehensive environmental planning is suggested. The most important tasks, in order to overcome the detrimental environmental repercursions of industrialization and urbanization, is to clearly analyse the problem. With a clearer understanding of the noise factors that adversely affect the society, it could be possible to organize laboratory research and work out programmes to disseminate knowledge thus gained and to suggest necessary preventive measures.

There must be proper environmental quality standards for cities so that the quality of the environment may be easily monitored. This may facilitate the formulation of plans for improving the surroundings. Necessary legislation regarding air, water and noise pollution in the cities may be enacted in order to ensure a better and healthier environment for the common citizen.

REFERENCES

Berry, Brian J.L. *et. al.*, *Noise Pollution, Land Use, Urban form and Environment*, 1972, p. 192.

Desh Bandu and Eklavya Chetan edited, *Current Trends in Indian Environment*, 1977, pp. 79-92.

Harry Rothman, 'Murderous Provinces, A Study of Pollution in Industrial Societies', *The Pollution Crisis in Cities*, London : Report Hart Davis, 1972, p. 192.

Harry Rothman, 'Murderous Provinces, A Study of Pollution Crisis in Cities', Report Hart Davis, London, 1972, p. 66.

Kumra, V.K., *Kanpur City : A Study in Environmental Pollution* (Ph.D. thesis, B.H.U., 1979).

State Health Institute, Lucknow.

SECTION EIGHT

POLITICAL GEOGRAPHY

TWENTY SEVEN

FRIEDRICH STANG

CYPRUS
Aspects of the Political and Economic Geography of the Partitioned Island

THE ROOTS OF THE "CYPRUS PROBLEM"

THERE are few countries whose fate has been so much influenced by their geographical location as Cyprus. Only 75 km off the Turkish coast, about 100 km from Syria, less than 400 km from Egypt, but 850 km from the Greek mainland, the island—with a population of 646,000 in 1982 and with its 9,251 km² the third largest in the Mediterranean—is not only of the greatest strategical importance but it has also been a centre of commerce and trade from the earliest time (Fig. 27.1).

Every power that wanted to dominate the eastern Mediterranean and its neighbouring regions has tried to establish itself in Cyprus or exert some influence on its affairs. So its history has always been determined by developments outside the control of the indigenous population. The list of occupants of the island in early history included the Mycenaeans, Phoenicians, Assyrians, Egyptians, Persians, Macedonians to the Romans. During its 300 years as a Roman province Christianity was introduced into Cyprus and later developed into an autocephalous "national" church. After the separation of the Roman Empire into West and East, Cyprus was part of the East Roman Empire of Byzantinum from 330 A.D. to 1191 A.D. Three hundred years of rule by the Lusignans, a French dynasty that came to power in connection with the crusades in Asia Minor, followed. Then the Republic of Venice took over the island in 1489 for a shorter period, using it

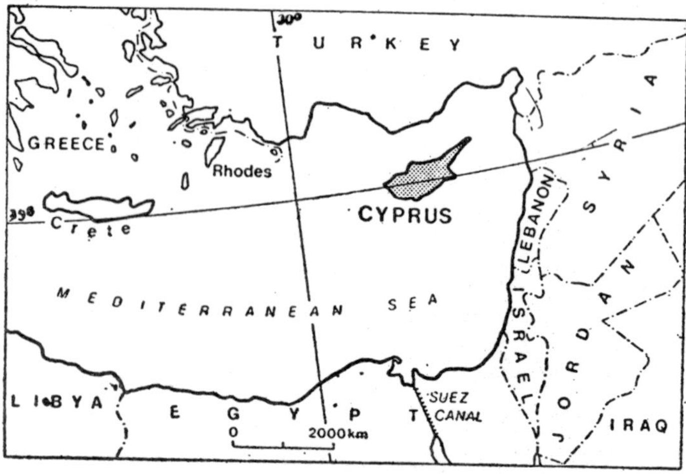

FIG. 27.1 : *Cyprus and the eastern mediterranean*

as an entrepôt for its eastern trade and as a bulkwark against Turkish attacks.

During all these centuries—since Greek settlers had come in the wake of the early Mycenaean traders and colonized the island in the 15th and 16th centuries B.C.—the population, its language and its culture has been Greek. Of all the invaders none had settled except a small ruling class. A new element was added to the population when Cyprus was conquered by the Ottoman Empire of the Turks : after 1571 about 22,000 Turkish settlers came to Cyprus.

The Turks were Moslems, but the Christian population was granted a large measure of autonomy. Their Archbishop was accepted as the speaker of his community and even given administrative powers. So the role of the Archbishop as the religious as well as the political leader, which today is still of such importance goes back to early Turkish times. Although the two communities lived as neighbours, they did not mix. This was not only owing to the difference of religion but also of language and a different cultural background. When the Turkish rule became more oppressive, there were serious risings between 1764 and 1821 but they were not caused by religious controversies but by the demand for more autonomy and less taxes.

The geopolitical situation of the 19th century in this part of the world was characterized by the decay of the Ottoman Empire and by the attempt of the British to consolidate their position in the world, particularly in the Mediterranean, which made Cyprus a focus of their interest. After the defeat of the Ottoman Empire in the Turkish-Russian War the Sultan concluded a treaty of assistance with Great Britain and in exchange handed over Cyprus to British administration in 1878. Having occupied Gibraltar in 1704 and Malta in 1800 Britain now extended its control to the eastern Mediterranean, which was since the opening of the Suez Canal the main route to India. During the First World War—which the Truks fought on the German side—Britain annexed Cyprus, gave its inhabitants British nationality and declared Cyprus a crown colony in 1925. Under British rule the economy, which previously had been depressed by high taxes and a number of natural calamities such droughts, epidemics and earthquakes, recovered and the population of the island grew from 186,000 in 1881 to 348,000 in 1931 and 574,000 in 1960.

For centuries the two ethnic and religious groups had lived together on Cyprus without serious problems in their relations. It was only in our century that the differences between them—which eventually led to the partition of the island—slowly started to build up. The two groups who before had considered themselves as Moslems or orthodox Christians, now began to speak of each other as Turks and Greeks. The idea of the European ethnic nation-state and of thinking in these categories—which usually went together with considering neighbouring nations as adversaries—had reached the island. Nationalism gained a new dimension after the First World War. The disintegration of the Ottoman Empire and the creation of a Turkish state resulted in a Greek-Turkish War, which also had repercussion on the relations of the two communities in Cyprus.

The spread of nationalist ideas on the island was supported by its school system. Under Turkish rule a strict separation of schools for Greeks and Turks had been established and this continued under British rule. The expansion of education made things worse, not so much because of a *divide et impera* policy behind it but simply because of the fact that more children now want to school. The administration for each group of schools was in close contact with the educational authorities in Athens or

Ankara, who supplied teachers, influenced the curricula, and sent the textbooks. The schools thus not only taught in different languages but they also had a different outlook on history and culture. They tended to educate children to consider themselves not as Cypriots but as Turks and Greeks who happened to live in Cyprus. Thus the co-operation of both groups in public life and in common organizations slowly broke up. The trade unions in the towns and the co-operatives in the rural villages, whose membership had included both communities, were accordingly reorganized.

The situation worsened after World War II with the spread of independence movements in the colonial empires in general and in Cyprus with the election of Makarios III as the new Archbishop and religious and political leader in particular. Makarios and the Greek Cypriots, who soon followed him blindly, fought for the end of British rule, but not in order to create an independent Cypriot state but to realize *Enosis*, the unification of Cyprus with Greece. For the Turkish Cypriots this would have meant living as a very small minority in a Greek Cyprus, so they now demanded *Taksis*, the partition of the island. Turkey which already felt threatened by the fact that its west coast was lined with Greek islands only a few miles off, considered a new Greek base near its important ports on the south coast as unbearable for strategic reasons.

In 1954 the Greek Cypriots with support from Greece founded a military organization to fight the British. As a result terror sm raged throughout the island. Finally, the British decided to grant Cyprus independence under the condition—to be guaranteed by Greece and Turkey together with Great Britain—that Cyprus as a whole or part of the island was not allowed to form a political or economic unit with any other state. Great Britain kept the sovereignty over its two military bases at Akrotiri and Dhekelia. There are, however, no formal boundaries between this British sovereign areas and the territory of the Cyprus Republic.

On 16 August 1960 the constitution was signed, with Archbishop Makarios as the new president declaring that it was forced on him and that *Enosis* was still the ultimate aim. The Greek Cypriots had 70 per cent and the Turkish 30 per cent of the 50 seats in the house of representatives. Important laws had to have the consent of a majority of both groups. The president of the republic (who had to be a Greek Cypriot) and the vice-president (a Turkish Cypriot) each had an additional veto power. Apart from the house of representatives there were two communal chambers, one Greek and the other Turkish Cypriot, to deal with the cultural and social matters of their respective communities. The percentage of both communities in public service and in the army was also laid down and led to the complaint of the Greek Cypriots that the minority of Turkish Cypriots was over-represented.

With much good will from all concerned the constitution might have solved the problems of the new republic. But this good will did not exist, and the constitution became a tool for each ethnic group to insist on its rights and to gain advantages over the other community. The constitution was not used to build a new state, but instead it played an important role in its disintegration.

THE PARTITION

Already during the fight for independence the conflicting interests of the two groups had reached such a pacs that any co-operation in a new state would have 'posed the greatest of difficulties. Therefore, since 1956 a partition of the island was under consideration. A British plan suggested a north-south partition with the western part for the Turkish and the eastern part for the Greek Cypriots. Whereas Turkey advocated an east-west line of partition along 35° lattitude which gave the Turkish Cypriots the north and the Greek Cypriots the south of the island. The problems of resettlement and the resistance of the Greek Cypriots to any partition finally led to the formation of the unpartitioned state. At the time of independence in 1960 of the 623 villages in Cyprus, 63 per cent were purely Greek, 19 per cent Turkish and 18 per cent had a mixed population. The percentage of pure Turkish Cypriot villages was particularly high in the regions of Paphos and Famagusta. It must be borne in mind, though, that the Turkish Cypriot settlements did not form one or a few continuous zones, but were dotted all over the island.

After only a few years of quiet, fighting broke out again

between the two groups by the end of 1963. When the Turkish representatives boycotted the sessions of parliament, the Greek Cypriot rest-parliament brought up the strength of the army, the 'National Guard', to about 15,000. Greece as well as Turkey, who were both entitled to keep a small contingent of soldiers on the island, increased their numbers against the treaty. In the course of the fighting and despite a UN peace keeping force sent to Cyprus in 1964, a spatial separation of the two groups took place. A demarcation line was drawn in the walled city of Nicosia. In the country side the Turkish Cypriots fled to Turkish villages for safety and concentrated in the districts north and west of Nicosia. By 1964 about half of the Turkish Cypriots lived in enclaves. Their establishment led to a complete isolation of the Turkish Cypriots. The Greek Cypriot government tried to break up the enclaves by economic sanctions and allowed only supplies of food to pass through. Though these blockades inflicted great hardship on the Turkish Cypriots, they failed to break their resistance. After renewed fighting had taken place in 1967, the tension eased. In 1968 the blockade was lifted but important roads leading through the enclaves were still only open to Greek Cypriots travelling with a UN convoy. Under these conditions any political and economic co-operation between the two groups could not be expected.

Since the spatial division in the form of enclaves had proved unworkable, the Turkish Cypriots—who all the time had to rely on Turkey for their support—kept insisting on partition or the establishment of federal state. On the other side, Greek interference in the affairs of Cyprus and differences between the military regime in Athens and the leftwing Archbishop in Nicosia made many Greek Cypriots more reluctant about a union with Greece. In July 1974 Makarios demanded from the government in Athens the withdrawal of its troups from Cyprus. In this situation the National Guard tried to depose Makarios, who was able to flee to England. Since the Turks now expected the Greek officers and the new president to unite Cyprus with Greece by force, they landed their troups at Kyrenia in July 1974 and extended this bridgehead till they occupied 37 per cent of the area of Cyprus in the northern part of the island. The result was an exodus that affected more than a third of the population: 160,000-180,000 Greek Cypriots fled to the south and 45,000-

54,000 Turkish Cypriots (the figures given by each side differ) went to the north. Of the island's total population of 621,000 in 1979, 473,000 lived in the south and, 148,000 in the north.

The Turkish cease-fire line runs over 180 km from the north-west to the south-east, separated from the Greek Cypriot territory by a safety zone of between a few metres and several kilometres width, being manned by UN forces (Fig. 27.2). For Greek or Turkish Cypriots it is next to impossible to cross this line, and also foreign tourists staying in the north cannot visit the south. Tourists coming from the south may enter the northern part in Nicosia but can only visit the towns of Kyrenia and Famagusta for a day via the two main roads leading there from Nicosia.

After the Turkish invasion the Greek Cypriots, who considered their government as the only legal representation for all of Cyprus, introduced an economic embargo on the north. They not only cut their own trading connection with the north but also insisted on other countries to do the same. So a ship coming from the port of Famagusta in northern Cyprus will not be allowed to enter a southern Cyprus port. International travel agencies or airlines going to the North would lose their licences for south Cyprus or their landing rights.

In 1975 the north proclaimed the 'Turkish Federated State of Cyprus' which, however, was not internationally recognized. Since then talks between the two groups have dragged on. The main points of conflict concerned :

1. The territorial question : The Greek Cypriots were not prepared to cede far more than a third of the island's territory to the Turkish Cypriots.

2. The status of the two "states" : The Turkish Cypriots wanted as much autonomy as possible, the Greek Cypriots a strong central government.

3. The question of property of refugees : the houses and the property of the Greek Cypriots who fled to the south were taken over in the north by Turkish refugees and settlers and a restitution to the original owners is very hard to bring about.

4. The equality of the two groups : The Turkish Cypriots demanded equality of rights for the two states and also a constitutional guarantee for equality in education, employment and pay, which the Greek considered impossible.

After many years of unsuccessful talks between the two

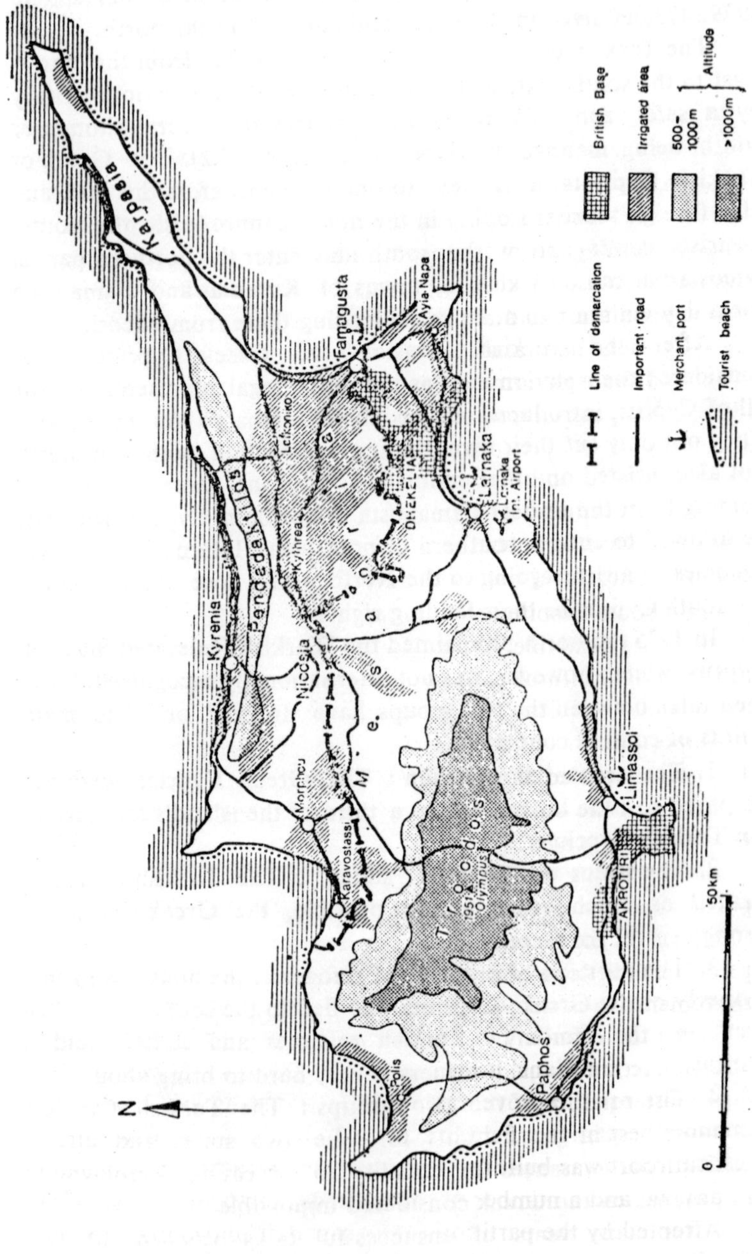

FIG. 27.2 : *Cyprus*

groups the Turkish Cypriots declared the North an independent state on November 15, 1983.

ECONOMIC DEVELOPMENT IN THE SOUTH

In spite of the political situation, the fighting between the two ethnic groups and the blockade of the Turkish Cypriots in the enclaves, the standard of living in Cyprus rose considerably in the decade 1964-74 and was higher than that of neighbouring countries such as Turkey, Egypt or even Greece. But the Turkish Cypriot population had no share in this new prosperity. The Greek Cypriots had always held trade and economy in their hands, and the isolation of the Turkish Cypriots in the enclaves widened the material gap. The partition of the island in 1974 with the concentration of Turkish Cypriots in the north and Greek Cypriots in the south gave a new spatial dimension to the economic disparities between the two ethnic groups. While in the north the economic potential could not be fully utilized and the economy, therefore, stagnated, the south experienced an economic boom despite the fact that here important resources were no more available. The north before 1974 supplied 70 per cent of the agricultural production ; two-thirds of the tourist beds were located here. The northern mountain chain was also the most important area of mining. The port of Famagusta which fell to the north handled 83 per cent of the movement of goods for the island and the only international airport near Nicosia could not be used any longer.

The following outlines on the restructuring of the most important sectors of the economy after partition will have to concentrate on the Greek Cypriot south, because on the Turkish Cypriot north little information is available and even the freedom of movement is restricted.

To put the economy of the south on a new basis 'Emergency Economic Action Plans' have been set up from 1975-76 onward. A very substantial improvement of infrastructure was necessary. This shows best in the field of communications : a new international airport was built in Larnaka, new seaports in Limassol and Larnaka, and a number of new roads substitute those that were disrupted by the partition of the island. The two main pillars of the Cypriot economy which were most affected by partition agriculture and tourism, received strong promotion, and in addition more industrialization was aimed at.

Agriculture

Of the main areas of agricultural production the citrus fruits growing district of Morphou, the olive plantations of Kyrenia and Kythrea and a substantial part of the grain producing central plain of Mesaoria fell to the North. The Greek Cypriots were left with the hinterland of Paphos and Limassol, where citrus fruits and olives are grown, and the plains around Larnaka and south of Famagusta which are to a large extent irrigated. Most of the vineyards are on the hillsides of the Troodos mountains, a region not affected by partition. The grapes are made into wine and sherry by big firms in Limassol which also take over the marketing and exporting. In the plain near Limassol dessert-grapes are grown on irrigated areas.

To make up for the loss of fruit trees in the north, lines of agriculture and horticulture were developed which allowed a quick adaption. Early vegetables and flowers for the European market are raised in simple hot-houses made of plastic sheets. Early potatoes have become one of the main export items.

Irrigation has been much extended. But in the mediterranean climate of Cyprus apart from the mountainous regions, only the winter months bring sufficient rain. So with the increasing use of irrigation, of industries, and also of hotels, water is getting scarce. New dams are built in the mountains, subsoil water is tapped, and the available water is used more efficiently. Dry farming in the central plain is improved because the south cannot meet the demand for wheat as the most important food, and for barley for feeding.

Apart from the efforts to raise agricultural output, regional programmes for the advancement of agriculture in backward areas which were already started before partition, were continued. They concern mainly the improvement of infrastructure in the Troodos mountains and better marketing facilities to improve incomes and so to stop migration to the towns.

A special problem arose with Britain's accession to the European Community. The partial loss of Britain as the biggest market for agricultural products from Cyprus made readjustments necessary by strengthening trade links with countries in the

Middle East and Eastern Europe. Cyprus itself is an associated member of the Community since 1972, but the establishment of a customs union, which was intended in two steps, has not yet been achieved—mainly owing to the unsettled political questions. Competition for the agricultural exports of Cyprus comes from the European Economic Community (EEC) states, France and Italy and lately from Greece ; it will grow when Spain and Portugal join the EEC.

Tourism

Cyprus is attractive for sun-hungry tourists from western and northern Europe because of 340 days of sunshine in the year. Apart from the sandy beaclies there are the resorts in the Troodos Mountains with a pleasant climate during the summer months and facilities for skiing in winter. The island also offers many interesting archeological sites from its long history.

After independence great efforts were made to develop international tourism and in economic planning priority was given to it. The importance of tourism as a foreign exchange earner grew as traditional sources of foreign exchange—the export of minerals and the income from British bases—were drying up. Another reason for fostering tourism was the employment it offered and the opportunities to advance regional development of the island through tourism. The number of visitors rose more than tenfold from 1964 (25,700) to 1973 when the income from tourism amounted to 40 per cent of all foreign exchange earned.

The warlike conditions of 1974 and the partition of the island caused severe setbacks. Famagusta and Kyrenia, the main centres of tourism, fell to the north. With much support from the government the Greek-Cypriots built new hotels in the southern part of the island catering for higher and middle income groups, concentrating on Limassol, Paphos, Larnaka and Ayia Napa—the latter in the extreme south-east as a sort of substitute for Famagusta. The hotel capacity which in 1973 stood at 10,800 beds in the undivided island and at only 3900 in the south, has risen to 12,600 in the Greek Cypriot part (1982) and the number of foreign visitors from 264,000 (all Cyprus) to 548,000 (south only). Foreign exchange earnings from tourism account for about a fifth of total earnings from exports of goods and services. But the rapid growth of tourism has begun to show disadvantages :

hotel staff is getting scarce and buildings and building sites have begun to spoil some parts of the island along the coast.

Industry

The development of industry in the southern part was not merely intended to balance the losses in agriculture but also to offer new employment to refugees from the north. With the support of the government almost 200 small units were established in 1976, a number of these had formerly operated in the north. Those industrial units which are mainly working for export are in the textile industry, in the manufacture of shoes and other leather goods and particularly in food processing and in the cement industry. Export orientation is a must, because the local market is too small. But international competition demands factories of bigger size which are able to use modern technology and can work more efficiently than most of the existing small and labour-intensive industries and cottage industries. In textiles and shoes Cyprus meets fierce competition from the low-wage countries of the Far East and tries to evade it by going into high-quality products. In the last years the economic boom in south Cyprus has overheated, resulting in higher wages, labour troubles, inflation and an unfavourable balance of trade.

Partition has also caused remarkable changes in the locational structure of industries. Earlier, the few industrial units—apart from some raw material orientated industries—were located at Nicosia as the biggest local market. To supply employment for refugees the existing industrial zone in Nicosia has been extended and south of it, near Strovolos, a new industrial area has been established together with a satellite township. The most striking expansion, however, took place in the new industrial areas of the port towns of Larnaka and Limassol. The industrial zone near Paphos on the other hand which was meant to boost the regional development of the area has not been such a success. Also, the establishment of export-oriented zones has not caught on well. The zone near the airport in Larnaka is still without a single factory.

DISPARITIES OF NORTH AND SOUTH

The influence of outside forces, which has for so long determined

Cyprus' political fate, can again be traced in its recent economic development after partition. The economic development in south Cyprus must of course first be attributed to the energy of the Greek Cypriots and their determination to carry on. But external factors have been of great help : financial aid from Greece, European states, and the USA has poured in, and, if effectively administered, comparatively small sums can have great results in such a small state. In addition, Greek Cypriots living in foreign countries were willing and able to invest in the south, and the government has given them every possible incentive. The advantageous location between Europe, the Middle East and North Africa has enormously helped the development of trade which gained an additional impetus with the reopening of the Suez Canal. The economic boom in the rich oil states opened new export markets, offered some employment in the first crucial years after partition, and now brings additional tourists from these countries. The conditions in Lebanon not only brought many wealthy refugees but also had the effect that part of the commerce which was formerly handled there was shifted to Cyprus.

The Turkish Cypriots—partly as a result of the development in the last hundred years and especially during the decade before partition—had less training and less economic power. Though some of the advantages which helped the south are valid for the whole island, they could not have a positive impact on the economy of the north because of the political situation. The Turkish occupation, international non-recognition of the Turkish Federated State of Cyprus and the embargo by the Greek Cypriots brought the north into political and economic isolation.

International tourism almost came to standstill and, is only very recently and slowly recovering. For years, 90 per cent of the visitors were Turks from the mainland who could not afford the high standard of the hotels but only came on a short excursion mainly for duty-free shopping. The hotels were formerly owned and run by Greek Cypriots and the Turkish Cypriots had little experience in this line of business. The difficulties are aggravated by the embargo of the Greek Cypriots who threaten tour operators with legal consequences if they use the requisitioned hotels in the north.

Agricultural exports meet with great difficulties : citrus fruits, the main export crop, are marketed in Europe via Turkey and a lot cannot be sold at all ; this gives farmers little incentive to improve production. Financial help from Turkey is substantial and keeps the north alive. But the Turkish Lira has been such a weak currency that imports from countries other than Turkey have to be paid for in currencies bought at black market rates. This has made raw materials too expensive and finished goods from the north uncompetitive.

The recent economic and political development gives the impression as if the Turkish Cypriots are now beginning to feel like the Greek Cypriots ten years ago, when they gave up their demand for 'Enosis' : The influence of the "mother country" is becoming too dominant. The declaration of independence in November 1983 by the Turkish north is certainly a reaction to the frustration resulting from international non-recognition of the Federated State and to the long unsuccessful talks with the Greek Cypriots, whose economic success has given them a strong bargaining position and may have made them—and the rest of the world concerned—complacent about the present political situation. But the declaration of independence of the north could also be a move towards more independence from Turkey, and so perhaps pave the way in the future for co-operation between the north and the south.

There is still some hope that the two communities could work together and thus solve the "Cyprus Problem" themselves. Then Cyprus might become not only a prosperous but also a more independent island.

REFERENCES

Bundesstelle für Aussenhandels information, Markt-information, Zypern—Wirtschaft in Zahlen, Köln, 1983.

Central Bank of Cyprus, Annual 1976ff, Nicosia, 1977ff.

Cyprus Geographical Association (ed.), *International Symposium on Political Geography*, Nicosia, 1976.

Cyprus Tourism Organization, *Annual Report*, 1974ff, Nicosia : 1975ff.

Drury, Michael, P., 'The Political Geography of Cyprus'. In John, I. Clarke and Howard Bowen—Jones, *Change and Development in the Middle East* London and New York : 1981, pp. 289-304.

Hahn, Barbara, Die Insel Zypern. Der wirtschafts—und politisch-geographische Entwicklungs—prozeB eines geteilten Kleinstaates-*Jahrbuch d Geogr Ges.* zu Hannover, Hannover, 1982.

Heinritz, Günter, Der griechisch-türkische Konflikt in Zypern und seine siedlöngs—und bevöl-kerungsgeographischen Konsequenzen bis zum Sommer 1974. In *Geographische Rundschau*, 27, 1975, pp. 93-99.

HMSO, *Cyprus*, Cmnd, 1093, London, 1960.

Karouzis, George, *Proposals for a Solution of the Cyprus Problem*, Nicosia : 1976.

Lee, Michael and Lee Hanka, *Cyprus* : Newton Abbot, 1973.

Patrick, Richard, *Political Geography and the Cyprus Conflict* : 1963-71, Publication Series, No. 4, Waterloo/Ontario : Department of Geography, 1976.

Public Information Office, *Basic Facts about the Turkish Federated State of Cyprus* (North-Nicosia : 1979).

Stang, Friedrich, Zypern-wirtschaftliche Entwicklung und Strukturveränderungen seit der Teilung der Insel. In : *Aachener Geographische Arbeiten*, 14 II, 1981, pp. 358-359, Aachen.

Government of North Nicosia, *Statistical Year-book 1979*, North-Nicosia : State Planning Organization, 1980.

Statistisches Bundesamt, Länder-kurzberichte Zypern 1975, 1978, 1981-Allgemeine Statistik des Auslandes, Stuttgart and Mainz.

TWENTY EIGHT

R.C. SHARMA

GEOPOLITICS BEYOND SHATT-AL-ARAB
Perspective on the Iran-Iraq Conflict

THE present Iran-Iraq conflict symbolizes the age-old spatial relationships of two contending political systems, with a varying *raison d'etre*, converging on a common border zone, where the major sources of friction and conflict lie in the general prevailing situation rather than in the boundary line itself.[1]

The present paper tries to investigate the issues involved between the two sovereign states in the context of their space-relationship in the Gulf region, where undoubtedly the desire for seeking complete suzerainty over the Shatt-al-Arab watercourse has been of prime importance through many centuries. The present situation is not very simple as it involves much more complicated questions than merely the question of boundary adjustments. In view of the complexity of the situation, it is not surprising that all attempts to resolve this conflict have so far failed.

AN OVERVIEW

The nations in question, are overwhelmingly rural with a very strong state set-up, has emerged out of a long history of adjustment in relation to the various internal and regional processes, both socio-economic and political in nature. The oil-based economy has further strengthened the manoeuvring capabilities of both the nations within the regional and international framework. The physical diversities are equally matched with the ethno-linguistic diversities, which are greatly magnified on the borders of these two nations—Iran having roughly 3.5 times more geographical area and 2.5 times more population compared to Iraq. This scale of distance puts certain limits on many aspects of these two countries, specially on the various strategic issues. The physical-ethno diversities on the common border zone place many stresses on the effectiveness of the central authority as well as create ambivalent framework which frequently comes under pressures of varying magnitudes resulting many a time in turbulent conflict situations.

Iran's demographic structure and composition has always been effective in favour of socially dominant classes comprised of tribal leaders, landowners, merchants, aristocrats and the *ulemas*. 'The states power remained conditional on the compliance of the pre-existing rural potentiates . . . . *in the form of landowners and the ulemas* . . . .[2] Specially after 1940s there have been significant changes in Iran, both in its economy and society but with serious spatial imbalances, resulting frequently in internally eruptive situations. Although there have been shifts in the power of *ulemas* but the processes of modernization and social transformation always got badly hit. Heikal and Stempel (1980) blame the increasing inability of the ruling institutions to respond to the revolutionary situation and they feel that the story of the present Islamic revolution is not over and Iran's future prospects are cloudy at best, disastrous at worst.[3]

Iraq being a typical product of hydraulic society has roots of social and economic transformation in its long history of man-environment-relationship on the great Mesopotamian plains. Recent history of Iraq tells of a sequential takeover of the Ottoman legacies by the European powers. However, it was in a slightly modified form, leaving many bilateral issues between these two countries as the basis for constant irritants. April, 1920 saw the British as the sole masters of these plains, ushering in an era of economic exploitation and military domination, which later culminated into the Baghdad Pact of 1955. Iraq remained in a state of turmoil for a very long period, ultimately paved the way for the July revolution. It enabled Iraqis to come out of the military alliances and the economic grip of the multinationals. Modernization and renewed cultural order have been the basic stems of contemporary Iraq. Socialism, as the fountainhead of

Arab nationalism, has been adopted as the cardinal principle to decide the internal developments and external linkages.

Oil production and oil-reserves are mostly concentrated on the borderlands of these two countries, largest reserves are to be found in the South, that is, beneath the swamps in the Shatt-al-Arab. The North is equally significant in terms of oil wealth. Most of Petro-chemical industrial development is in the strategic south. Ethnicity further complicates the matters in the south as well as in the north, where the two significant minorities of Iran, that is, the Kurds in North and the Arabs and Bahais in the South merge smoothly in the cultural landscape of the adjacent areas of Iraq.

Shatt-al-Arab has been the most strategic region on the Gulf, as it provides the lone outlet to the ocean for the Mesopotimian plains. It has always been the target of the contending power in the region to have a complete suzerainty over this water-course, which fluctuated seasonally and intermittantly during the course of history. No doubt the water-courses have certain advantages as inter-state boundaries but they have seldom proved satisfactory, as they unnecessarily divide communities and create problems in water-utilization and management. The shifting courses bring in confusion and disruption in the life of the border residents. The case of this water-course has undergone severe tests in the recent history of the region.

The conflict on Shatt-al-Arab is unique in the sense that it has survived all treaties and border agreements, signed from time to time during a very long history of about four centuries. It has emerged repeatedly in one form or another.[4] It has belied all efforts for a permanent settlement. Territorial transgressions and abrogations over the agreed upon details symbolise the essence and ethos of the quest for sovereignity over the border region. The Accord of Algiers is explicit in the sense that it proclaims sanctity of the borders and non-intervention in the domestic affairs of the contracting parties. 'The Accord of Algiers embodied a settlement in which the legal and political aspects are so balanced that an encroachment on any of its elements is bound to upset this balance and consequently lead to the virtual nullification of the Accord itself.' It is widely believed that the Shah of Iran forced upon Iraq this agreement with the help of the western alliances. The rule of Thalweg line was adhered to resulting into territorial gains to Iran.

Developments and forces

Where do the roots of Iran-Iraq conflict lie is a baffling question and needs better understanding than a simple explanation. One has to comprehend the developments which have taken place, both internally and regionally. One has to critically examine the role of others to understand the conflict in the area.

The development in Iran, especially since the early 1960s, brought to the forefront the pronounced role of the country in the region. The military buildup of 1970s was to exercise military pressure, probably to support expansionist designs in the region. The basic premise in the mind of the Shah was possibly a significant role for his country in the Gulf and for this he could get patronage and encouragement of the US and her allies. The exit of the Shah and the capture of power by the fundamentalists in Iran brought in a more complex situation. The sudden upsurge of Islamic fundamentalism in Iran and its intended export to the neighbouring areas precipitated the crisis faster than it was expected. The fundamentalist reassertion triggered off Shia-Sunni differences in the region and internally damaged the emerging progressive forces. It could be compared with the Arab nationalism, and it brought to the forefront the traditional Persian-Arab enimity.[5]

In an interview with Mousavi Tabrizi, the Chairman of the Revolutionary Court of Iran states that the present conflict should not be viewed in a limited sense as it symbolises the efforts of the imperialists and reactionaries to check our progress towards a better society . . . it is the super-power game. It is not the simply a fight with Iraq but with all exploiters, imperialists and reactionaries. Bani-Sadr, the former President now in exile in France, had justified the fundamentalists' approach towards the necessity of this conflict and still condemns the present regime in Iraq. The desire to annexe the neighbouring areas, either territorially or otherwise, could be taken as an application of the basic principle of geopolitics, i.e., the might of the nation grows at the expense of the adjoining territory. However, the present exercise may simply be to find adherents for a particular ideology beyond the national borders.

No one doubts the strategic significance of the Shatt-al-Arab as a lone outlet for Iraq and this perspective figures very prominently in the calculations of the nations around it. The Shah's desire to emerge as the over-lord in the Gulf was matched with Iran's geographical strength in terms of its space and resources, both natural and human, *vis-a-vis* the other Gulf countries. The power-politics syndrome was apparently favouring this design in the region. The succeeding regime of the fundamentalists seems to have taken crude lessons from the power-politics game over the Gulf, which really figure very high in their strategic perspective and therefore there is a direct clash with the national interests of Arab countries like Iraq, Kuwait, and Saudi Arabia. Making an assessment of the causes which might have led to the present conflict, it would not be unrealistic to add that the strategic perspectives on the Gulf and the basic need of having a secured outlet for a nation like Iraq must have figured very high in the minds of both the regimes.

As stated elsewhere, the industrial complexes concentrating on the borderlands of both the countries as well as the vast oil reserves in the swamps of the Shatt-al-Arab and adjacent Khuzistan and further north in the forelands of the Kurdish area of both the countries explain the genuine concern on the part of the combating nations. Oil as a source of revenue has no equivalent replacement, it is, therefore, but imperative on the part of these two nations to have full security in these areas. No one likes to be deprived of this gift of nature and therefore, it figures very high in any strategic consideration. Time and again these borderlands have always been under great stress, atleast for the last four centuries, partly because of the ethnicity and partly due to the resources. Demographic structure in these borderlands is such that it has always caused worries and problems to the regimes in Iran and Iraq, and therefore, it is necessary to search for an amicable settlement of issues involved here. These borderlands have always been the weaker zones of the suzerainty of the respective regimes of the countries.

Frequency of convulsions in the internal set up of these countries, the legacies of the alien domination, the juxtaposed situation of Islamic fundamentalism and Arab nationalism and socialism, the quest for establishing hegemoney over the Gulf and the resources on the borderlands have provided the basis for the interplay of the various regional and international forces culminating in the form of the present conflict. One single premise seems to be inadequate to explain the course and conduct of the war during the last three years. International forces have always had an important say in the develpments in the region in order to seek an assured supply of oil from the region, in order to have access to the ready markets as well as to fullfil some of their strategic considerations. The Gulf and the countries like Iran, Iraq and Turkey have prominently figured in the super-powers' strategic consideration and manipulation. The geographical space here is amply littered with strategic military bases and the daployment of sophisticated arms.

Responses and Prospects

The consequences of this prolonged war are grave and the situation calls for a concerted effort by the international community to seek a peaceful settlement. But is it possible to have a lasting and durable peace and security in the region ? 'The governments usually make more stringent demands of a settlement for ending a war than they imposed upon the relationship during the prewar period.'[6] This is precisely the exercise involved in the present conflict. It has defied so far all moves for a peaceful settlement, primarily because the conditions laid down by the combatants seem to be unrealistic specially by Iran.

In an interview Ali Akbar Hashmi Rafsanjani,[6a] the President of the Islamic Republic Party and Chairman of the Iranian Majlis, openly confess that 'the war has rendered about 2 million people homeless, oil refineries are not working, many industries are closed, youngmen are fighting on the front instead of working in the factories, a lot of money is being wasted in purchasing weapons abroad and . . . we are not able to stand completely on the face of this war.' President Khomenei in a recent interview[6b] said that 'we are ready for peace but peace with dignity and honour as well as with definite promises.'

Iraq faces severe damages to its centres of vital economic importance, the threat of closure of the Strait of Hormuz and the loss of man-power. Ongoing development projects show strains of the war. Fighting the war involves high cost in blood and money and the country has to undergo tighter discipline and rigid organization as it gears up all its resources to meet the challenge in spite of the facts of geographical disadvantages, that is, the limited scale of distance leaving little room for retreat if necessary, the limited quantum of available man-power to man the front and possibly lesser resources at hand to cover the emergent situation.

Since its beginning, the Iran-Iraq Conflict has attracted various peace moves and efforts of international and regional organisations.[7] Iraq declared unilaterally a ceasefire from 5-8 October, 1980 and expressed a desire to come to the negotiating table. But these moves were spurred by the Iranians. The fresh armed attacks and counter attacks continued. The situation continues to be serious, gravely endangering international peace and security. No one doubts the gravity of the consequences of the prolongation of armed conflict between Iran and Iraq. It is evident that there is a wide consensus to end the war, simultaneously at the UN, the Non-Aligned Countries and the Organisation of the Islamic Conference.

Then what is it that prevents solution of the Conflict and why does the conflict persist ? Opinions differ. However, one may evaluate the whole situation in the light of the basic assumptions in the context of realities obtainable in the region. In the words of the former Iranian President Bani-Sadr and in the light of the comments of the former minister of foreign affairs, Qotub Zadeh, the regime in power in Iraq has no right to stay. Essentially it was an indirect provocation. In an interview on 26th June, 1983 in Paris,[8] Bani Sadr emphasised on the sinister plan of the regime in power in Iraq to dismember Iran by attacking in the oil-region of Khuzistan and by creating rebellion in Kurdish territory of Iran. Similar statements and overtures are being made by other Iranian leaders. No doubt such overtures resulted into intense provocation on the other side of the border, whereas Iraq blames Iran of having expansionist designs in the region.

A raw conclusion which one may like to draw here is that the strategic significance of the space seems to be a point of reference where one may label responsibility for the genesis and continuation of the hostility. Or is it just an exercise to keep oneself in power by diverting the anti-regime upsurge by the progressive elements of the country, as it has been the practice of many a fundamentalists regimes. Although it is generally happening in the decolonised world, that in case of war, one of the contending party refuses to come to the negotiating table and continues to ignore the various peace moves, both at the regional and international levels. The pertinent question here is on the genuineness of such efforts, as both the developed and the developing nations of the world with their set national interests try to gain out of such a situation. The case of Iran-Iraq conflict is not much different in this context.

FOOTNOTES

1. S.B., Jones, *Boundary Making*, Washington, D.C., 1945, pp. 13-14.
2. F. Halliday, "*Iran Dictatorship and Development*, New York, 1979, p. 14.
3. M. Heikal, *The Return of the Ayatollah*, London, 1980 ; Stempel J.D., *Inside the Iranian Revolution*, 1980.
4. Chronologically a total of about eighteen treaties have been concluded since 1520 A.D., e.g. treaty of Zahab, 1639 ; Kardon Treaty 1746 ; first Treaty of Erzrum, 1823 and the second in 1847 ; the Protocol of Constantinople 1913 ; the Frontier Treaty of 1937 ; and the Accord of Algiers 1975 etc.
5. Recent interviews of some of the Ayatollahs and the ministers in the present regime in Iran very clearly bring out inherent traditional hatred towards the other regimes and this further clarifies their designs for regional hegemony.
6. Ikle, F.C., 'The purpose of Fighting (and of Ceasing to Fight)' in Dawid S. Mclellan, et. al. (ed.), *The Theory and Practice of International Relations*, New Delhi 1977, p. 312.

   6a and 6b Interviews by Nasira Sharma (Tehran), Nov. and Dec., 1980, respectively.
7. The Security Council Resolution, 28th Sept., 1980 which called upon the two parties to ceasefire and settle the conflict by peaceful means ; Third Summit meeting of the Islamic Conference held in Saudi Arabia, Ja., 1981 ; the Ministerial Conference of Non-Aligned Countries, New Delhi, Feb., 1981, the Security Council resolution, 4th Oct., 1982 ; the General Assembly resolution, 22nd Oct. 1982 ; and the Security Council statement on 21st Feb., 1983 on the serious situation etc.

   Interview of Mr. Bani Sadr in Paris on 26th June, 1983 was taken by Nasira Sharma New Delhi.

INDEX